Catholic Physics

Catholic Physics

Jesuit Natural Philosophy in Early Modern Germany

MARCUS HELLYER

University of Notre Dame Press
Notre Dame, Indiana

Notre Dame, Indiana 46556

www.undpress.nd.edu

Published in the United States of America

Library of Congress Cataloging-in-Publication Data
Hellyer, Marcus.
Catholic physics : Jesuit natural philosophy in early modern Germany / Marcus Hellyer.
p. cm.
Includes bibliographical references (p.) and index.
ISBN 0-268-03071-5 (cloth : alk. paper)
1. Physics—Study and teaching—Germany—History.
2. Jesuits—Education—Germany—History. I. Title.
QC47.G3H45 2004
530'.07'1143—dc22
2004024688

∞ *The paper in this book meets the guidelines for permanence and durability of the Committee on Production Guidelines for Book Longevity of the Council on Library Resources.*

CONTENTS

Part 3. The Eighteenth Century

FIGURES

ACKNOWLEDGMENTS

My foremost intellectual debts are to my co-advisors, Robert S. Westman and Luce Giard. I owe thanks also to Professors Martin J. S. Rudwick and John Marino, whose seminars I attended and who served on my dissertation committee. I enjoyed many discussions, both inside and outside seminars, with my graduate school colleagues, in particular Benjamin Bertram, Margaret Garber, Rene Hayden, Margaret Meredith, and Andrew Zimmerman. Along the way, I have also benefited from the friendship, collegiality, and intellectual input of Alix Cooper, Charlotte Methuen, Cees Leijenhorst, Rolf Decot, and Patrizia Conforti. My colleagues at Brandeis, especially Govind Sreenivasan, Sylvia Marina Arrom, Sabine von Mering, and John Schrecker, have been very supportive of this project and my work in general. The work of scholars such as Roger Ariew, Dennis Des Chene, Peter Dear, and Mordechai Feingold has been particularly influential, and I would like to recognize them more explicitly than by merely citing their work in a footnote. I also would like to acknowledge the example of scholarship and of being a scholar set by Allan Mitchell. In addition I thank the two anonymous readers of the manuscript of this book both for their immensely helpful suggestions and for performing an unremunerative and generally thankless task, yet one that is vital for the health and intellectual honesty of our profession.

All historical works need a strong supporting crew, and this one is no exception. Mary Allen at the University of California, San Diego (UCSD), and Judy Brown and Dona Delorenzo at Brandeis have provided not only administrative expertise but also their friendship. Eliot Wirshbo at UCSD taught me Latin early in my graduate training;

without it this project would have been impossible. I have benefited from the help, knowledge, and support of many archivists and librarians, in particular Doug Stewart of UCSD and then the Dibner Institute, Rosemary Repperger of the Stadtbibliothek Mainz, Joseph de Kok, S.J., of the Archivum Romanum Societatis Iesu, and the Interlibrary Loan crew at Brandeis University. I particularly appreciate the enthusiastic support of my editor at the University of Notre Dame Press, Jeff Gainey.

Material included in this book has appeared earlier as "'Because the Authority of My Superiors Commands': Censorship, Physics and German Jesuits," *Early Science and Medicine* 1 (1996), and "Jesuit Physics in Eighteenth-Century Germany: Some Important Continuities," in *The Jesuits: Culture, Sciences, and the Arts, 1540–1773*, edited by John W. O'Malley et al. (Toronto: University of Toronto Press, 1999). I thank Brill Academic Publishers and the University of Toronto Press for their kind permission to republish.

The dissertation research that was the starting point of this project was funded by a doctoral dissertation fellowship from the National Science Foundation and a research fellowship at the Institute for European History in Mainz, Germany. I have since benefited from the generous support of a visiting fellowship at the Jesuit Institute at Boston College and a Brandeis University Bernstein Junior Faculty fellowship.

Finally, I thank my long-suffering family: my parents (particularly for learning not to ask when it was going to be done); my wife, Belinda; and our daughters, Isabelle and Madelyn.

ABBREVIATIONS

AHSI *Archivum Historicum Societatis Iesu*

APUG Archive of the Pontifical University Gregoriana, Rome

ARSI Archivum Romanum Societatis Iesu

BHSA Bayerisches Hauptstaatsarchiv, Munich

BSAW Bayerisches Staatsarchiv, Würzburg

BSB Bayerische Staatsbibliothek, Munich

Const. Ignatius of Loyola, *The Constitutions of the Society of Jesus,* trans. with introd. and commentary by George E. Ganss, S.J. (St. Louis: Institute of Jesuit Sources, 1970).

dB/S Augustin de Backer, Aloys de Backer, and Carlos Sommervogel, eds., *Bibliotheque de la Compagnie de Jesus* (Louvain, 1960).

Denz. Heinrich Denzinger, ed., *Kompendium der Glaubensbekenntnisse und kirchlichen Lehrentscheidungen,* 37th ed. (Freiburg i. Br.: Herder, 1991).

EABP Erzbischöflische Akademische Bibliothek, Paderborn, Archiv des Paderborner Studienfonds

GA Otto von Guericke, *Neue (sogenannte) Magdeburger Versuche über den leeren Raum: Nebst Briefen, Urkunden und anderen Zeugnissen seiner Lebens- und Schaffensgeschichte* [aka Große Ausgabe], ed. and trans. Hans Schimank (Düsseldorf: VDI-Verlag, 1968).

ILPW	Ignatius of Loyola, *Personal Writings,* ed. Joseph A. Munitz, S.J., and Philip Endean (Harmondsworth: Penguin, 1996).
ISJ	*Institutum Societatis Iesu* (Florence, 1893).
Mon. Ig.	*Monumenta Ignatiana, ex autographis vel ex antiquioribus exemplis collecta. Series prima. Sancti Ignatii de Loyola Societatis Jesu fundatoris epistolae et instructiones,* 12 vols. (Madrid: Typis G. Lopez del Horno, 1903–11).
Mon. paed.	Ladislaus Lukács, S.J., ed., *Monumenta paedagogica Societatis Iesu,* 7 vols. (Rome: Institutum Historicum Societatis Iesu, 1965–92).
MüUB	Universitätsbibliothek München
Pachtler	G. M. Pachtler, ed., *Ratio Studiorum et Institutiones Scholasticae Societatis Jesu per Germaniam olim vigentes collectae concinnatae dilucidatae,* 4 vols. (Berlin: A. Hoffman, 1887–94).
PCE	Peter Canisius, S.J., *Beati Petri Canisii, Societatis Iesu, epistulae et acta,* 7 vols., ed. Otto Braunsberger, S.J. (Freiburg i. Br.: Herder, 1896–1923.
PSGWL	Gottfried Wilhelm Leibniz, *Die philosophischen Schriften von Gottfried Wilhelm Leibniz,* vol. 2, ed. C. J. Gerhardt (Berlin, 1879).
RS	"The Ratio Studiorum of 1599," in *St. Ignatius and the Ratio Studiorum,* ed. Edward A. Fitzpatrick (New York: McGraw-Hill, 1933).
S.J.	Society of Jesus
Spindler	Max Spindler, ed., *Electoralis Academiae Scientiarum Boicae Primordia: Briefe aus der Gründungszeit der Bayerischen Akademie der Wissenschaften* (Munich: C. H. Beck, 1959).
ST	Thomas Aquinas, *Summa theologiae,* vol. 58, *The Eucharistic Presence* (3a. 73–78), trans. William Barden, O.P. (New York: Blackfriars, 1965).
StAK	StadtArchiv Köln
StAMz	Stadtarchiv Mainz
SBD	Studienbibliothek Dillingen
StBMz	Staatbibliothek Mainz
WüUB	Universitätsbibliothek Würzburg

INTRODUCTION

THE FIRST JESUITS TO TRAVEL THROUGH THE GERMAN LANDS WERE shocked by what they found there in the 1540s. Coming from countries that were still largely Catholic, they were unprepared to witness the success of the Reformation, which appeared to most Catholics to be unstoppable. Jerónimo Nadal wrote back to Ignatius of Loyola, the founder of the Society of Jesus, that "[t]here is the very grave danger that if the remnant of Catholics here are not helped, in two years there will be not one in Germany." But the magnitude of the task confronting them did not lead to despair. For these men, anxious to do God's work, the situation offered unparalleled opportunities. Nadal insisted that "in no part of the world is the Society, supported of course by God's grace, more needed; in no part of the world would the Society be more helpful." And, Nadal was aware, if the Jesuits did not aid the Catholic cause, no one else could. This would be its special mission: "I think that the task of helping Germany in its religious life is reserved to the Society."[1]

As in many other lands, the Jesuits' efforts to save the Catholic cause in Germany centered on educating future Catholic leaders. Catholic higher education had virtually collapsed in the wake of the Reformation and its associated upheavals. The Jesuits filled this void so effectively that by the early seventeenth century they held a virtual monopoly over the faculties of arts at the Catholic universities, a monopoly that was to last until the suppression of the Society in 1773. One consequence of great interest here is that virtually every professor of natural philosophy or mathematics in Catholic Germany was a Jesuit. Consequently, if a person was fortunate enough to belong to the small yet influential minority in the Catholic lands of Germany who had received a university education and had gone on to run the bureaucracies of church and state, then he (for women were not admitted to German universities at the time) in all likelihood would have learned all he knew of the various branches of natural philosophy from the fathers of the Society of Jesus.

This book tells the story of how this monumental educational enterprise taught natural philosophy. It is a history that spans a vast period; from the middle of the sixteenth century to 1773, Germany moved from the beginnings of the Catholic Reformation through the Thirty Years' War and the baroque period to the Age of the *Aufklärung*, the German Enlightenment. It involves dozens of institutions, hundreds of professors, thousands of texts, and tens of thousands of students. Beginning the account from the point at which the Jesuit colleges were firmly established, around 1600, limits the material only marginally. What unifies this vast subject matter is that throughout this long era, the history of natural philosophy at Catholic universities in Germany was necessarily a history of *Jesuit* natural philosophy. This book examines what that was, how it was taught, and how it changed.

Although the Jesuit colleges are fascinating in their own right, this account goes beyond an aggregation of institutional histories.[2] Examining the natural philosophy taught at the Jesuit colleges reveals much about the cultural and intellectual world of early modern science. The German provinces of the Society of Jesus produced few great natural philosophers, particularly after the middle of the seventeenth century. Most of the professors of natural philosophy at the Jesuit colleges were not creating significant new natural knowledge; this task was not yet part of the self-conception of the German academic. Yet these colleges provide an excellent opportunity to study the dissemination of scientific knowledge in early modern Europe. In the peculiar cultural and institutional environment of the Jesuit colleges, one can observe how, when, and in what form many of the crucial changes that occurred in physics, those developments that constituted the core of what has been traditionally called the Scientific Revolution, entered the classroom. By examining how natural philosophy fit into the larger framework of the Jesuit pedagogical mission, we can also identify factors that hindered or encouraged the dissemination of both knowledge about nature and methods of studying it. This study shows that although Jesuit natural philosophy was transformed in many ways between 1600 and 1773, there were also remarkable continuities as the New Science of the seventeenth century was fused onto a core of peripatetic natural philosophy in a synthesis that endured until the middle of the eighteenth century.

Reassessing Jesuit Science

To study the Society of Jesus one must peel back layers of distortion and misrepresentation, much intentional, much the result of merely recapitulating the laziness or bigotry of earlier historians.[3] Virtually from their inception in 1540,[4] the Jesuits were the targets of hostile criticism. In Germany this came not only from Protestants, who saw their advances halted in large part by the Jesuits, but also from other Catholics who came to resent the influence the Society gained at princely courts and universities throughout Catholic Germany. During the *Aufklärung,* the German Enlightenment, the Society was among the chief targets of educational reformers in both the Protestant and Catholic territories due to its prominent role in education. The jurist Friedrich Carl von Moser, for example, wrote in 1767 that Catholic education would be bad as long as the Jesuits were in charge of it because "for them, everywhere is Paraguay."[5]

After the Society was reestablished in 1814 it was once again subjected to polemical attacks, particularly in the latter part of the century; in an age of nationalism, an international and ultramontane entity such as the Society of Jesus was an attractive target. The Bismarckian age witnessed Germany's undisputed position at the pinnacle of natural science, but it was also a period marked by rampant nationalism, militarism, and the anti-Catholic *Kulturkampf,* during which the Iron Chancellor expelled the Jesuits from Germany in 1872.[6] The same year, Carl Prantl published a history of the Bavarian University of Ingolstadt—where the Jesuits had taught for two centuries—claiming that "the intervention of the Jesuit Society was an immeasurable misfortune for the university, for here we see the effects of an institute dangerous to the public good that consciously or unconsciously injected into every one of its members in greater or lesser degree an element of evil." The Jesuits were a "poison" that exhausted the university through its efforts to fight them off.[7] Other historians simply ignored the Jesuit educational system, which in effect meant ignoring Catholic higher education in its entirety.[8]

The late nineteenth century also saw the flowering of works alleging the inherent, inevitable incompatibility of science and religion, a theory now conveniently referred to as "the conflict thesis." Much historical literature explicitly linked scientific progress to Protestantism, while deeming Catholicism and science to be inherently incompatible. In America John William Draper, in his *History of the Conflict between Religion and Science* (1874), claimed that the Catholic Church's persecution of scientists had left its hands "steeped in blood."[9]

Fortunately more recent works have, on the whole, restored a modicum of decorum to discussion and reaffirmed the importance of the Jesuit contribution to German education, although few have examined the place of science at Jesuit institutions.[10] Laetitia Boehm, for example, has provided some much needed perspective, revealing how even twentieth-century historians have unconsciously adopted many of the sentiments shared by the Jesuits' eighteenth-century critics. She has also noted that much modern criticism of the Jesuits rests upon a conception of *Forschung* (research) that became widespread only in the nineteenth century and that was not—and could not be—shared in the previous century by the Jesuits themselves.[11]

We now know that Jesuit scholarship and Jesuit science throughout Europe did not resemble the colossal disaster that generations of critics depicted.[12] On the contrary, the later sixteenth and early seventeenth centuries can be described as a Golden Age of Jesuit science. Not only did many of the traditional heroes of the Scientific Revolution have significant relationships with Jesuit scholars—Galileo is a prime example—but, as active participants in the intellectual life of Europe, numerous members of the Society were themselves significant contributors to natural philosophy and mathematics.[13] Many Jesuit mathematicians and natural philosophers achieved prominence well beyond their own order, among them the mathematicians Christoph Clavius and Christoph Scheiner. A broader definition of science that included scholastic textbook natural philosophy also would encompass such influential philosophers as Francisco de Toledo, Francisco Suárez, and the Coimbran commentators (Conimbricenses). Furthermore, this first century or so of the Society's existence included the development of the Jesuits' educational institutions and the formulation and implementation of the model for the Jesuit educational system realized in the *Ratio studiorum* of 1599.[14]

What is still obscure is the path of Jesuit science after the middle of the seventeenth century. Assumptions about the repercussions of the trial of Galileo for science in Catholic lands have created the impression of a caesura around 1633.[15] This has led to a further assumption that Jesuit science declined significantly after this point or even died.[16] But while one may grant that the way the Jesuits did science differed after the condemnation of Galileo, they were nevertheless still studying, discussing, and writing about nature.[17] Jesuit zeal for publishing scientific works did not flag in the eighteenth century but actually increased considerably.[18] In fact, in Germany the number of surviving Jesuit academic texts increased virtually exponentially from the start of the eighteenth century. Jesuit professors produced masses of material in numerous fields of natural science, ranging from huge, multivolume textbooks through disputations and lecture manuscripts to single-page thesis sheets. Such texts number in the thousands; in light of their teaching monopoly, it would be more surprising if the German Jesuits had *not* produced such a huge corpus.[19]

The Jesuit colleges and universities were vibrant places. Traditional institutional histories that rely on foundational and normative documents such as charters and statutes do not do them justice. In particular, we cannot accurately determine the content of classroom instruction from such sources. Statutes governing the curriculum stayed in effect for many years after their inception, during which time the actual curriculum could change considerably. Furthermore, such documents generally give very few specific instructions on the content of instruction. Also a decree or statute insisting that a certain subject be taught may not have introduced anything new but rather may have merely confirmed what was already the practice in the classroom.[20] Studies that merely search through the sources to find the first mention of Newton or some other great scientist, theory, or device in order to celebrate the university's role in the advancement of science are hopelessly colored from the outset by anachronism.[21] So for all their importance, we still know surprisingly little about the teaching of science at the Jesuit colleges in Germany.

The Aim of This Study

By 1773, the year of the Suppression, the natural philosophy taught at the Jesuit college differed dramatically from that taught around 1600. The subject of this book is the transformation of Jesuit natural philosophy from a largely scholastic body of knowledge and discourse into an experimental, mathematized science. This was a long process. Following the pioneering work of Charles Schmitt, historians have become increasingly aware of the longevity of Aristotelian philosophy, including the branches of natural philosophy.[22] Schmitt recognized that scholastic philosophy based largely on Aristotle was extremely elastic, capable of absorbing numerous elements from other philosophical systems while maintaining a core of scholastic principles and methods. We can see the remarkable flexibility and adaptability of Aristotelian natural philosophy particularly clearly in the case of the Jesuits. The Jesuits' *Ratio studiorum* of 1599, which governed the Society's educational mission, specified that Aristotle was to provide the foundation of Jesuit philosophical instruction. Furthermore, St. Thomas Aquinas's thirteenth-century synthesis of Aristotelian philosophy and Christian theology, which formed the foundation of the theology taught in Jesuit colleges, was comprehensible only to students who had mastered Aristotelian-scholastic philosophy. Significant aspects of this philosophical system survived in the Society's German colleges well into the eighteenth century.

This does not mean that one must concede the charge leveled by the Jesuits' Enlightenment (and later) critics that the Society remained steadfastly committed to scholastic philosophy until the Suppression in 1773 and failed to appreciate any of the developments in the natural sciences made over the preceding two centuries. Such attacks do not recognize that the Jesuits were quite capable of integrating novelties into their curriculum, despite their commitment to many aspects of Aristotelian philosophy. It is crucial, however, to determine what those novelties were and under what conditions such integration occurred.

Little has been done to examine the Jesuits on their own terms and to situate the German Jesuit natural philosophers in their own traditions and institutions. This book corrects that situation by studying Jesuit physics in Germany over the long term, from 1600 until the Suppression in 1773. An approach that traces the course of Jesuit instruction over the long term, rather than focusing on a single individual or scientific controversy, can identify significant traditions—in this case traditions of interaction between natural philosophy and theology, of the rules and practices governing what could be said and done in natural philosophy, of the place and purpose of natural philosophy within the Society, and of Jesuit dialogue with practitioners and scholars outside the Society. One of the goals of this study is to identify the major characteristics of Jesuit physics and to map out continuities into the eighteenth century.

Clearly it is much too large a task to study all fields of the physics taught at Jesuit universities and the changes they underwent over a period of almost two centuries. To identify both continuities and innovations in practices and classroom content, three themes will be pursued through the seventeenth and eighteenth centuries. The first of these is the constraints placed on Jesuit natural philosophers by their order's efforts to maintain uniformity in its teaching during an era marked by rapid developments in science. Therefore, practices of censorship of science, which were established early in the Society's history and which were continued throughout its existence, will receive much attention. But this is not simply a story of repression; Jesuit professors used several strategies to exploit the spaces left by the structures and practices of censorship to teach and publish novelties. Despite the supervision exercised by the Society's hierarchy, its members were capable of appreciating many of the developments that occurred in physics over the course of the seventeenth and eighteenth centuries and integrating them into their teaching and publication.

The second theme is the relationship between theology and natural philosophy within the Society. This theme is related to the first, for the hierarchical standing of theology over philosophy, which required that philosophy conform to the truths of the

Catholic faith, was one of the principal constraints on the teaching of physics in the Society. Perhaps the best example of how theological concerns could limit what natural philosophers were able to say was the question of how the transubstantiation of the bread and wine into the body and blood of Christ could be explained physically—what we can term the physics of the Eucharist. The Society believed that the Council of Trent enshrined an Aristotelian-scholastic theory of matter as the only one compatible with the orthodox account of transubstantiation. As a result, the Society and its members stubbornly rejected alternative matter theories such as Cartesian atomism, relenting only after the middle of the eighteenth century.

In contrast to this example of dogged resistance to the "New Science," the German Jesuits showed considerable openness to experimental philosophy. Their adoption of experiment forms the third theme of this work. The treatment will focus on the pneumatic experiments conducted with the Torricellian mercury tube and the air pump. Not only did a Jesuit mathematician, Kaspar Schott, play a vital role in the dissemination of the pump, but in many ways these two instruments can be seen as emblematic of the new experimental philosophy. They have also attracted the attention of modern historians of science.[23] The air pump is featured in the work that established the importance of examining the production of scientific knowledge as a social enterprise, Steven Shapin and Simon Schaffer's *Leviathan and the Air-Pump.*[24] But because Shapin and Schaffer were examining the relationship between the production of scientific knowledge and a particular social group at a particular political conjuncture, they devoted relatively little attention to the pump's earlier history in a social setting different from the one they were considering. This neglected "prehistory" is a story in which the German Jesuits played a prominent role. A study of the German Jesuits' use of pneumatic experiments also adds to our understanding of the diffusion of the experimental philosophy in the early modern period.[25]

Conflicting traditions existed within the same order and even within the same individual. While ongoing censorship was expressed in a discourse that altered little over two centuries, Jesuit natural philosophers sought to subvert this censorship with the same methods in the early seventeenth as in the mid–eighteenth century. While the Jesuit polemics against Descartes and his matter theory were virtually identical in 1650 and 1750, in many other ways the structure, form, and content of the Jesuit natural philosophy curriculum underwent fundamental changes between these two dates as the result of a tradition of integrating novelties despite the censorship. In the quarter-century before the Suppression in 1773, the last vestiges of scholastic natural philosophy were removed.

A Note on Scope and Terminology

All historians know it is difficult to speak of "Germany" in this period. The Holy Roman Empire consisted of hundreds of territories possessing widely differing degrees of independence from the emperor. Some populations in the empire did not speak German and some outside it did. The Jesuits' own administrative hierarchy did not map neatly onto this already chaotic picture. The Society was divided into the assistancies of Italy, Spain, Portugal, France, and Germany, but they were not necessarily coterminous with the nations or linguistic communities of the same name. The assistancies were further divided into provinces. The provinces of the German assistancy stretched from Belgium to Lithuania. Its German-speaking provinces were Lower Rhine (which corresponds roughly to the northwest of the modern Federal Republic of Germany), Upper Rhine (central Germany, including the Mosel and Main Valleys), Upper Germany (southern Germany and Switzerland), and Austria (including many of the Hapsburgs' non-German territories). The Bohemian province also had a large number of German speakers.

Although some provinces had close associations with particular states (such as the Upper German province with Bavaria), all Jesuit provinces in Germany covered numerous political entities. This meant that the Jesuits had to deal with numerous jurisdictions, as does any historian who studies them. So while the Jesuits had a centralized hierarchy and a single document governing their pedagogical mission, the famous *Ratio studiorum* of 1599, their educational institutions were scattered across scores of territories, each of which had its own policies on education. Consider some of the universities in Germany where the Jesuits taught. Ingolstadt was in the Duchy (later Electorate) of Bavaria, while a short trip up the Danube Dillingen was in the territory of the prince-bishop of Augsburg. On the Main River, Würzburg was in the territory of the prince-bishop of Würzburg, while Mainz, downstream at the confluence of the Main and the Rhine, was in the territory of the elector-archbishop of Mainz. Sometimes one person governed both these sees, other times not. The University of Cologne was in the city of Cologne, officially the seat of the elector-archbishop of Cologne, but the elector-archbishop had virtually no power in the city itself; it was the town council that set educational policy. In addition there were the universities in the Hapsburg lands at Vienna, Prague, Innsbruck, Graz, Olmuc, and Freiburg im Breisgau, widely scattered and with different traditions but all subject to the Hapsburgs' increasingly absolutist pretensions.

The Jesuits for their part were permitted by their own statutes considerable flexibility in dealing with local conditions; indeed, this was one of the hallmarks of the Jesuit "way of proceeding," to use their own term. Thus, while it is possible to identify clear patterns in Jesuit education, doing justice to the historical specificity of every individual institution would be a mammoth task, well beyond the reach of one book. However, as this is not a study of particular institutions but rather an attempt to sketch out major facets of the natural philosophy practiced at the Jesuit universities and colleges in German provinces, a detailed study of every university is not necessary.[26] This work focuses on three universities—Mainz, Würzburg, and Ingolstadt. Both Mainz and Würzburg were in the Upper Rhenish province. Neither was particularly prominent, but both were typical examples of the small universities that predominated in Germany.[27] The Bavarian University of Ingolstadt in the Upper German province was a somewhat more significant institution for Catholic Germany in this period.[28] This book also draws freely on materials from other Jesuit institutions in the German assistancy when they shed light on the problems being discussed, including numerous texts from the Universities of Dillingen and Freiburg im Breisgau,[29] both, like Ingolstadt, in the Upper German province; the Tricoronatum, the Jesuit college at the University of Cologne, and the University of Paderborn, both in the Lower Rhenish province; and the University of Prague, in the Bohemian province. These materials reveal a high degree of uniformity in educational practices and content throughout Germany. This is not to say that Jesuit education was by any means identical in every institution. Local practices and the influence of secular governments prevented this. Thus not all general claims made here are necessarily valid for every institution throughout the German assistancy, let alone the entire Society.

Since the subject of this book is natural philosophy, we should spend a few moments addressing it, although we shall be spending considerably more time on it in coming chapters. *Natural philosophy* is not used in this work as a synonym for science. Natural philosophy was not science. Careless use of language effaces the distinctions between them and misrepresents natural philosophy.[30] While many subjects addressed by the modern discipline of physics can be found in natural philosophy, an equal number have been long banished from "science." Dennis Des Chene has described natural philosophy as "a kind of clearinghouse in which physics, metaphysics and theology could meet and negotiate their claims, much less needed now that those disciplines have gone their own ways."[31] Although this book traces one process by which those disciplines came to part, at our starting point the Jesuit curriculum was still an active clearinghouse, or perhaps better said, wrestling ring.

The Jesuits often used the term *natural philosophy* synonymously with *physics; physica generalis* and *physica particularis* together formed the natural philosophical curriculum. But at the Jesuit colleges the term *physics* had several meanings. The second year of the three-year philosophy course taught at Jesuit colleges was called *Physica.* Second-year students were then *Physici.* In this year natural philosophy was taught, following the order, if not the content, of Aristotle's books of natural philosophy. The first part of the year dealt with the subjects covered by Aristotle's *Physics,* which addressed the most general questions relating to all natural bodies; the second part dealt with particular bodies—heavenly bodies, atmospheric bodies, the elements, and living bodies.

At grave risk of being overly pedantic, in this study "Physics" refers to the second year of the curriculum, "physics" refers to the discipline of natural philosophy, and "*Physics*" italicized refers to Aristotle's book. The adjective *physical* is used to make an explicit contrast to mathematical or metaphysical methods, claims, or subjects. We shall do our best to be consistent.

PART 1

Discourses and Institutions

CHAPTER 1

Managing Philosophy in the Society of Jesus

FROM ITS INCEPTION, THE SOCIETY OF JESUS WAS AN ORDER OF THE learned. The ten companions that Ignatius of Loyola (1491–1556) gathered around him in 1532 were masters and advanced students from the University of Paris, a major center of learning in sixteenth-century Europe. Throughout the century, the Society's members included some of the most prominent humanists, theologians, mathematicians, and philosophers of their time. Its colleges, where the branches of philosophy were granted great esteem, were widely acknowledged to be among the best in Europe. But the Jesuits did not pursue worldly knowledge purely for its own sake. While philosophy had an important role to play in the Society's larger mission, distinct boundaries had to be placed around its practice. This chapter examines the role of philosophy in the Society of Jesus and the Society's methods of establishing and maintaining boundaries within which the teaching and writing of sound philosophy could be ensured.

A chronological approach, such as tracing the evolution and adoption of the final, 1599 edition of the Society's pedagogical blueprint, the *Ratio studiorum,* could not do this story justice in the space available and, moreover, has been done already.[1] Rather, this chapter will adopt a thematic approach. It will show how the esteem that the Society granted to philosophy led to an emphasis on the maintenance of uniformity in its philosophical teaching. The chapter will then trace the development of the methods by which that uniformity was defined and enforced over the second half of the sixteenth century. The constant tension within the Society between its commitment to uniformity in philosophy and its impulse to grant professors considerable liberty in their teaching and writing will thereby become clear.

The Purpose of Philosophy

It was by no means unusual for a religious order to concern itself with the teaching of the worldly sciences. The view that philosophy was necessary to inspire and prepare students for the study of theology can be traced back at least as far as Church fathers such as Justin Martyr, Clement, and Origen.[2] At medieval universities, the liberal arts were taught as training preparatory to theology, and completion of studies at the faculty of arts was generally required before a student could progress to the higher faculty of theology. Many of the professors were clergy, often belonging to the mendicant Dominican and Franciscan orders.[3] The first Jesuits were located firmly in this tradition.

While Ignatius and his companions were aware that they would have to provide training for future members of the Society in order to maintain a supply of well-educated recruits to replenish their own ranks, they did not originally envisage that the primary mission of the Society would be the administration of schools for the laity. But once the Jesuits began teaching, they were met by such a demand for education across Europe that already by the end of the 1540s, at the urging of princes and cities, the Society was opening schools for external students in addition to their own scholastics.[4] Confronted by questions concerning the goals of education and the curriculum necessary to meet them, the first Jesuits adopted the pedagogy of the University of Paris, where they had been educated.

From the outset of the Society's teaching enterprise, Ignatius and his followers regarded philosophy as an essential prerequisite for theology.[5] Sometime between 1565 and 1570, as part of the first attempt to systematize Jesuit teaching practices, one of the main architects of the Jesuit educational system, Jerónimo Nadal (1507–80), defined the purpose of philosophy: "Since the arts or natural sciences dispose the intellect to theology and serve the perfect cognition and use of it, and in themselves help towards this end, let them be treated with the appropriate diligence by learned teachers seeking sincerely in all matters the honor and glory of God. Logic, physics, metaphysics, moral science, and also mathematics are to be treated insofar as they pertain to our announced ends."[6] Virtually the same wording was retained thirty years later in the final, 1599 version of the *Ratio studiorum* in the preamble to its section "Rules for the Professor of Philosophy."

Ignatius was so convinced of the necessity of a sound basis in classical languages and philosophy for theology students that he refused requests put to the Society to establish theological schools until the necessary classes in humanities and philosophy had been securely established. He insisted upon this in 1551, when King Ferdinand of Germany asked that the Society reestablish the faculty of theology at the University of Vienna. The king hoped that uneducated students would come from the provinces to study theology in Vienna. Ignatius countered that "such students, even though they be well disposed, will not have the proper foundation in logic and philosophy, or even perhaps in languages. Such a foundation is indispensable."[7] The situation was similar somewhat later in Würzburg, where Peter Canisius (1521–97), the Society's provincial in Germany, and General Francisco Borja (1510–72, general from 1565) were unwilling to commit themselves to reviving the theology faculty at the university there until the necessary material and pedagogical foundations were in place.[8]

The most influential Jesuit philosophy textbook expressed the relationship between theology and philosophy almost exactly as Nadal had. Francisco Suárez (1548–1617), in the proemium to his *Disputationes metaphysicae,* the most important metaphysics textbook of the seventeenth century, wrote that philosophical questions frequently arose during the treatment of theological mysteries. As it was too confusing for students to deal with such questions in a theology class, he declared that they had to be treated separately as the preparation for theology. For Suárez it was impossible to become a perfect theologian without first building a solid foundation in philosophy.[9]

The Jesuits also inherited the long tradition stretching back to antiquity that held that the arts and sciences were not only an essential tool for theology but also salutary in themselves if studied correctly. Ignatius was convinced of the usefulness of the liberal arts and philosophy for the Society's wider apostolic activities. His secretary, Juan Alfonso de Polanco (1516–77), informed Urban Fernandes, master of novices at the Jesuit college at Coimbra in Portugal, in June 1551 that Ignatius

> heartily wishes that all be well grounded in grammar and the humanities, especially if age and inclination are in one's favor. Then he excludes no kind of approved learning, neither poetry, nor rhetoric, nor logic, nor natural philosophy, nor ethics, nor metaphysics, nor mathematics, especially, as I said, in the case of those who are of the proper age and ability. For the Society is happy to be provided with

> all possible weapons, on condition that they who wield them do so for edification and are ready to use them or not as shall be judged best.[10]

The various branches of philosophy could be usefully applied to the vast number of tasks that the Society of Jesus assigned its members around the world as they worked in missions, princely courts, and colleges.

For the first Jesuits, drawing on the scholastic tradition they had imbibed at Paris, there was no inherent contradiction between faith and reason. This acceptance of the legitimacy of philosophy and reason was wholeheartedly shared by subsequent generations of Jesuits in Germany. In contrast to many Protestant theologians, who, as part of a reaction against the medieval Church and its scholastic learning, had denounced philosophy in general, the German Jesuits defended the powers of reason and their application in philosophy.[11] The Jesuit controversialist theologian Jakob Gretser (1562–1625) sharply condemned Luther's attacks on reason and philosophy. Luther, he claimed, had denounced reason as Satan's whore, who could do nothing but blaspheme. On the contrary, asserted Gretser, natural reason inclined one toward the good—that is, toward God. God and the dogmas of faith were not contrary to reason but above reason. Reason could reach them but only if it received the light of faith. Luther might insist that the speculative sciences such as natural philosophy were erroneous and sinful, but in reality, Gretser wrote, they made men better.[12]

Philosophy and Morals

But if philosophy could make men better, it could also make them worse. The practice and teaching of philosophy was a morally charged field. Bad philosophy not only caused theological problems but, the Jesuits were convinced, could also have deleterious moral effects. Diego Ledesma (1519–75), another early Jesuit influential in the formulation of the *Ratio studiorum,* writing in 1574 to support the introduction of a list of approved philosophical opinions for use in schools, argued that

> these [philosophical opinions] greatly help that which pertains to morals; for they are full of religion and, as experience shows, preserve the devotion and simplicity and sincerity of the mind and turn the tranquil mind to God, humble and subject to Him, and excite hope and loyalty and love of God and the action of grace. The contrary doctrines and contrary opinions, however (as we have, alas, sufficiently, and beyond sufficiently, experienced, which we do not say without pain

> in our soul), dry up devotion, and supply many causes of temptation against the faith and religion, and often turn the mind toward doubt and ambiguity and vacillation, etc.[13]

While philosophy could prepare the student intellectually for training in theology, it could also turn the student to or away from God and an upright life.

An incident featuring the most influential of the early Jesuits in Germany, Peter Canisius, is particularly illuminating in this respect. Canisius had played a crucial role in establishing the Society of Jesus in Germany and restoring Catholicism in the face of awesome hurdles. It was he who with the most meager of resources secured the foundation of Jesuit colleges at universities that had either foundered or gone over to Protestantism. From his early studies at the University of Cologne he had devoted his life to struggling against the Protestant heresy and reviving the old Church. He was the acknowledged champion of Catholicism in Germany. People in Augsburg, for example, expected that he "would expel the heretical preachers with the rod if he stayed there long."[14] His awareness of the tenuous position of the Society and Church in Germany made him particularly sensitive to dangers within his own ranks. In September 1567 Canisius's brother, Theodoric, the rector of the newly established Jesuit university in Dillingen, wrote to General Borja reporting the apostasy of a Jesuit brother, Balthasar Zuger. He blamed Zuger's fall on two masters, Anton Klesel and Simon Damer, "who, praising Averroes beyond measure, spoke extremely contemptuously of Christian philosophers such as St. Thomas. And it is amazing how rudely galling Master Antonius was, in that having abandoned his public duties and theology, he thought it was right to spend all his time in reading and scrutinizing philosophy."[15]

Averroes (Ibn Rushd, 1126–98) was the twelfth-century Arab philosopher who insisted on the rationality of creation and the ability of reason to comprehend it through the study of philosophy. For Averroes, at least in many medieval and early modern Christians' reading of him, nothing was beyond reason, not even faith and revelation.[16] Although the Jesuits were convinced that reason and faith were compatible, many feared that too high an estimation of the powers of reason could lead philosophy to question the tenets of revealed religion, which were ultimately beyond reason. The suitability of Averroes for use in the classroom was, therefore, hotly disputed within the Society.

As far as Theodoric Canisius was concerned, the two masters had already succumbed to the Averroist temptation. The study of Averroes had led to the abandonment of both their religious obligations and their study of theology. Furthermore, he claimed that a group of students was too close to these professors and could not be separated from them. Some of them had developed "effeminate manners" and enjoyed the "reading of profane poets." The accusation of moral decay was clear.[17]

These masters were not the only professors who had suffered moral lapses that Theodoric Canisius believed were caused by Averroist philosophy. He reported on 10 December 1567 to General Borja the case of Joannes Vicus Anglus, who had arrived at Dillingen the previous June.[18] Vicus's superiors soon realized that he was committed to Averroist opinions and was diligently teaching them. Thinking he should not teach philosophy, they demoted him to prefect of the school of rhetoric. But still he refused to follow the prefect of studies' directions. According to Theodoric Canisius, Vicus criticized his superiors and brothers and tried to win them over to his opinions. Although Vicus had been forbidden to praise Averroes, he did so anyway in the presence of the students, stating that Averroes was preferable to St. Thomas.[19] Eventually Vicus's behavior and sarcasm became unbearable to his brothers, so the provincial, Peter Canisius, ordered him to perform penances. Vicus obeyed, but as he afterwards remained just as stubborn in his opinions, Peter Canisius and his advisors finally decided to dismiss him from the Society.[20] Another master, the Spaniard Alfons Pisano, wrote that Vicus said he was performing his punishment but did not care about the judgment of anyone other than God.[21] The prefect of higher studies at Dillingen, Gerard Pastelius, wrote that Vicus feigned devotion to superiors when present, slandered them when absent, spoke badly of St. Thomas, praised Averroes, granted himself privileges, and hardly did the punishments given to him.[22] Pisano recommended to Peter Canisius that Vicus be expelled or imprisoned.[23] After Peter Canisius took the offender away, Pisano rejoiced that "through the grace of God our college has been freed from the spirit of infidelity" but warned that "the spirit of schism and sedition continues to ravage the vineyard of the Lord."[24]

The aberrant masters had been charged with a lengthy catalogue of offenses: the neglect of religious obligations, the abandonment of theology, immorality, arrogance, quarrelsomeness and sarcasm, disrespect and disobedience toward superiors, schism and sedition, and, ultimately, apostasy. Peter Canisius, the provincial, identified the cause of this decay in a letter to General Borja: the cases of apostasy were caused by too much contact with Averroist professors sent from Rome who followed reason alone and respected no authority. Nothing was a surer road to heresy in Canisius's view. Averroist philosophy had taken deep root in the Society and had to be extirpated. Canisius's solution was twofold: Rome should send men strong in their faith and vocation, and, if possible, all members should be prohibited from voicing their own opinions against the consensus of the schoolmen. He urged the Society's censors to supervise everybody's speech and report anything suspicious. Averroism had to be torn up by the roots, and the provincial had to be given the power to cut out the pestilent parts of the body.[25]

The next year, 1568, the provincial congregation of the Upper German province passed a resolution requesting firm action from the Society's general congregation, stating that "we greatly desire that a severe law be promulgated throughout the whole Society, so that the opinion of either Averroes or any other philosopher that fights with the Christian faith or with the common doctrine of the schools in any respect may not be defended or confirmed."[26] In short, precisely because of its importance, philosophy was something that had to be closely watched. Despite its utility, the potential problems that could arise from the unfettered use of reason were considerable.

Uniformity and Obedience

While danger did not confront all the early Jesuits as immediately as the Canisius brothers, it was clear to them that there were good and bad philosophies. The task confronting the Society was to identify which was which and to ensure that the good was taught and the bad excluded from its schools. From the outset the Society considered good philosophical instruction to be synonymous with uniform instruction. Ensuring uniformity was easier said than done. There was not only the problem of developing effective measures to promote uniformity but also the question of how to preserve the liberty theologians needed to theologize and philosophers to philosophize. The Society grappled with the problem of how to ensure uniformity (*uniformitas*) of correct instruction throughout the second half of the sixteenth century until the final draft of the *Ratio studiorum* was sent out to the provinces in 1599. Yet even after this—indeed, until the middle of the eighteenth century—the twin issues of defining philosophical correctness and enforcing it were of constant and pressing concern.

While the emphasis on uniformity may seem quite contrary to the aims of higher education today, there were several reasons for it: doctrinal, pedagogical, and existential. The first of these was critical. As the Society became increasingly aware of the inroads made by heresy, it recognized that one of its prime tasks was to fight the ignorance they held to be its cause. Nowhere was this more pressing than in Germany. The desperate straits of the Catholic Church in Germany were reflected in Peter Canisius's unflagging efforts to root out false doctrines in his Society. Sound doctrine was an essential part of the Jesuits' efforts to defend the central tenets of the Catholic faith and refute the Protestants' assaults on them. The Society could not allow its members to teach false and dangerous doctrines. All Jesuit discourse on the issue of theological and philosophical opinions was unanimous in stating the need to uphold and spread sound Catholic doctrine.

Second, the Society was deeply concerned with issues of sound pedagogy. The structure and method of the Society's teaching enterprise were designed to allow students to master a set body of knowledge as efficiently as possible. Squabbling among professors over details undermined the authority of the teacher, wasted time, and confused and bored the students. Letters to Rome from the provinces requesting the introduction of measures to enforce uniformity frequently justified their requests with accounts of the inefficiency of professors contradicting each other.

Beyond these factors, there was also an existential reason for the insistence on uniformity. The Society's emphasis on obedience is well known, but it is perhaps the most misunderstood of all elements of the Jesuit way of proceeding. The Society of Jesus as Ignatius conceived it was to be radically different from existing orders. Jesuits would not live apart from the world in monasteries but roam through it. Theirs would be an active life, engaging in a vast range of tasks in all social groups and cultural environments. They would often be acting independently, far away from a superior. To maintain coherence in such a extended enterprise, Ignatius recognized that it was crucial that Jesuits be properly formed. Part of this formation included the development of the virtue of obedience. Just as the older orders excelled in more external forms of piety, Ignatius wanted the Society to excel in this internal virtue.

His best-known explication of the concept of obedience is his letter of 26 March 1553 to the members of the Society in Portugal. The letter became a canonical text throughout the Society.[27] Citing St. Gregory, Ignatius stated that obedience was "the only virtue which plants all the other virtues in the mind, and preserves them once they are planted."[28] He outlined the three stages necessary for complete obedience. The first degree of obedience was simply to do what one was told. This was easy, for it could be done either willingly or unwillingly. More difficult was the second degree, which "goes beyond the mere performance of an order to the adoption of the superior's wishes. Rather, it is a question of stripping oneself of one's own wishes and putting on the divine wishes interpreted by the superior."[29]

The third and most perfect degree of obedience was the most difficult. Whoever aimed at perfect obedience "must not stop at their wishes but also include their thoughts." Without this highest form of obedience, the first two were, in the long term, impossible. Displaying his scholastic training, Ignatius wrote that "the appetitive powers of the soul naturally follow the apprehensive and, in the long run, the will cannot obey without violence against one's judgment." He continued, "When one acts in opposition to one's judgment, one cannot obey lovingly and cheerfully as long as such repugnance remains."[30] By requiring that the individual members of the Society understand why they were being commanded and make their superior's will and opinion their own, Ignatius wanted something far beyond the unthinking obedience of popular stereotype.

Ignatius justified obedience in religious terms; one obeyed the superior not because he was himself good or just or capable but because he held the place and authority of Christ. Thus, by submitting oneself to the will of the superior, one submitted oneself to the will of Christ. One's will was consumed in a holocaust, as Ignatius put it, leaving nothing. As St. Thomas had written, obedience was the most important virtue because it was the sacrifice of man's most prized possession, the will.[31] To prefer one's own will to that of the superior was to deny God's rule in the universe.

However, Ignatius also recognized the crucial role of obedience in maintaining social coherence. Ignatius had learned from the decline of the older orders that had decayed since the time of their foundation in the Middle Ages through abuses and laxity. In stating the importance of obedience for the maintenance of any order's rule, he wrote that "so much the better is the government where such subordination is safeguarded, and if this subordination is faulty in any society, its failings become all too obvious."[32] Ignatius would institute measures to ensure that this did not happen to his order. He argued, "[T]o encourage the unity needed by every religious congregation to stay alive, St. Paul is so insistent that 'all should have one mind and one voice' [1 Cor 1:10]. It is union in thought and will that preserves them."[33] Ignatius was not unusual in this conviction; the early modern view of society was not pluralistic. A society where everybody acted together was considered strong. It could resist being torn apart from within or being overwhelmed from without. On the other hand, as the Society's *Constitutions* noted, diversity of opinions was the mother of discord.[34] Thus Ignatius's insistence upon adherence to uniformity was not a manifestation of tyranny but the articulation of a conviction considered necessary for the survival of any society and for the unity of Christians.

Obedience, however, was not mindless submission. Subordinates in the Society were certainly not denied the opportunity to speak. Ignatius encouraged them to make "representations" to their superior. If they felt the superior was not making the proper decision they could offer advice, provided they remained "indifferent," not preferring their own opinion but willingly accepting whatever the superior decided. Superiors were not to make decisions without consulting both those expert in the matter and those to be affected. Even after his superior had made his decision, a Jesuit could wait and then make another representation if he thought the decision was ill advised.[35]

Furthermore, the Society's respect for the value of experience meant that all measures were to be tested in practice and that suggestions from subordinates were to be sought. A prime example can be seen in the development of the *Ratio studiorum.* Two trial versions were circulated and put into practice in 1586 and 1591. Suggestions from the provinces in response to this trial were then incorporated into the third and final version, which was officially implemented in 1599.

Aristotle and the Society of Jesus

Since the Jesuits acknowledged the need for uniformity and obedience in the Society, they needed to stipulate a body of knowledge that would constitute the basis of this uniformity in its philosophical teaching. From the earliest days of the Society this need was fulfilled by Aristotle. The Society's *Constitutions* specifically stated that Aristotle was to be followed in philosophy.[36] Although a pagan, his works had been Christianized over the preceding centuries since the reintroduction of his corpus to the Latin West in the twelfth and thirteenth centuries and had become the basis of the university curriculum.[37] The Third General Congregation of the Society in 1573 further specified that in philosophy Aristotle had to be taught in such a way as to support scholastic theology.[38] For many Jesuits this was not difficult, since they did not perceive any major conflict between Aristotle's philosophy and the tenets of the Christian faith, as Jakob Gretser's defense of Aristotle in the face of Luther's hostility shows. Gretser wrote that Luther tried to exterminate philosophy because St. Paul had taught that one must be wary of philosophy. But, Gretser responded, St. Paul clearly meant, not Aristotle's philosophy, but only empty, false philosophy that leads away from God. In Christian schools, he assured his readers, any dangerous parts of Aristotle were rejected. In fact, Gretser wrote, in contrast to Luther's claim that innumerable things in Aristotle's *Physics* were inconsistent with divine matters, there were actually very few such passages, and those had come from a misreading of Aristotle. Doctrines such as the mortality of the soul and the eternity of the world, both abhorrent to orthodox Catholic theologians, had been imposed on Aristotle by later commentators.[39]

But the problem was that by the late sixteenth century there were many Aristotles. There were vast numbers of ancient, medieval, and contemporary translations and commentators to choose from, many with conflicting views, not to mention the growing numbers of critics of Aristotle. So vague injunctions such as that of the *Constitutions* to follow Aristotle were not much help for professors in the classroom, who had to decide which of Aristotle and his interpreters' opinions were acceptable. Was Aristotle to be followed literally, or could he be challenged on details while the essentials were maintained? What, then, were the essentials? Or was it enough merely to follow some kind of Aristotelian method? What, then, was this method? Which commentators could be used? Not all were equally acceptable. For example, while Averroes was particularly influential at sixteenth-century Italian universities, many Jesuits considered him to be extremely pernicious.

The Jesuit response to these problems developed throughout the sixteenth century and culminated in the 1599 *Ratio studiorum,* although even after this the issue continued to be contested, as later chapters will show.[40] The 1599 *Ratio studiorum* was considerably more detailed than the *Constitutions* in its instructions: "In matters of any importance let [the professor of philosophy] not depart from Aristotle unless something foreign to the doctrine which academies everywhere approve of; much more so if it is repugnant to the orthodox faith, and if there are any arguments of this or any other philosopher against this, he should strenuously seek to refute them according to the Lateran Council."[41] Aristotelian commentators hostile to Christianity were to be avoided, and the use of Averroes received only grudging permission: "[I]f anything good is to be cited from him, let [the professor] bring it out without praise and, if possible, let him show that he has taken it from someone else."[42] Since the *Ratio* built the philosophical curriculum on the works of Aristotle, the professor of philosophy was instructed to put great effort into interpreting his corpus well. He was to "persuade his students that their philosophy will be but very partial and mutilated unless they highly esteem this study of [Aristotle's] text."[43]

While the 1599 *Ratio* confirmed Aristotle's position as the Society's preeminent authority in philosophy, it was still not particularly detailed; it contained no lists of specific propositions that should or should not be taught, and it did not define exactly the "matters of any importance" in which professors could not depart from Aristotle. Nor did it present in detail measures that would ensure a uniform adherence to Aristotle (or divergence from him when necessary). However, the Society did have such measures, which it developed in a long process of refinement through practice.

Lists of Propositions

Uniformity was a desirable thing, and Aristotle was to provide the foundation upon which it was to rest. The problem facing the Society was how to grant adequate liberty to its scholars while maintaining uniformity in a time of great intellectual foment that produced increasingly large numbers of alternatives to Aristotle on virtually every philosophical question. The desire for uniformity was not simply imposed from above; calls came from provinces asking the hierarchy to impose restrictions on excessive liberty in philosophy and theology. Fernando Perez reported to General Claudio Acquaviva

(1542–1615, general from 1581) from Spain in 1587 that professors were teaching "not only new things but also less safe and received things" and implored "by Christ's guts" that the members' freedom to choose opinions be constrained.[44] But constraining freedom without crushing it would be a difficult matter.

The Society adopted three main methods of fostering uniformity: lists of banned opinions, textbooks, and general guidelines for assessing the acceptability of theological and philosophical opinions. None was exclusive of the others, and very often the three were used in conjunction with differing emphases. The first of these methods was simply to publish a list classifying opinions as either banned or approved. The first of these lists was drafted in response to Averroism. Reports that Benito Pereira (or Pereyra, 1535–1610) had taught Averroist opinions at the Collegio Romano sparked a controversy that lasted for two decades at the college and prompted a rash of measures aimed at controlling the choice of both theological and philosophical opinion in the Society. In 1565 General Francisco Borja published a list of propositions explicitly banning the teaching of certain Averroist opinions, such as the mortality of the soul.[45]

Ledesma, one of Pereira's bitterest opponents in Rome, affirmed the need for at least a list of approved and obligatory opinions in a memorandum to Borja's successor, Everard Mercurian (1574–80, general from 1573), in 1574. Experience showed, Ledesma claimed, that general guidelines on the choice of opinions were by themselves insufficient to maintain uniformity because professors could circumvent them. The responsibility for choice was put on the shoulders of local superiors, who were not always the best judges of such matters, and this led to daily squabbles when professors disagreed with their superiors. Liberty of doctrine, Ledesma stated, was no different from liberty of religion and faith. As the Society had no intention of permitting the latter, why should it even consider allowing the former? The only way to stop people following "impious Arabs" such as Averroes was a thorough list of approved propositions. He concluded that the specific propositions on Borja's 1565 list were the best because these were the propositions at the heart of the disagreements at the Collegio Romano and that they had to be retained to keep such squabbles under control.[46]

Circumstances changed, however. Borja died in 1572, Ledesma in 1575, while Pereira continued as professor of theology at the Collegio Romano. In 1582 the newly elected general, Claudius Acquaviva, asked the professors of the Collegio Romano to evaluate Borja's list.[47] They rejected virtually all the opinions on it without being able to agree on a list to replace it.[48] Acquaviva had asked the professors to suggest how novel opinions in theology and philosophy could be avoided. While many responses referred specifically to theology and St. Thomas, their authors often added that their remarks applied equally well to philosophy and Aristotle. They were in complete disagreement on the question of whether particular philosophical opinions ought to be proscribed. Some of the professors thought a few propositions should be banned if they caused problems, but none of the professors suggested that detailed lists should be maintained.[49]

Robert Bellarmine (1542–1621), the Society's most prominent theologian, recommended that two lists of theological propositions should be drawn up: one of forbidden opinions and one of more or less approved ones. But he reminded the general that Ignatius had not wanted absolute uniformity, only uniformity as far as it was possible.[50] Alfons Salmerón (1515–85), who, as the last survivor of the founding members of the Society, had considerable authority, argued for a large measure of liberty. He wrote that the Society should follow the truth rather than any one man because the latter would set a man up as a god. Rather, the best doctrine should be accepted. A list had been tried, he said referring to Borja's 1565 one, and it did not work. If the Society must have a list, it should be kept as brief as possible to avoid unduly restricting the mind.[51] In 1591 Bellarmine opposed a list of forbidden propositions, writing, perhaps with an eye on General Borja's 1565 list, that it was more dangerous to list propositions wrongly than not to list any at all.[52]

Extensive lists of banned opinions had fallen out of favor, although the congregations of individual provinces still passed local measures and occasionally the general banned a particular proposition on the advice of the Society's revisers general. Although a short list of approved philosophical opinions was included in the 1586 version of the *Ratio studiorum,* it drew such criticism from the provinces, including Germany, that no list was included in the final 1599 version.[53] It was not until 1651, in response to a whole new set of circumstances and threats, that a lengthy list of forbidden philosophical theses again received official sanction and was circulated throughout the whole Society.

Textbooks

A second means that promised the hope of uniformity in philosophical doctrine was the standard textbook, a project required by the *Constitutions* but never executed.[54] Certainly, in keeping with the usual university pedagogy, Jesuits scholars published commentaries on parts of the Aristotelian corpus that professors could draw upon to craft their lectures.

In 1585, however, the professors of the Collegio Romano considered whether the Society should produce a textbook that all of its members would be bound to explicate and defend in the classroom. Although their discussion focused on the feasibility of a theology textbook, the final paragraph of their recommendation stated that their reasoning was also valid for philosophy, as it formed the foundation of theology. Considering the numerous advantages and disadvantages of producing such a text, the professors first suggested potential difficulties. They acknowledged that a textbook could become a "harsh yoke" for future teachers, "especially since neither opinions nor the method of teaching agree at all times and places."[55] Limiting the Society's professors to one accepted set of opinions and forbidding new ones could limit the effectiveness of the Society's intellectual and ministerial efforts: "Why should access to new inventions be denied to our members, with which they could illuminate the Church, as do the doctors of other orders? Especially as many parts of theology that have not yet been raised to perfection could firmly cut down the generation of new heresies if they were gradually perfected."[56] Also, if they could not alter the textbook, teachers would become lazy. They would cease to consult the sources themselves, and the Society would soon be without "excellent theologians and inventors of enlightening things."[57] The Roman professors implicitly recognized that the sciences were not static bodies of knowledge but constantly changed to meet the demands of the time. "Fixing" knowledge in a text failed to acknowledge the dynamic nature of the sciences.

But despite these significant problems, the professors unanimously agreed to recommend to the general that a commentary on St. Thomas be published at Rome by seven "excellent" theologians. All members would be bound to explicate and follow it. The reasons in its favor, they claimed, far outweighed the reasons against. First, it would preserve uniformity, inherently a worthwhile aim, as the *Constitutions* stated. Not only should the same opinions be taught, but they should be taught in the same way. A textbook would also provide many pedagogical benefits. Students learned primarily not by writing down lectures but by engaging in frequent exercises such as repetitions, disputations, and interrogations. A textbook would allow the professor to concentrate on such activities rather than on preparing classes. It would also preserve the health of both the students and teachers. In addition, students could refer to the textbook for those questions the professor did not cover in class, and professors could study it for material with which they were unfamiliar and to refute specific objections. Furthermore, such a book would stop the spiritual dangers that arose when people immersed themselves in theology and philosophy, neglecting piety and duty; the older orders had declined through a lack not of learning but of piety. A textbook would liberate Jesuits from the distraction of excessive study.[58]

The Roman professors maintained that such a textbook would still allow for some diversity. The textbook would not be the opinion of any one professor, since it would be written by seven professors, who would not have to agree on every point. Professors could disagree with opinions expressed in the textbook, but they could not teach their own opinions. After all, the Society would not make them teach absurd opinions. Nonetheless, there would still be variations among professors, for while they would not be permitted to differ in important matters, they would vary in "memory, use of Latin, readiness, sharpness, affability, clarity." Some would use many arguments, others few.[59] In sum, the professors of the Collegio Romano regarded textbooks as the best way to negotiate the compromise between uniformity and liberty, ensuring the former without unduly restricting the latter.

Despite the Roman professors' recommendation, the Society never did commission or adopt one standard philosophical textbook. Rather, its members wrote many textbooks, which were widely used both inside and outside the Society.[60] Nevertheless, there were some early signs that in Germany textbooks would achieve what the Collegio Romano professors had intended them to do. In May 1575 Adrian Loeffius (1520–?), the rector of the Jesuit college at the University of Trier in the Rhenish province, wrote to General Mercurian, reporting that the philosophy professors had begun to use Francisco Toledo's logic and physics textbooks with much success.[61] He noted that the textbooks saved the professors much dictation and the students much writing. Great license in teaching and defending various opinions had been suppressed, in marked contrast to the squabbles that had resulted from different things being taught in successive years.[62] This was exactly what the professors of the Collegio Romano had wanted from textbooks. Yet only two years later the rector of the Jesuit college at the University of Mainz, Georg Bader (d. 1612), told the general that the professors at Trier had returned to the harmful practice of dictating from their own notes.[63] Bader lamented the change but provided no explanation. Although there is no record of the exact reason why the professors in Trier found Toledo's texts unsatisfactory, it is possible to identify the sources of dissatisfaction with textbooks more generally.

In 1602 the general sent a number of instructions to the provinces to assess the progress of the implementation of the 1599 *Ratio studiorum* throughout the Society. To this end, visitors of studies were appointed to examine the schools in their province. One of their tasks was to consult with professors on the question of which authors were to be taught in order to limit philosophy professors' "liberty and license in opinions" and prevent the spread of dangerous opinions.[64] Pedro Ximénez (1554–1633), the visitor of studies for the Austrian province, found that the professors at the college in Graz had begun to use the logic textbook of the Jesuit Pedro da Fonseca but had later returned to the old custom of explicating the *Isagoge* of the Neoplatonic philosopher Porphyry (c. 234–c. 305) because they felt that it was the most suitable text.[65] In fact, the visitor reported to the general that most of the professors at Graz, Vienna, and Olmuc did not think that there was any textbook fit to be read that covered the whole curriculum.[66]

When the provincial of Austria summoned a committee to consider the issue, its report revealed that the misgivings of the professors at the Collegio Romano about textbooks in 1582 had been justified. The Austrians acknowledged the benefits of a textbook: it spared the students the labor of writing, it helped both advanced and new students, and it "restrained the desire for new opinions."[67] But they identified numerous pedagogical problems that textbooks created. Students became bored and professors lazy. Teachers needed to write their own lectures in order to comprehend the material properly and be able to defend sound doctrine. Also, particular textbooks had their own weaknesses. Toledo had written too long ago and did not consider more recent authors. The Conimbricenses were also unsatisfactory,[68] and no one could be found within the province who could write a text that could withstand the professors' scrutiny.[69] Furthermore, the textbook was not an effective method of limiting excessive liberty of opinion. Professors could still disagree with the text if the prefect of studies allowed, and this led to constant arguments between professors and prefects, especially as prefects were often no more knowledgeable than professors.[70]

On the other hand, the report acknowledged that too much uniformity was undesirable; it was possible to stand on opposite sides of many philosophical opinions without showing any sign of novelty. God had intended some things to be left uncertain, so why should one constrain the human mind by leaving no freedom to philosophize? Therefore, the Austrian professors recommended that if the Society wanted to limit new opinions, it should produce a catalogue of opinions to be avoided. If a textbook were used, then the Society should also produce a list of opinions in which it was permitted to depart from the text rather than leave it up to individual professors and prefects to decide.

The problems predicted by the Roman professors in 1582 had occurred in actual practice: no textbook could ever be completely satisfactory. There were too many questions on which disagreement was possible without harming the faith, and professors would rather explore these questions and teach their own opinions than adhere to any single view. The Austrians were certainly not advocating complete freedom to philosophize, but to them a list of illicit propositions seemed more effective than a textbook in maintaining the balance between liberty and uniformity. Rather than assuming that everything was banned except those things in the textbook, it seemed better to assume that everything was permitted except those things on the list. The discussion had virtually come full circle back to 1565, when General Borja had drawn up the first list of forbidden propositions.

The Jesuits produced many philosophy textbooks that their professors drew upon for their lectures, but never did a single textbook achieve canonical status. When the Seventh General Congregation of 1615 was asked to consider the issue of standard theology and philosophy textbooks, it deferred making any decision on the matter.[71] Jesuit professors continued to prefer to write their own lectures, as the large numbers of lecture manuscripts scattered in libraries throughout Europe testifies.

General Guidelines

A third method in the quest for uniformity was provided by general guidelines on the choice of theological and philosophical opinions. Like lists of propositions and textbooks, the issue of guidelines was long debated in the Society. Prompted by the controversy over the teaching of Averroist doctrines at the Collegio Romano, General Borja codified twenty-five years of discussion on the matter.[72] Along with his 1565 list of banned opinions, he supplied five criteria for choosing opinions. First, nothing could be taught that harmed the faith; second, nothing could be taught against the "received axioms of the philosophers," such as the four types of causes, the four elements, and the three principles of natural bodies; third, nothing could contradict the most common opinion of the philosophers and theologians; fourth, nothing could be defended against the common opinion unless the prefect of studies was first consulted; and fifth, nothing new could be introduced into philosophy or theology unless the superior or prefect was first consulted.[73] Although later commentators rightly noted that the criteria were repetitive and could be edited, they were to remain in some form or other a standard part of the decrees issued by the Society's hierarchy over the next two hundred years.

The Society explicitly acknowledged that "most common" meant there was potential for change as opinions gained or lost widespread acceptance. Therefore it was important to determine what was actually meant by "most common"—in particular, most common among whom? This was open to interpretation. "Most common" was originally intended to mean that sense accepted by the majority of scholastic doctors such as Aquinas.[74] But this was not self-evident to all Jesuits, including Paul Hoffaeus (1524–1608), Canisius's successor as provincial of Upper Germany, who wrote to General Mercurian in 1578 asking for clarification on what Borja had meant by "common opinion." He asked Mercurian to exercise moderation in defining it, since the guidelines implied that anything outside the more common opinion was forbidden.[75] In the following century some Jesuits came to see "more common opinion of the philosophers" as referring not to scholastic doctors but to contemporary philosophers, many of whom were not Aristotelians at all.

Despite these problems, Borja's general guidelines were remarkably long-lived and mostly preferred to lists of particular banned or approved opinions. While General Acquaviva dropped Borja's list of banned propositions in 1582 at the recommendation of the Collegio Romano professors, he explicitly reconfirmed Borja's guidelines on the choice of opinion.[76] The following year Acquaviva wrote to Georg Bader, the new provincial of the Upper German province, that the Society would not be formulating a catalogue of banned opinions, since "such a catalogue would be nearly infinite and certainly useless." Rather, a "general formula" would be circulated to guide local superiors in evaluating individual opinions.[77]

Local Practice: Hoffaeus and the Canisius Brothers

Despite all the measures implemented to maintain a balance between the two conflicting needs of uniformity and academic liberty, often the degree of liberty granted to professors in the classroom depended largely on local circumstances and the personal foibles of individual superiors. Some superiors exercised close supervision over their subordinates. A superior with his own philosophical and theological commitments could make life very unpleasant for a professor who insisted on teaching opinions that differed from the superior's. Other superiors allowed considerably more freedom.

An example of differing styles of leadership can be seen in Germany in a comparison of the Canisius brothers and Paul Hoffaeus. The effects of repressive supervision were reported by Antonio Balduine, a young logic professor who arrived at the Jesuit university in Dillingen from Rome soon after the controversies over Averroism in Rome and Dillingen mentioned earlier. Benito Pereira, the Collegio Romano professor who had been accused of spreading Averroism, had been Balduine's professor. Balduine complained to the general in 1570 that Rector Theodoric Canisius would not let the professors give any lectures without their first being vetted by the prefect of studies. For Balduine the situation was particularly intolerable: "He seems particularly annoyed at Father Benedict Pereira, my teacher, who, he recently said, was the author of all apostasies and atheists in the Society. And he said that he detested his doctrine, because of which I was also suspect." Canisius wanted Balduine to write all of his lectures for the coming year in advance so that he could check that Balduine would not teach any of Pereira's opinions. The young professor wondered how this was possible and complained: "From this Your Reverence can gather how much confusion ensues, if any rector according to his own will proscribes to his lecturers those conclusions that are to be held in the schools and that they are to approve of. And what is worse, he does all this and things like it without the advice of the consulters. He has brought it about that throughout this province Benito is considered to be a curse and all his students are suspect."[78] Balduine asked that the general rein Canisius in and restore Pereira's reputation. Balduine was, however, mistaken in his claim that Canisius was acting according to his own will and without advice. Canisius had the support both of the provincial, who was his brother Peter, and of the provincial congregation, which had just asked the general to act against Averroism throughout the Society.

Such perennial controversies prompted Paul Hoffaeus, Peter Canisius's successor as provincial of Upper Germany, to write to General Mercurian in May 1578, asking that the liberty of professors not be excessively restricted. While he granted that the *Constitutions* forbade "dissonant opinions," to maintain this stricture would be impossible and counterproductive.[79] The mind should be given room to exercise, as long as it stayed within the limits of faith and morals, for it was the nature of good minds to say novel things. However, he believed, good minds were frightened away from discussing novelties if they constantly had to defend themselves against the charge of novelty. Furthermore, as the teachers at the Society's schools would not have the same opinion for perpetuity, any attempt to stop differences of opinion would be in vain. Two months later he wrote again, imploring that "in speculative matters in which it is possible to speak without danger to the faith and without scandal, . . . an honest liberty be granted to professors, lest we cheat them of their talents."[80]

Hoffaeus and the Canisius brothers appear to have had very different views on philosophical liberty, but one should be wary of painting the contrast too strongly. Peter Canisius regarded Hoffaeus highly and saw him as a worthy successor to the position of provincial.[81] Hoffaeus was himself prepared to curtail too much novelty. At the urging of the general, Hoffaeus wrote a circular to his province in 1577, the year before the letter cited above, reminding censors to be alert for novel opinions. All academic theses had to be checked before publication, and any novelties were to be sent to Rome, where their utility and necessity were to be checked. One of the four censors Hoffaeus appointed to this task was in fact Theodoric Canisius.[82] From what we have seen of Theodoric, it is not surprising that he took his job seriously. He soon complained that members were skirting the censorship regulations.[83] Perhaps it was his zealotry that provoked Hoffaeus's letter the next year to Mercurian, asking for greater leniency in the interpretation of uniformity.

These men embody the two poles that marked the range of Jesuit attitudes to the issue of liberty and uniformity. But both ends of the spectrum agreed on much:

philosophy must be in accordance with the Catholic faith, and it must be taught so as to prepare the student for scholastic theology. Moreover, both sides agreed that Aristotle was the preeminent but not infallible and sole authority on matters of philosophy; Canisius opposed Pereira and his students not because they departed from Aristotle but because he saw their philosophy as leading away from the Catholic faith.

Diego Ledesma presciently summed up the dilemma when he wrote in 1564–65 that "in teaching philosophy a twofold abuse should be avoided: first, too much liberty, which indeed harms the faith, as experience shows in the academies of Italy; second, that people are bound to the doctrine of just one or another author. This has produced hateful and contemptible things in Italy."[84] By the end of the sixteenth century, the Jesuits had formed the consensus that Aristotle was to be their preeminent authority in philosophy but that because Aristotle was not infallible Jesuit philosophers were not to be bound to his authority alone. The Society of Jesus had developed various techniques to assist Jesuit philosophers in navigating the path between excessive adherence to Aristotle and excessive liberty in the choice of opinions. However, in the early seventeenth century, this attitude toward Aristotle and uniformity underwent a marked change.

A Shift in the Balance

Toward the end of the long generalate of Claudius Acquaviva, which lasted from 1581 until 1615, the balance between uniformity and liberty in philosophy swung increasingly in favor of uniformity. At the beginning of his generalate, Acquaviva actually moderated the Society's emphasis on Aristotle, merely confirming Borja's very general rules governing the choice of opinion.[85] This was in keeping with Acquaviva's conviction that the spiritual life of the company required, not further rules and prescriptions, but a better understanding and assimilation of the spirit of the Society as formed by the Society's foundational documents.[86]

However, in the early seventeenth century, this climate changed. Despite the implementation of the final version of the *Ratio studiorum* throughout the Society, letters were still reaching Rome from the provinces complaining about excessive license in philosophy that created divisiveness. Also, Acquaviva and the general of the Dominican order had both been chided by Pope Paul V in 1607 for their orders' involvement in a long and often unseemly squabble over the role of grace in salvation. As his theologians had decided that a range of opinions was permissible in the matter, the pope instructed both orders to show more decorum and to stop their mutual accusations of heresy.[87] The very thing that the policy of obedience and uniformity was designed to prevent had come to pass: an embarrassing reprimand from the pope. And in 1610 the Jesuit theologian Leonard Lessius threatened to stir the matter up again with a new work on the controversial issue.[88]

In May 1611 Acquaviva wrote to the Society's local superiors, expressing his concern over the lack of uniformity in the choice of opinions. It was not enough merely to express the same conclusions, he wrote. Rather, they had to be reached in the same way. Firm boundaries had to be set; otherwise, new dangers would constantly arise. Although Acquaviva was concerned primarily with theological questions and did not specifically address philosophy, his letter was expressed in broad terms encompassing all higher studies. The extent of the problem was made clear by the flood of responses from the provinces suggesting remedies.[89]

Two years later, in December 1613, Acquaviva circulated another letter on uniformity of doctrine. Again, his chief concern was with theology, in particular scholastic theology, yet in this letter he specifically mentioned philosophy, stating that "the provincial should watch diligently and ensure that opinions taught in philosophy are subordinate to theology and that our philosophers follow one Aristotle, wherever his doctrine does not dissent from the Catholic truth." This was a very conservative interpretation of the *Ratio studiorum* that departed from the tradition dating back to Borja's 1565 criteria and even beyond. Rather than recommending the "more common" opinion of the doctors who interpreted Aristotle, Acquaviva wrote that Aristotle was to be followed in all philosophical matters.[90] Whether Acquaviva was aware that he thereby exceeded all previous rulings and raised the authority of Aristotle to hitherto unmatched heights is unclear. But his instruction severely restricted the potential for change in philosophy that was built into the older view. That his ruling on philosophy was understood in this way in the German provinces can be seen from the 1614 response of Heinrich Scheren, provincial of the Rhenish province, to several queries regarding Acquaviva's letter posed by a college rector. Scheren explained that "the letter decrees that in merely physical matters Aristotle is to be followed wherever he teaches nothing contrary to the faith."[91]

Although he had written at the start of his office that general guidelines made a list of specific propositions unnecessary, in 1613 Acquaviva circulated a lengthy list of theological propositions taken from Thomas's *Summa theologiae* that had to be taught or, since various scholastic doctors disagreed with Thomas, on which a range of opinion was possible.[92] However, Acquaviva did not go so far as to publish a list of approved or prohibited philosophical opinions, stating in his 1611 and 1613 letters that the existing rules, such as the *Ratio studiorum* and the decrees of the general congregations, were sufficient to maintain uniformity.

Around the time that Acquaviva was attempting to assert some control over the liberty permitted to Jesuit natural philosophers and mathematicians, the range of opinions available in their sciences was blossoming, and many Jesuits held distinctly un-Aristotelian opinions. This was particularly the case in the area of astronomy. During the first decades of the seventeenth century, many Jesuit mathematicians and, to a lesser extent, natural philosophers were enthusiastically adopting many of the developments occurring around them. This created problems for received Aristotelian views, but these Jesuit scholars did not abandon Aristotle wholesale. Rather, they generally sought to integrate recent developments into a basically Aristotelian framework. In adopting a moderate path, they were prepared to question Aristotelian opinions such as the immutability of the heavens, but they maintained a firm commitment to geocentrism. At the same time, some Jesuit scholars called for greater freedom in intellectual questions, particularly where no matters of faith were at stake.[93] The mechanisms designed to create uniformity were coming under greater pressure than ever before.

CHAPTER 2

Censorship and Its Limits

MEMBERS OF THE SOCIETY OF JESUS TAUGHT AT HUNDREDS OF COLLEGES around the world. This mission included publishing works on all branches of philosophy. But these were a small part of a much larger publishing enterprise. From its first days the Society placed great emphasis on publishing, and in many ways it was a community of scholars and authors. The thousands of works listed in the nine volumes of de Backer et al.'s *Bibliothèque de la Compagnie de Jesus* attest to its fecundity. Jesuits wrote in a stunning range of fields, genres, and languages for audiences of all ranks and regions. They came from many different nationalities and cultural and intellectual traditions. Not surprisingly, keeping order in this vast undertaking was no easy task. The Society of Jesus established elaborate and extensive mechanisms to cultivate uniformity in teaching and publication. However, if these mechanisms—the culture of obedience, lists of propositions, textbooks, and guidelines governing the choice of opinions—ultimately failed to guarantee uniformity on issues for which the Society deemed it necessary, then a further mechanism was needed, namely censorship in the narrower sense: the review and correction of all books prior to publication.

Censorship played an extremely important role in the practice of natural philosophy and mathematics in the Society throughout the seventeenth century and indeed well into the eighteenth. It clearly placed limitations on what Jesuits could publish in some areas of natural philosophy. Numerous Jesuit philosophers and mathematicians encountered difficulties with their censors and were forced to rewrite important sections of their texts against their wishes. Despite this, censorship, like the other measures designed to foster uniformity, did not create complete uniformity. Considerable space always remained in the system for authors to express opinions that the Society had forbidden.

This chapter will examine how censorship functioned within the Society. It will consider what censorship was meant to do and how effectively it fulfilled its function. It will also examine whether Jesuit natural philosophers and mathematicians saw the strictures of censorship as overly constraining and, if so, how successfully they could circumvent such constraints.

The Workings of Censorship

The Society did not suddenly create the mechanisms of censorship in the first decades of the seventeenth century in response to the rise of the New Science. Censorship was required by the *Constitutions* and was practiced from the Society's earliest days.[1] General Claudio Acquaviva codified the rules governing censorship in 1601 and established a committee of five revisers general, with one from each of the five assistancies at the Society's flagship college, the Collegio Romano.[2] The revisers general were to consider queries from the provinces concerning the admissibility of theological and philosophical opinions. Furthermore, all manuscripts intended for publication were to be sent to Rome for the revisers' approval after first being approved in the author's home province by three censors or revisers.[3] As it became increasingly unworkable to send all manuscripts to Rome, works were read by provincial censors who sent only their reports to Rome where a decision would be made on the basis of those reports. If necessary, the revisers general could have the manuscript sent to Rome for further examination.[4]

Ensuring uniformity and orthodoxy in theology and philosophy was not the only reason for establishing reviews of texts; another vital concern was to protect the reputation of the Society and avoid scandal. Nowhere was this of greater concern than in Germany. Since Peter Canisius's earliest efforts to shore up the Catholic faith in the German lands, the Jesuits had been anxious to use the medium of print to refute the errors of the Protestants and spread correct doctrine. But they sought to avoid unseemly quarrels with both "heretics" and other Catholics.[5] Works that could embarrass and alienate princes and the nobility were very closely scrutinized. General Muzio Vitelleschi (1563–1645, general from 1615) admonished censors not to approve anything that seemed to criticize the princes or indeed anything that dealt with the state of nations at all.[6] He also ruled in 1632 that nothing was to be published that would bring "harm to princes, prelates, religious, or any kingdoms or nations."[7] Historical studies and chronologies could easily raise embarrassing questions about the lineage of noble houses and their dynastic claims, so great care was taken in Germany lest princes and nobles be offended.[8]

Reviews also served the important function of ensuring that only good books were published. Revisers were to examine all aspects of the book: that its style and spelling were correct; that it did not stoop to flattery, bragging, or insults; and that the author was well versed in his subject. The book had to be useful and of benefit to the Society's members; if it merely repeated what others had written on the subject, it could not be published.[9] Local and Roman revisers' reports often commented on all of these factors, offering suggestions on how the book could be improved; spelling mistakes, errors in mathematical calculations, inaccurate quotation or attribution of sources, and weak reasoning were all corrected. Manuscripts could be and were rejected for being inadequate on any of these grounds, although more commonly the revisers granted qualified approval; once the corrections had been carried out, the book could be published. This constructive aspect was an important part of Jesuit censorship that should not be ignored; however, of principal interest here is the role of censorship in maintaining uniformity in natural philosophy and mathematics.

Acquaviva's 1611 and 1613 decrees requiring adherence to Aristotle were reflected in the revisers' rulings. In 1614 the Jesuit mathematician Giuseppe Biancani (1566–1624) attempted to publish a work examining those passages of Aristotle's works that dealt with mathematics. One of the revisers, Giovanni Camerota (1559–1644), was unhappy with Biancani's criticism of Aristotle, in particular his support for Tycho's views on the mutability and the fluidity of the heavens and the motion of the planets by their own motive forces.[10] On Biancani's treatment of floating bodies, the censor wrote: "The addition to Biancani's book about floating bodies moving in water should not be published, since it is an attack on Aristotle and not an explanation of him (as the title indicates). Neither the conclusions nor the arguments to prove it are due to the author but to Galileo. And it is enough that they can be read in Galileo's writings. It does not seem to be either proper or useful for our members' books to contain the ideas of Galileo, especially when they are contrary to Aristotle."[11] Biancani's text was unsatisfactory because it was an attack on Aristotle. Camerota made no comment concerning the correctness of Aristotle's opinions; he simply assumed that Aristotle was not to be criticized. This was indeed the logical extension of Acquaviva's 1613 letter. In the published version, Biancani toned down his criticisms and contented himself with merely reporting modern views without granting them explicit approval. This was a strategy widely used by Jesuit natural philosophers and mathematicians, but one that had to be used with precision lest the revisers perceive that the author actually approved of the opinion he claimed he was merely reporting.

Camerota's criticism of Biancani was symptomatic of changes occurring within the Society. While the Jesuits were among the first to acknowledge Galileo's telescopic discoveries, they became increasingly wary of the cosmological implications Galileo drew from them. With the placement in 1616 of Copernicus's and other heliocentrists' works on the Index of Prohibited Books, no Jesuit could overtly support heliocentrism. While the role of the Jesuits in the trial and condemnation of Galileo is still disputed, by the time of his trial in 1633, Galileo had been abandoned by his former supporters in the Society.[12]

The *Ordinatio* of 1651

In the first decades of the seventeenth century, the balance constantly renegotiated throughout the first sixty years of the Society's existence between allowing liberty and enforcing uniformity in philosophy was clearly beginning to swing away from liberty. Anything that was novel, in the sense of departing from Aristotle, or at least what the general or the revisers felt was the correct interpretation of Aristotle, earned their suspicion. Acquaviva's successor, Muzio Vitelleschi, reaffirmed his commitment to the mechanisms of censorship and also insisted that Aristotle's opinions were to be followed.[13] Furthermore, the adherence to Aristotle and the medieval scholastics that Camerota expected of Biancani was consistently imposed by later revisers general. They all ruled that philosophical opinions that dissented from Aristotle and the common opinion of the schoolmen could not be taught in the Society's schools, even if they had no bearing on theological matters.[14] Although many in the Society had recognized in the late sixteenth century the futility of establishing a list of banned opinions, the revisers general repeatedly warned the general of a crisis in the Society's school and recommended the implementation of such a list.[15] The generals and revisers were not the only ones concerned with the spread of novelties; many Jesuits throughout the provinces were disconcerted by too much openness toward novelties and complained to Rome. They sent to the revisers general examples of dubious opinions from the whole of the natural philosophical curriculum that were being taught in schools.

Novel opinions concerning the possibility of the vacuum, the motion of bodies, the functioning of perception, and the composition of the heavens were among the numerous natural philosophical opinions that came before the revisers general. But the crucial area in which dissent could not be tolerated was the physics of the Eucharist—that is, those physical concepts used to explain the transubstantiation of the bread and wine into the body and blood of Christ with only the species of the bread and wine remaining.[16] This question will be explored in considerable detail in chapter 4. Sufficient for the present discussion is that the physics of the Eucharist touched upon a large number of philosophical opinions from many parts of the curriculum. For example, any opinions that attempted to situate the substance of matter primarily in quantity or extension, that stated that the continuum was not infinitely divisible but composed of indivisible points, or that equated substance with matter, quantity, or extension were judged by the revisers general to run the risk of undermining the orthodox account of the physics of transubstantiation.

In 1648, perhaps influenced by the revisers' constant warnings of crisis, General Vincent Carrafa (1585–1649, general from 1646) ordered them to begin to solicit opinion from the provinces for a potential list of banned theses, but before he could take any action he died.[17] When the Society's Ninth General Congregation assembled in Rome in late 1649 to elect the new general and discuss the most pressing issues confronting the Society, the issue of uniformity in studies was among them. Earlier that year, the Collegio Romano had been the scene of a bitter dispute when the revisers general sought to prevent Sforza Pallavicino, one of the Society's most learned members, from publishing a number of theological works. In response Pallavicino included in his treatise *Vindicationes Societatis Jesu* his view of the ideal cultural and intellectual environment and then presented copies of the work to the congregation's delegates.[18] The revisers general also warned the congregation of a crisis, claiming that they were being ignored and that Aristotle was being laughed out of the Society.[19]

The congregation received numerous queries or *postulata* from provincial congregations concerning scholarly matters that put paid to any stereotypes about monolithic conformity and obedience within the Society of Jesus. While the provinces were concerned about professors who taught dangerous opinions, they were just as concerned about pedagogical effectiveness, lamenting professors who wasted time in endless disputes about useless questions and never got through all the required material. Professors argued with each other and undermined the authority of the prefects of studies.[20] The congregation instructed one of its "subcommittees," the Deputatio pro Studiis, to examine the complaints. The Deputatio's response was curiously ambiguous.[21] It stated that no new measures were needed beyond enforcement of the existing provisions in the *Constitutions* and the *Ratio studiorum,* yet it also ignored the Society's traditional aversion toward lists of banned opinions and produced an extensive list of philosophical and theological opinions that could not be taught in Jesuit schools or proposed in theses or disputations. The opinions on the list were taken largely from the rulings of the revisers general on the opinions sent from the provinces in 1648 at Carrafa's request.[22] It was promulgated under Carrafa's successor, General Francesco Piccolomini, as part of the *Ordinatio pro studiis superioribus* (1651), which reaffirmed the Society's commitment to maintaining the "solidity and uniformity of doctrine."[23]

Clearly, not every possible question in philosophy was covered by the list; for such cases the Ordinatio recalled the guidelines governing the choice of opinion that had existed since Borja's 1565 decree: the preeminence of Aristotle and Thomas, adherence to the most common opinion, suspicion toward novelties. The list of sixty-five philosophical opinions paid particular attention to those that endangered the physics of the Eucharist, such as atomism and the identity of matter and quantity. But many opinions on the list did not have an immediately obvious theological connection; rather—and this was not explicitly stated in the Ordinatio, but it is clear from the revisers' records from which the opinions were taken—they differed from Aristotle or from the more common opinion of the schools. As it was to attract considerable controversy over the next century, proposition 41 is particularly noteworthy: "Heaviness and lightness do not differ in species, but only in regard to more or less." That is, one could not teach as true that lightness was a relative quality and merely "less" heaviness. As with all the opinions included on the list, the Ordinatio did not attempt to justify its inclusion or specify what was to be taught in its place, but the peripatetic view was that heaviness and lightness were distinct qualities; lightness existed absolutely, not relatively.[24]

The Ordinatio may appear to mark the triumph of the conservative party within the Society over those who favored a more open integration of novelties into Jesuit natural philosophy. Yet there was still room for change in the system. For example, the guidelines governing the choice of opinion had always allowed for the possibility that a previously unacceptable opinion could be taught if it became the more common. For example, on the fluidity of the heavens, an opinion that contradicted Aristotle's theory of solid spheres, General Vitelleschi had written in 1631 that "[i]f at some time that opinion, put forth by others, becomes common and receives the approval of the majority, then we will easily allow our members to follow what is seen to be more probable in that matter."[25] This was in fact what happened, and the doctrine of the fluidity of the heavens was not included in the Ordinatio. This avenue for change was left open by the 1651 Ordinatio's reaffirmation that previously unacceptable opinions could later be approved. The revisers responded in 1674 to the query whether it was permitted to teach that rarefaction and condensation occurred through the intromission and emission of corpuscles. They wrote that regardless of the truth or falsity of the opinion, or whether it was in conformity with Aristotle, it might now be taught, "since now it seems to be commonly taught by sufficiently many people."[26]

The Ordinatio of 1651 did not end the teaching and writing of natural philosophy in the Society of Jesus anymore than the trial of Galileo had eighteen years earlier, but it did pose problems for Jesuit natural philosophers. To assess the true impact on natural philosophy of the various measures that together created the apparatus of censorship, it is necessary to look beyond normative statements and study closely the actual practices of censorship in the Society. In fact, censorship was contested at many levels within the Society, and it was still possible for professors to discuss virtually any opinion in natural philosophy.

Interpreting and Applying the Ordinatio

It would be misleading to suggest that all Jesuits were unhappy with the Ordinatio and that the rules of censorship were imposed by a "reactionary" center on an unwilling periphery; not only the revisers but also members throughout the provinces wanted greater uniformity. Furthermore, not all Jesuits were attracted to the changes occurring in mathematics and physics in the seventeenth century. In fact, many Physics professors, educated entirely within the Society, had had little exposure to them. There were virtually no "professional" Physics professors in the Society in Germany until well into the eighteenth century. Rather, philosophy professors were junior faculty who generally taught the three-year philosophical curriculum only once or twice before moving on to the theological faculty, the missions, or other duties. Therefore they had little chance to investigate the subject matter in much depth beyond what they had learned themselves as students or had taken from standard textbooks. They had little cause to disagree with the Ordinatio or with their more senior colleagues who occupied the office of revisers or prefects of studies.

However, many in the Society were dissatisfied with the Ordinatio. Among them were members who had adopted a more open position toward novel doctrines before 1651. These scholars attempted to offer an alternative reading of the Ordinatio that would create enough room for them to continue their work. There was sufficient ambiguity within the Ordinatio to allow them to make cogent criticisms of it. Even before the Ordinatio was promulgated, the problems with lists that had been identified in the 1580s resurfaced. Pierre Le Cazre (1589–1664), provincial of Champagne, who had been at the Ninth General Congregation but not on the committee that drew up the list of propositions on the Ordinatio, voiced dissent in a memorandum in which he pointed out the contradictions and inadequacies of the proposed list.[27] Le Cazre was well informed on developments in natural philosophy: for example, he had written a text on the fall of natural bodies and letters to Gassendi questioning both his and Galileo's theories on the subject.[28]

Le Cazre's view of the criteria governing selection of opinions was fundamentally different from that of the authors of the Ordinatio. Rather than taking conformity to Aristotle as the chief yardstick, he suggested that some of the banned opinions were supported by "many experiences" [multis experientiis].[29] Furthermore, while Le Cazre used one of Borja's 1565 criteria, namely whether an opinion was the most common, he was using "most common" in a different sense from Borja's. For Le Cazre, the most common opinion was that favored not by a majority of scholastic doctors but by contemporary natural philosophers, including those who were not Aristotelians. Many of the opinions placed on the banned list, he claimed, were in fact the more common opinion or were necessary corollaries of the more common opinion. Of proposition 41 he wrote: "Thus far, it is the more common opinion that lightness is nothing other than lesser heaviness, and experiments [experimentiae] are very convincing on this."[30] Thus, by appealing to a very different source of authority in natural philosophy and a different community of natural philosophers, Le Cazre offered judgments on the acceptability of several opinions completely at odds with those of the Ordinatio. None of Le Cazre's recommendations were adopted in the final version of the Ordinatio, but the questioning of both the content of the list and the criteria it was based upon by a respected provincial shows that the foundations of the Society's system of censorship were contested at a senior level.

Another area of contestation was the question of what it actually meant for a proposition to be forbidden. Since the Ordinatio offered no explanation for why a particular proposition was placed on the list, the intent of the prohibition was not always clear. Many propositions consisted of several sentences or clauses, but it was not apparent whether the whole proposition was banned or merely particular parts of it. In addition, if a proposition was banned, did this mean that its opposite had to be taught, and if this was the case, what exactly was the opposite? Another difficulty was raised by the passage of time; many of the propositions on the list were placed there to settle arguments over points of scholastic philosophy, but as Jesuit natural philosophy moved away from its scholastic and peripatetic origins over the course of the eighteenth century, the Jesuits' familiarity with those debates declined, and the sense of the prohibition became increasingly opaque. Furthermore, as Le Cazre pointed out in his objections, while the list included several propositions that were the consequence of other propositions, the propositions from which they were derived were not banned, which was quite inconsistent.[31] Thus there was ample room for disagreement over how the Ordinatio should be interpreted and applied.

An incident involving Athanasius Kircher's (1602–80) *Iter exstaticum* reveals how two German Jesuits believed the Ordinatio should be read. In 1656 Kircher published in Rome with the Society's imprimatur an account of an imaginary journey through the cosmos, in which he discoursed on its nature and structure. Kircher sent several copies to his friend and former pupil Kaspar Schott (1608–66) in Würzburg, where it was very favorably received by the Jesuits, particularly by Melchior Cornaeus (1598–1665), the college's rector, who was working on a philosophical textbook.[32] Schott undertook to publish a second edition of the work in Würzburg so that it would be available to audiences in Germany.[33] However, before the Würzburg edition appeared, the original work was subjected to an anonymous attack that Kircher forwarded to Schott. Its author was very familiar with the Ordinatio, for he cited three propositions from that list that he believed Kircher violated. The critic also claimed that Kircher made six assertions against the faith. In the Würzburg edition, Schott reprinted the critique but included refutations of its charges by himself and Cornaeus.[34]

After responding to the individual charges, Schott concluded that "for any of our authors or doctors to be considered to have sinned against the said Ordinations, it does not suffice that from his sayings some prohibited proposition seems to be deduced, even by means of a probable consequence, but he must have taught it either in the terms themselves [of the Ordinatio] or certainly in something so connected with them that the consequence cannot be avoided by any probable reasoning."[35] Schott's argument is clear: even if a forbidden proposition could be deduced from an opinion expressed by a Jesuit author, this was not sufficient grounds to censor it. For an opinion to be classified as unacceptable, the connection had to be explicit. Schott saw the danger of allowing opinions distinct from but connected to those on the list to be censored and sought to preserve room to speak in natural philosophy.

Cornaeus also appealed for a moderate application of the Ordinatio, writing that "the words of the Ecstatic Author [Kircher] should be interpreted benignly and should not be censored so rigidly." It was unfair to cite single sentences out of context, especially when Kircher explained himself at length elsewhere in the work. Nor should one read the work in a sense other than that in which it was meant. In response to the critic's claim that Kircher assigned nature a soul and said it was alive, Cornaeus wrote, "The locution of the Ecstatic Author is clearly tropical and metaphorical, from which you cannot rightly infer anything. Since his real opinion is well established, what is the point of calling his words into disrepute?"[36] Like Schott, Cornaeus tried to open up space for authors; the censors should restrict themselves to judging the fundamental sense of a work and not take isolated statements out of context.[37]

The Probable and the Hypothetical

Regardless of how Schott, Cornaeus, or any other Jesuit authors thought the Ordinatio should be interpreted and applied, they still had to submit their works to the provincial censors or even the Roman revisers if they wanted the Society's imprimatur. The effectiveness of censorship depended on the willingness of the provincial censors to enforce its provisions. Such willingness was not always present. Lapses were frequent enough to prompt General Goswin Nickel (1584–1664, general from 1652) to write to the provinces in 1654, threatening to restore the requirement that all manuscripts be sent to Rome for approval if the provincial revisers did not improve.[38] But the grumbling from Rome notwithstanding, the revisers do appear to have been on the whole conscientious, so authors who wished to publish propositions that were proscribed had to come to terms with censorship.

One method of publishing without following the Ordinatio was simply to circumvent the whole process and publish without having the work read by the revisers. This was clearly forbidden but occurred anyway. Kaspar Schott did it; his *Joco-seriorum* appeared in print in 1662, even though General Oliva had ruled that it could not be published, as it did not meet the standards expected of the Society's authors. The *Joco-seriorum,* a collection of natural magical tricks and anecdotes, such as how to nail a chicken by the head to a table without killing it, was indeed a frivolous book. Oliva ordered that it be suppressed. The incident earned Schott a reprimand from the general but no punishment.[39] There are few cases of Jesuits publishing natural philosophical texts without the Society's imprimatur; such behavior could result in removal from teaching offices. Jesuits did publish numerous anonymous texts with the Society's tacit permission in order to distance the Society from potentially ugly disputes.[40]

A more effective way of gaining the Society's imprimatur for forbidden or dubious opinions was through the judicious use of terms such as *true, false,* and *probable* as well as *hypothesis.* It was possible to teach or publish virtually any opinion if one correctly identified its degree of probability. This mode of discourse had a long history. Jesuit philosophical and theological texts had assigned degrees of probability to opinions long before 1651. Indeed, the use of *hypothesis* to comply with lists of banned opinions but still conduct natural philosophy dates back at least to the aftermath of the 1277 condemnation of Aristotelian theses by Bishop Tempier of Paris. A classic example is the treatment of the earth's motion by the Parisian professor Nicole Oresme (ca. 1325–82), which has been termed "the most startling argument of the fourteenth century [with] its equally startling disappointing conclusion."[41] Oresme provided weighty arguments to show that one could not determine whether the heavens moved around the earth diurnally or the earth turned on its axis while the heavens stood still. One could not demonstrate that the heavens moved, Oresme argued, and he suggested that it was more probable that the earth moved. Nevertheless, he concluded that the earth did stand still while the heavens moved, as God had established the world in that way and as stating the contrary was against natural reason and the articles of faith.[42]

The same form of argumentation can be seen the Jesuits' treatment of the question of the motion and position of the earth in the cosmos, which was in virtually every Jesuit textbook and lecture course on natural philosophy. The heliocentric view that the earth both rotated around its axis and revolved around the sun at the center of the universe had been treated by some Jesuits before 1633. Although it was unequivocally condemned in that year at the trial of Galileo, this did not silence the Jesuits on the issue. After the condemnation, Jesuits in Germany still discussed heliocentric cosmology at considerable length in lectures, dissertations, and textbooks.[43] Some Jesuits devoted scant attention to the Copernican system, but many discussed it in some detail. Cornaeus, for example, called Copernicus a man of "incomparable genius" and devoted ten pages to his system in his philosophy textbook.[44]

Yet Jesuit discussions of Copernicus until at least the 1750s had one thing in common: they all insisted that Copernicus's system could be treated only as a hypothesis. The term did not have its current meaning—a provisional explanation that does not yet have the status of a fully-fledged theory—but rather was considered to be a mathematical construct that could account for the observed phenomena but had no physical reality.[45] Cornaeus left no doubt that Copernicus's system was false and contrary to Scripture. Similarly, Kaspar Knittel in Prague wrote that Copernicus's system was "*pura hypothesis.*" Its incompatibility with a literal interpretation of Scripture meant it could not have any status beyond that.[46] In contrast, the Jesuits were unanimous in stating that the Tychonic system was a "thesis"; it accounted for the observed phenomena, conformed to physics, and agreed with Scripture and therefore could be defended as true.[47]

In effect, the Jesuits followed the instructions that Cardinal Bellarmine had given to Galileo in 1616 but that the Florentine astronomer and philosopher had neglected in his *Dialogue Concerning the Two Chief World Systems.* Galileo failed to distinguish the view he taught from the view he held to be true, employing the device of a fantasy, which the Inquisition and its consultants regarded as transparently implausible. The Jesuits, however, convincingly stated that while they might report the views of the heliocentrists, they did not hold them to be true. Galileo's unwillingness to compromise his commitment to Copernicanism led to his unfortunate encounter with the Inquisition in 1633, which prevented him from addressing the issue of Copernicanism ever again. By following the "rules," Jesuit astronomers and natural philosophers were able to disseminate knowledge of the competing cosmological systems throughout the seventeenth century.

Just as the use of the term *hypothesis* allowed Jesuit authors to treat "incorrect" opinions in great depth, the application of the term *probable* could introduce a degree of uncertainty sufficient to get a novel philosophical opinion into print, particularly if there was nothing explicitly forbidding it. In Jesuit philosophical and theological texts authors frequently assigned degrees of probability: "probable," "more probable," "greatly probable," and "most probable," for example. In many questions, both philosophical and theological, there was no officially sanctioned "correct" answer. Provided Aristotle or St. Thomas's opinion was granted a degree of probability, other opinions could be openly discussed.

Authors had to exercise care when dealing with dangerous opinions such as atomism and the Cartesian view of body; whatever the arguments in their favor, they had to be declared false. The corresponding orthodox opinions, such as the composition of bodies from matter and form and the existence of absolute accidents, had to be described as true. But again, if one declared atomism and heliocentrism to be false and provided the arguments against them, one could discuss them in considerable detail. Whether such statements were sincere is irrelevant here.[48] Whether Jesuit professors agreed with Copernicus or Descartes, by classifying their systems as false, they were able to provide students with a detailed treatment of their opinions both in published texts and in the classroom.

Local Circumstances

Another way the provisions of censorship could be circumvented was, in effect, to go over the revisers' heads and appeal to the general for a special exception to the rules. The Jesuits were renowned for their flexibility and ability to adapt their rules to local circumstances; individual cases could be granted exceptions. One example from the German assistancy shows how particular favor from the general allowed questionable opinions on a potentially dangerous issue to be published. Rodrigo de Arriaga (1592–1667), a Spanish Jesuit, had taught in Prague for many years with great distinction.[49] His work was, however, sometimes controversial. One of the catalysts of the 1651 Ordinatio was that a professor at the Collegio Romano, Sforza Pallavicino, and an unnamed German professor had been caught teaching the Zenonist doctrine of quantity and had been forced to recant. This doctrine, which asserted that quantity consisted of points, had been repeatedly and strenuously rejected by the revisers general as incompatible with the orthodox account of Eucharist. In a letter from General Vincenzio Carrafa, Arriaga was named as the German professor's source.[50] The text in question was doubtless his philosophy textbook, *Cursus philosophicus,* which enjoyed wide circulation throughout the German assistancy. Its first edition, published in 1632, was followed by several later editions, the last of which appeared posthumously in 1669, well after the 1651 Ordinatio.[51] Several of the doctrines that Arriaga had taught and actually published before 1651 were banned by the Ordinatio. Sometime before his death and the appearance of the final 1669 edition, Arriaga successfully asked Vicar General Paul Oliva for permission to include these proscribed propositions in the forthcoming edition. In the preface to this edition Arriaga wrote:

> Having been humbly petitioned by me, [General Oliva] conceded mostly gracefully, partly because I originally published those few opinions in good faith, . . . partly because they are completely accepted here at the University of Prague, partly because they pertain purely to philosophical matters and not at all to matters of faith and morals, and finally partly to honor me. However, he did not want to concede to any other through this favor the license to teach those opinions in our schools, let alone publish them in print.[52]

Arriaga listed here a range of valid reasons for why the 1651 Ordinatio could be suspended. The general certainly had the authority to do this. But why were they applied in Arriaga's case on such a potentially dangerous issue, one specifically forbidden and for which other professors had already been disciplined? The section in the 1669 edition of his *Cursus philosophicus* that deals with the continuum and its composition from Zenonian points is identical to the one in the pre-1651 editions. Arriaga's claim that this was purely a philosophical matter was somewhat disingenuous, as he included a section in his chapter on Zenonian points entitled "Argumenta theologica," in which he refuted the claim that Zeno's opinion was among those of Wyclif condemned at the Council of Constance. Yet Arriaga did not explicitly construct any links between Zenonian points and the Eucharist. Elsewhere in his textbook, he upheld those elements of physics necessary for the orthodox view of transubstantiation. He certainly did not say that Zenonian points were atoms or that bodies consisted solely of points without substantial form—two opinions explicitly forbidden in the 1651 Ordinatio.

Oliva probably thought it would cause more damage to draw attention to the issue than to ignore it. Arriaga, as he himself acknowledged, had a great reputation in the Society because of his services. He had arrived in Prague at the height of the Counter-Reformation purge in Bohemia after the Battle of White Mountain and was instrumental in establishing Jesuit control over the Bohemian educational system. He had served for many years as dean of the faculty of arts and as rector of the Jesuit college at the University of Prague and had twice represented the province at the general congregation in Rome. His reputation as a philosopher extended well beyond the Society—"to see Prague is to hear Arriaga," the saying went. Issues of prestige and reputation were of considerable concern to the Society. It would have been a source of some embarrassment if somebody, particularly a "heretic," asked why a textbook that had circulated in so many editions for so long was now deemed unsatisfactory. Also, Oliva made it clear that this case did not establish a precedent; permission to publish these opinions was granted to Arriaga alone. No further editions were printed after that of 1669, which appeared two years after his death.

A concern with local circumstances was clearly a factor at work in the actual application of censorship. Rather than adhering blindly to the rules in all cases, superiors were willing to allow publication if it meant preventing scandal and sparing reputations.

Patronage and Censorship

The Jesuits' flexibility as well as their concern to supervise philosophical freedom can also be seen in the Society's use of patronage. The Society was aware of the possibilities offered by influential patrons, and its members consciously dedicated works to patrons to win and cultivate their friendship. As was the custom of the time, virtually all Jesuit works of natural philosophy were dedicated to patrons. The prestige of an influential patron could grant a Jesuit some leverage in dealing with the Society's internal review procedures.

In 1658 Schott seized the occasion of a visit by Emperor Leopold I to the Jesuit college in Würzburg to dedicate to him the encyclopedia of mathematics he had been instructed to write—an offer that the emperor, a self-described "mathematicus," enthusiastically accepted.[53] Although General Vincenzio Carrafa had banned the teaching of military engineering ten years earlier as being inconsistent with the *Constitu-*

tions and tasks of the Society, Schott claimed that it would offend the emperor to offer him an incomplete work that lacked a chapter on one of the most useful fields of practical mathematics. He included a chapter on military engineering and was allowed to publish.[54] Although the patronage offered by the emperor did give Schott a certain leverage, one should be wary of reading too much into this. If military architecture had been banned for reasons of political expediency at the end of the Thirty Years' War, the ban could be lifted for the same reason in order to retain the goodwill of the emperor. Furthermore, Schott's work was mathematical, not natural philosophical, and certainly did not deal with potentially dangerous opinions such as those linked to the physics of the Eucharist.[55]

The Society was in fact anxious to ensure that the authority of a patron supported only the correct version of such important opinions. Joseph Pfriemb, professor of philosophy in the Jesuit philosophy faculty at the University of Mainz, was quick to respond in 1748 to the claim of the Benedictine Andreas Gordon that Prince Max Joseph of Bavaria, a student of the Jesuits, defended "novel" opinions in a disputation in the presence of his father, the elector of Bavaria.[56] According to Pfriemb, the prince had not defended any novel opinions or any that were "unsuitable for perceiving the mysteries of the Catholic faith."[57] The prince had rejected Cartesian principles and atomism, affirmed the centrality of the earth, and stated in opposition to the Cartesians that brute creatures had an animate soul. His mathematical theses consisted only of those "that all scholastics either claim as their own or do not think should be condemned."[58] The Society might not censure the philosophical views of a prince, but it could try to ensure that such notables did not seem to approve of dangerous opinions.

Many factors, then, influenced the actual practice of censorship of physical texts in the Society of Jesus in Germany. The 1651 Ordinatio, while of great significance, did not necessarily have the impact on Jesuit natural philosophy that one might have expected. There was still considerable room for those who wanted to discuss novelties and "dubious" opinions, as the next example shows.

Melchior Cornaeus and the Problem of Absolute Lightness

Melchior Cornaeus's attitude toward the Ordinatio of 1651 was moderate, yet while he was willing to call for a reasoned application of the Ordinatio, he was not prepared to openly disobey the Society's decrees when his own work was censored. For several years before 1651, he had been working on a philosophical textbook. Unfortunately for

Cornaeus, several opinions he held and wished to publish were directly forbidden by the Ordinatio, including the theory that lightness was merely a lack of heaviness.

In 1653 Cornaeus wrote to Athanasius Kircher, an old acquaintance from their student days, stating his concerns. He declared, "I am neither willing nor able to repudiate those [banned opinions], partly because I consider them to be truer than any others, partly because I taught and wrote them before the censure of the [Ninth General] Congregation."[59] Cornaeus had written to General Piccolomini right after the Ordinatio appeared, asking for permission to publish his own opinions, "but in a matter so recent, I was able to accomplish nothing." Now with things "cooling off"—not to mention that Piccolomini had died in 1651—he was trying once again at the urging of "many people." Therefore, he asked Kircher to intercede on his behalf with the new general, the German Goswin Nickel, so that he might be granted permission to publish his original opinions.

Cornaeus assured Kircher that in obscure matters his mind was "held captive to truth and obedience" but that this was not one of them. Cornaeus was himself convinced of its truth: "This opinion on lightness appears certain to me, and so manifestly supported by the weight of ancient [authority] and reason that it is hardly to be eliminated from our schools." While he insisted, "If I am not permitted to write what I think, then I will never write anything at all," he moderated his brusqueness with the humble conclusion, "If I err, correct, instruct, and teach me; you will have an obedient student."

Cornaeus also petitioned Nickel directly, but the general's response specifically denied Cornaeus permission to print opinions banned in the 1651 Ordinatio: "What Your Reverence asks me—that you may publish several propositions that were prohibited by the sacred authority of the Ninth General Congregation in the Ordinatio for Higher Studies to be taught in our schools—I can in no way grant. The way Your Reverence thinks we can guard against others' displeasure is not at all acceptable to us and is greatly adverse to our integrity and customary way of doing things."[60] It would seem that Cornaeus did not have the same standing in the Society that Arriaga did.

Despite the negative response from the general, Cornaeus upheld his promise to Kircher to write what he believed, and, in his *Curriculum philosophiae peripateticae*, published in 1657, he specifically denied that there was positive lightness. He cited numerous ancient sources such as Aristotle, Plato, Epicurus, and Archimedes and recounted several hydrostatic experiments that he interpreted as showing that all upward movement was violent (i.e., not from an inherent, natural principle of upwards motion) and was caused by the downward motion of denser bodies taking the place of lighter bodies and forcing them upwards. Absolute lightness was an "empty, superfluous" concept, and the arguments against it, he claimed, were very convincing.[61]

But after this treatment Cornaeus included a section entitled "What Ought to Be Thought about Heaviness and Lightness." He opened it by saying, "What I have just taught about heaviness and lightness according to the opinion of learned men, I myself have openly taught and held for many years. Now because the authority of my superiors commands something else, I say that it is probable that heaviness and lightness are two positive qualities, really distinct in substance and mutually contrary just as heat and cold are. . . . And because authority commands that we subscribe to this opinion, I subscribe and I approve of it."[62] He then listed authorities favoring this view, including Aristotle (who had also been one of his authorities opposing it), and described several experiments that he interpreted as showing that air is naturally light. For example, he noted that bubbles ascend to the top of a tube of water due to the inherent lightness of air. He then repeated the experiment with wood, which in turn ascended to the top of the water. "Therefore," he wrote, "there is in wood a positive lightness, which is the principle of upwards motion."[63] The obvious question of why wood does not display this same lightness in the medium of air was not discussed. Rather, Cornaeus limited himself to saying, "Indeed, I can see how the opposing view can respond to this argument, but I prefer to be obedient rather than the victim of a curious mind."[64]

Several of the factors influencing the practice of censorship and natural philosophy identified earlier appear in this example. There is Cornaeus's appeal, albeit unsuccessful, for a special exemption. There is also the dispute over where authority in natural philosophy should lie; Cornaeus preferred that it lie with reason and observation but accepted the Society's ruling that it lay with certain human authorities. His behavior may seem openly duplicitous; after dealing in great depth with a position he believed was correct, he then declared it to be false because he was instructed to. Such behavior, however, becomes more comprehensible when seen in the context of Jesuit physical discourse, which used hypotheses and degrees of probability. Cornaeus's treatment of the question of lightness clearly fits into this ongoing and long-lasting tradition. His disapproval of the official view was quite explicit and blunt, and his acceptance of the official opinion ungracious, but there was nothing unusual in presenting a detailed treatment of an opinion that he ultimately judged to be incorrect. There seems little doubt that Cornaeus did prefer his original view, but by saying that the official opinion was more probable, he was able to get his own view past the censors and into print, and, moreover, in a form that made it considerably more convincing than the official one. Furthermore, Cornaeus did not publicly adhere to the official opinion to any degree greater than necessary; he limited himself to saying that the official opinion was merely probable, not true.

Cornaeus continued the Jesuit tradition of integrating new doctrines into a traditional framework. He was certainly no convinced enemy of Aristotle. Indeed, he saw Aristotle as particularly useful: he published a polemical dialogue in which Aristotle returns to life as a convinced Catholic and employs his philosophy to refute Lutheran and Calvinist errors.[65] While the demands of censorship were frustrating, Cornaeus had enough room to work within them and integrate much of the new physics.

Despite the formidable array of mechanisms that the Society of Jesus established to define and enforce the uniformity and orthodoxy in physics required by its foundational texts, reports of the demise of Jesuit physics after 1633 or 1651 are greatly exaggerated. The documents governing censorship should not be taken at face value. The study of a much wider range of sources reveals that the application of such normative documents was mediated by numerous factors that created considerable room for maneuver in the system, even in the decade immediately following the promulgation of the Ordinatio of 1651. Perhaps the most important of these were the possibilities inherent in the nuanced usage of terms such as *true, probable,* and *false.* On the other hand, censorship was not completely insignificant. The censors could be remarkably stubborn. Open adherence to doctrines such as atomism or Copernicanism was impossible. Yet Jesuit professors and authors enjoyed considerable room to explore a wide range of opinions—many questions in natural philosophy were left virtually untouched by the Ordinatio—and many authors took advantage of this room to maneuver.

CHAPTER 3

The Colleges

The Founding of the Colleges

The debates around the place and purpose of philosophy did not, of course, occur in an institutional vacuum, but developed as the size of the Society of Jesus' teaching enterprise grew. Although the Society was not founded to teach, once it began to open colleges demand soon outstripped its ability to staff them. The fact that the Jesuits did not charge their students tuition made their schools particularly attractive. Education soon became the most prominent of the Society's apostolic missions in Europe. By the end of the sixteenth century the Society had opened hundreds of colleges across Europe.

Despite another popular stereotype, the Society of Jesus was not founded specifically to combat the spread of Protestantism. At the University of Paris the first Jesuits were only dimly aware of the religious turmoil in Germany.[1] Nevertheless, the first Jesuits to travel through the German lands in the 1540s and 1550s, among them Pierre Faber, Jerónimo Nadal, and most importantly Peter Canisius, soon recognized the true extent of the threat posed by the Protestants to the old faith and the role the Society could play in averting it. These early missionaries to Germany realized that the ignorance and laxity of clergy were major factors in the rapid spread of the Reformation. The central solution that the Jesuits adopted in Germany was no different from that adopted in other regions where the faith was in decline, such as Sicily and southern Italy: improving the training of the clergy so that they could competently teach correct Catholic doctrine and live as examples of it. As Ignatius wrote to King Ferdinand in April 1551: "[A]mong other remedies which it is proper to use against the widespread disease which afflicts Germany, that one should be sought which is to be found in the presence at the universities of men who, because of the example of their religious life and the soundness of their Catholic teaching, will endeavor to help others and to lead them to what is good. This seems to be not only a prudent and very practical thought, but one that is necessary and even inspired by God."[2]

This plan required an educational system that could impart such training. When the Jesuits arrived in the Holy Roman Empire in the mid–sixteenth century, however, those universities that had not already gone over to the Reformers were in a shambles. At the University of Vienna, for example, the teaching of theology had virtually ceased, and the other faculties were little better. Most of the professors were Protestants. They did not lecture, attendance had plummeted, and the university did not even make the effort to collect and award the funds for endowed scholarships.[3] At the Bavarian University of Ingolstadt, many of the chairs in the faculty of arts lay vacant.[4] In other cities the decline in civil order made the functioning of institutions of education impossible.[5] In such desperate times, ecclesiastical and lay princes loyal to the old Church soon realized that as teachers and professors the Jesuits could bring considerable aid to the defense of the faith and presented the Society with numerous requests to found colleges in Germany.[6]

Despite the splendor the colleges acquired during the seventeenth and eighteenth centuries (and the fantastical wealth the Jesuits' critics claimed they hoarded), the process of establishing the colleges was not easy.[7] Even though Ignatius refused to commit the Society to any project that was not adequately endowed by a prince or other patron, the colleges' early days were marked heroic achievements based on few resources. Paul Hoffaeus, the Upper German provincial, reported in 1594 that the college in Würzburg had "many debts, virtually no provisions, sour wine in small quantity, very bad beer."[8] The colleges often began in abandoned, dilapidated monasteries. Nevertheless, the German colleges multiplied rapidly, and the Society received more requests than it had the personnel to fulfill. Ideally, the Jesuits began by teaching the lower studies of Greek, Latin, and rhetoric. At selected colleges, the curriculum was expanded to include the philosophy triennium once students had completed the humanistic curriculum.

By the early seventeenth century the Jesuit colleges in Germany had overcome problems of funding and staffing shortages and were institutionally well established. But the Swedish invasion of southern Germany in 1631 during the Thirty Years' War resulted in the complete closure of many universities and colleges. The war brought troops who demanded billets, exacted contributions, closed down colleges, vandalized their property, and were generally disruptive. Libraries were particularly attractive forms of booty to princes and generals wanting to instantaneously improve their cultural cachet. Many colleges, such as Würzburg, lost their entire collections during the Thirty Years' War. As well as being plundered and vandalized, the library of the college at Paderborn suffered the indignity of receiving a direct hit from a cannonball in 1646. The Society was able to reopen the colleges after several years as the imperial faction's military fortunes recovered following the death of King Gustavus Adolphus in 1632, but in some cases it was not possible to attain the prewar staffing levels until the 1640s.

Even though the Thirty Years' War ended in 1648, the Peace of Westphalia did not end wars on German territory. In fact the activities and even the existence of some Jesuit colleges were repeatedly threatened by wars, including the Nine Years' War (1689–97), a succession of wars of succession—Spanish (1701–14), Polish (1733–38), and Austrian (1740–48)—and the Seven Years' War (1756–63). The Bavarian phase of the War of Austrian Succession from 1743 to 1745 was particularly hard on the colleges in southern Germany. In addition to wars, plagues forced the temporary closure of colleges, and, of course, the two pestilences often went together. Nevertheless, the massive dislocations of the Thirty Years' War were not repeated.

The appendix lists the colleges in the German assistancy that taught Physics and mathematics and demonstrates the significance of the Jesuit educational enterprise in Germany. While many produced virtually no philosophical texts in any genre, and while even some colleges with university status produced little, their contribution to the diffusion of natural philosophy through classroom instruction should not be underestimated. Most philosophical publications came from the colleges that were part of universities, in particular those integrated into the older, preexisting universities. These were the most prestigious institutions and were assigned the Society's best professors (figure 1).

College, University, and Prince

The amalgamation of the Jesuit college with the university distinguished Jesuit education in the empire from many other states in Europe. In France and Italy, for example,

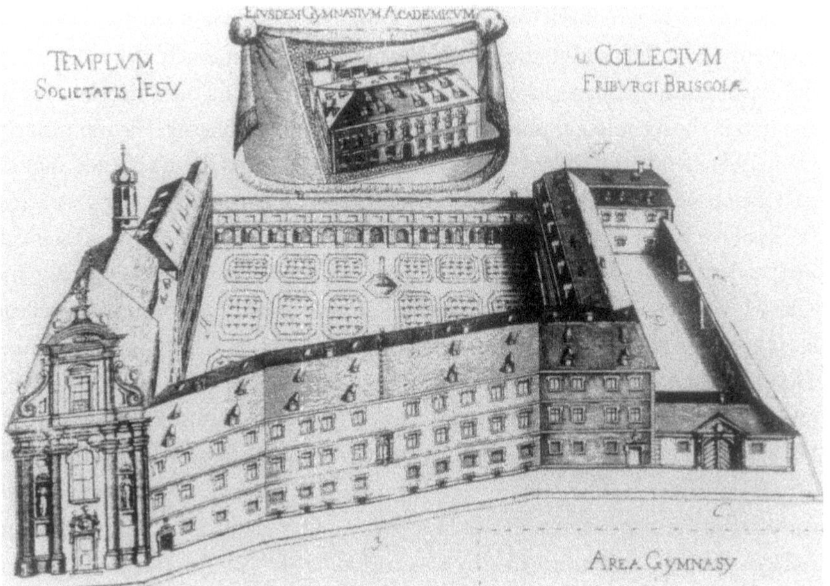

FIGURE 1. The Jesuit college at Freiburg im Breisgau, founded in 1620. The Jesuit colleges and universities were often among the most imposing buildings in their towns. Courtesy of the Universitätsarchiv Freiburg; UAF D49/928.

as a result of the opposition of university professors to the Society's attempts to teach at universities, the Jesuits colleges remained, on the whole, separate from the established universities. In Germany the Jesuits set up their own colleges, independent of existing educational institutions, but they were also invited by princes to take over chairs and even whole faculties at existing universities, or indeed entire universities.[9] However, the Jesuits' educational philosophy, reinforced by experience, prevented them from assigning professors of theology to university chairs without first establishing the necessary foundation of humanistic and philosophical education. The necessity of this policy was confirmed by the failure of the Jesuits' first attempts to teach theology at Ingolstadt in the early 1550s. The Society suspended its educational activities there and did not return until Duke Albrecht V of Bavaria promised sufficient support for a Jesuit gymnasium that would teach the prerequisites for theological studies.[10]

While the first Jesuits in Germany were wary about infringing on the customs, rights, and privileges of the faculties of the existing institutions, the Society eventually insisted that it be allowed to implement its own educational system rather than being obliged to conform to existing structures. Thus the Jesuits brought their own rules and practices to faculties that in some cases had been following their own statutes since the fourteenth century. This often led to serious friction with the existing universities, which jealously guarded their institutional independence.[11] As the Jesuits had the support of princes and bishops anxious to revitalize their universities, the generally moribund arts faculties could not put up serious resistance. But if conditions were unacceptable, the Society could, albeit reluctantly, withdraw its services. When the Jesuits encountered spirited resistance to their presence at the University of Ingolstadt from its faculty in 1573, they moved their college to Munich and did not return until 1576, when Duke Albrecht V supported their claims. In 1585 his successor, Wilhelm V, decreed that the entire arts faculty would be entrusted to the Jesuits. The duke was swayed more by the Society's promise of solid training for clergy and bureaucrats than by the university's appeals to academic liberty, and he handed over all the chairs in the faculty of arts to the Society, saving himself the professors' salaries in the bargain.[12]

The usual outcome of such tensions in Germany was that a Jesuit college consisting of a humanistic gymnasium, a philosophy faculty, and a theology faculty was incorporated into the university, replacing the universities' arts and theological faculties (or at least most of the latter).[13] Humanistic subjects, such as classical languages and rhetoric, traditionally taught in Germany by the arts faculty, were, according to Jesuit practice confirmed in the *Ratio studiorum,* taught in a five-year curriculum of lower studies at the Jesuit gymnasium before students embarked upon philosophical studies.[14] In general the professors of the Jesuit philosophy faculty constituted the university's faculty of arts in its entirety. This was the case at Ingolstadt and from 1561 onward at Mainz.[15] Similarly, at Vienna, the Jesuit college grew rapidly after its founding in 1550. In 1588 the college had over eight hundred students while the university had less than one-tenth this number.[16] Finally, according to the *Sanctio Pragmatica* of 1623, the Jesuit college replaced the university's moribund arts and theology faculties.

There were variations on the pattern. In 1556 at the University of Cologne, the Jesuits were entrusted with one of the existing three *Bursen,* or colleges, that constituted the faculty of arts. The other two continued to exist while the Jesuits refashioned theirs, the Tricoronatum, to conform to the *Ratio.*[17] At other universities the Jesuits had a freer hand, as there was no established institution to deal with. At Würzburg, where the original university had foundered, the Jesuits established a college at the invitation of the prince-bishop in 1567. This college then constituted the theology and philosophy faculties of the university when it was revived with the full four faculties in 1582.[18] At Dillingen a different model was followed: the university was founded by the bishop of Augsburg in 1551 specifically to be a two-faculty university (i.e., theology and philosophy). After initial difficulties the whole institution was handed over to the Jesuits in 1563. The university acquired the faculties of law and medicine considerably later.[19] The two-faculty model was followed in other places such as Bamberg, Paderborn, and Molsheim.[20] Often the Society first established a gymnasium to teach humanities and began philosophy classes only when enough qualified students had completed the humanities curriculum.

Of crucial significance is that in its dealings with prince and university, the Society insisted that the Jesuit provincial have the exclusive right to appoint the professors to the faculties it administered.[21] Without this, it would have been impossible to implement the *Ratio studiorum.* This was a privilege that princes, anxious to have Jesuits at their universities, willingly granted. Whatever other changes occurred in the seventeenth and eighteenth centuries, this privilege was essentially still in effect at the suppression of the Society in 1773. It ensured that the Society was able to determine the curriculum and staff of the philosophy faculties. This did not mean, however, that Jesuit teaching continued to adhere rigidly to the *Ratio* in all respects. For example, in the first decades of the eighteenth century, at the prompting of the princes, most Jesuit philosophy faculties established chairs for history, a subject not mentioned at all in the *Ratio.*

Despite its privileges, the Jesuit faculty of philosophy was subject to pressures from outside the Society. As a graft onto an older, larger institution, the university, its precise status was open to debate. The question of how a Jesuit college legally and intellectually fit into the university was constantly disputed by the Society and the universities. These conflicts were manifested in continual petty squabbles over order in processions, seating at ceremonial occasions, proper forms of address, the relationship between the university chancellor and college rector, and eligibility for membership of the university senate.

But more substantive issues were also at stake. While the role of philosophy was defined in the Jesuits' normative documents as a preparatory stage for theological studies, its role in a four-faculty university that also taught law and medicine was not quite so clear. Arts had always been considered the preparatory faculty for the three higher faculties, but university statutes had not always required that students complete or even attend classes in the faculty of arts before enrolling in one of the higher faculties. Also, many students had traditionally attended more than one faculty simultaneously. The Jesuits, however, insisted that all students going on to higher studies first complete their three-year philosophy curriculum, the triennium. The Jesuits also discouraged students from even attending classes at the other faculties while they were still enrolled in philosophy. Professors and future students of law and medicine increasingly regarded this as burdensome and unnecessary, so continuous strife resulted among the faculties.

Where the Jesuits enjoyed the favor of the prince, however, they were able to insist successfully that all students first complete the triennium in philosophy.[22]

The princes themselves exerted considerable influence over the Jesuit colleges, for they held ultimate authority over the universities. They could decree whatever changes they liked, although enforcing them was another matter. For example, from the later seventeenth century, rulers themselves began to question the value of teaching philosophy as a three-year preparation for theology to students who had no intention of undertaking theological studies. Consequently they forced the Jesuits to implement a two-year philosophical curriculum, the biennium.[23] Thus the need to win and maintain the support of the princes who ruled the myriad territories throughout Germany in which the Jesuits administered colleges was a constant concern for the Jesuits; their *litterae annuae,* or annual reports, always explicitly acknowledged the prince's generosity.

In summary, by the early seventeenth century, the Society of Jesus had largely replaced the old faculties of arts at universities throughout most of Catholic Germany with the humanistic gymnasia and philosophy faculties of their own colleges, in which they had the power to implement the *Ratio studiorum,* appoint professors, and set the curriculum.

The Size of the Institution

Jesuit colleges were very small by modern standards. A representative example of the staffing of a full Jesuit college that taught the entire curriculum from lower studies through philosophy to theology is provided by the roster of the Jesuit college at the University of Würzburg in 1657, a year in which Kaspar Schott and Melchior Cornaeus were both there.[24] The roster includes only those who were members of the Society of Jesus:

Persons in the Würzburg College	43
Priests	19
Those who teach theology and philosophy	8
Masters who are not priests	5
Students of theology who are not priests	11
Coadjutors	7
Scholastic who is not a priest	1

"Those who teach theology and philosophy" comprised four professors of theology (two for scholastic theology, one for casuistry or moral theology, one for Scripture) and four for philosophy (one each for Logic, Physics, and Metaphysics and a mathematician who was also responsible for ethics and sometimes biblical languages). The five "masters who were not priests" taught the five humanities classes of Latin, Greek, and rhetoric. They usually had completed their philosophical studies and after several years of teaching humanities would go on to complete their theological education. The "coadjutors" were brothers who performed services essential to the daily running of the Jesuit community, such as cooks, gardeners, or doormen. The staffing of the college at Würzburg was quite typical of German colleges that taught the full curriculum from the humanities through philosophy to theology.

The Jesuit universities were among the midsized institutions in Germany with around two to three hundred students (excluding pupils in the humanities classes, who could total one thousand).[25] This figure, however, includes students of the higher faculties as well as of philosophy. The number of students in particular classes can be difficult to determine. Numbers of graduates are better recorded, as graduation ceremonies were often accompanied by engraved sheets listing the graduating masters and bachelors. A 1713 list of all masters who had graduated from the University of Würzburg after completing the philosophical triennium from 1618 to 1713 shows that numbers averaged in the mid-twenties each year, although with substantial deviations.[26] But numbers of graduates are a poor indicator of enrollments. Innsbruck is one of the few universities that have reasonably reliable records of both enrollments and graduates. From the academic year 1671/72 to 1696/97, enrollment averaged eighty-three Logici, sixty-three Physici, and only thirty-one Metaphysici. Few masters actually graduated: an average of eighteen in this same period.[27] After Innsbruck switched to the biennium in 1735/36, the discrepancy between enrollments and graduates was even greater; the sixteen years between 1735/36 and 1754/55 for which numbers of enrollments and graduates are available witnessed an average of 108 Physici but only twenty-nine graduating masters.[28] Since the higher faculties did not require incoming students to have formally graduated from arts or philosophy, many students simply did not graduate because of the high fees required.

Students came from a wide range of backgrounds. At Innsbruck around a quarter were nobles. The sons of the Wittelsbachs, the Bavarian ruling dynasty, often studied at Ingolstadt. Universities often had fencing, riding, and dancing masters to provide nobles with a complete education. Regular clergy, from the Society of Jesus and other orders, and secular clergy numbered among the philosophy students. Universities also admitted students without any means, who had to sign themselves "paupers" in the university register. In the eighteenth century, absolutist rulers who sought to keep their subjects in their ordained places discouraged the Jesuits from accepting students from the lower classes, but they did so anyway.

The Jesuits' students went on to occupy positions as the elites of early modern Catholic Germany. Many nobles spent several years at a university before returning to govern their territories or seeking military careers. Other former students went on to gain positions in the developing bureaucracies of the German states without going on to the higher faculties. But since completion of the triennium or at least its Logic and Physics components was generally required for admission to the higher faculties, the vast majority of the empire's bishops, abbots, canons, judges, privy councillors, and court physicians had studied under the professors of the Society of Jesus.

Town and Gown

The number of priests (eleven) at the Würzburg college who did not teach is striking—it is in fact larger than the number of those who did. Yet there was a similar ratio at all Jesuit colleges in Germany. Nonteaching priests undertook a wide range of tasks within the colleges, serving as rectors, prefects of studies, confessors, preachers, and prefects of the various congregations that fostered the students' piety. But the college also served as a base from which Jesuit apostolic influence could be projected into the city and surrounding regions.[29]

The colleges themselves were physical statements of Jesuit intellectual and cultural preeminence throughout Catholic Germany. Once the Jesuits had secured the resources to replace their initial, shabby buildings, the colleges were often the largest structures in the city after the cathedral and the prince's residence. Inevitably they were urban institutions, usually located in the heart of the city.[30] The colleges' splendid churches, open to and attended by the public, were adorned with the rich iconography developed by the Society, presenting the local college as part of a pious enterprise that spanned the globe. Often ceiling frescos, such as those in the college church at Dillingen, depicted Jesuit missionaries proselytizing in the four continents of the world: Europe, Africa, Asia, and America. Prominent personages including princes, bishops, and emperors, were invited to the Jesuit dramas performed at the colleges, but these were also open to the inhabitants of the city. Outside the college, Jesuits performed charitable works throughout the city among the poor and criminal. Missions were sent to remote areas of the countryside, where priests were few and where both heretical and folk beliefs encroached upon the purity of Catholic faith and practice.[31]

In fact the annual reports sent by the colleges to the provincial and then collated and forwarded to Rome record all kinds of activities *except* those with academic content. Accounts of miraculous conversions, bleeding hosts, visits of notables, and extortion by occupying troops fill the reports. Precise count was kept of the numbers of Protestants converted and hosts given. The 1703 report from Upper Germany records that the Eucharist was given 87,980 times at the college church at Ingolstadt that year (and 1,559,565 times in the entire province), but it contains virtually no information on the content of instruction, disputations, or publications.[32]

However, these reports had to follow a standardized format and do not accurately portray the college's intellectual activities and their significance. The college functioned as the intellectual focus of the town. In addition to the regular curriculum, events in the academic calendar were opportunities to display Jesuit learning and cultural preeminence to society. These included public disputations, to which, again, local dignitaries were invited. Exercises in patronage, the disputations were accompanied by sheets of theses that the student defending the theses dedicated to a patron who paid for the sheets' publications. Masterpieces of the engraver's art, these large sheets were elaborately illustrated and replete with emblems. The original engraver left small blanks into which the name of the defender, his professor, and his patron could be inserted along with the theses, generally taken from the entire philosophical curriculum, that he would be defending (figures 2 and 3).[33]

Festive events such as graduations or the visits of notables gave professors the opportunity to give addresses that also touched upon natural philosophical topics. While these occasion were not primarily pedagogical but served to reinforce the bond between the college, the prince, and society and to display the prestige that each conferred upon the other, they did spread Jesuit natural philosophy to an audience well beyond the students actually enrolled in the colleges. Unfortunately the texts of such speeches have been lost, but the titles of many survive on the sheets announcing such events or the programs accompanying them.[34] They reveal a broad range of interests. At Würzburg, discussions of the validity and accuracy of astrology were perennial favorites even into the 1760s, as were lectures on the workings of natural magic, such as a 1650 oration pondering whether sympathetic powder could cure a wound if sprinkled on the swords that caused it or the linen that bandaged it.[35] Topical physical questions were also considered, such as "Whether and how the moon caused the tides?" in 1667. Such orations were often meant to be light in tone, and indeed many had jocular titles, such as "Why did Aristotle throw himself into the sea?" and "Whether and which wine, Franconian or Rhenish, is suitable for philosophers?" However, the titles do reveal changes occurring in Jesuit natural philosophy: for example, the relevance

PLURIMUM REVERENDO AC PERILLUSTRI DOMINO

DN. FRANCISCO GEORGIO DE Schönborn
L. BARONI METROPOLIT. ET CATHEDRALIUM ECCLESIARUM
MOGUNTINÆ, BAMBERG. ET HERBIPOLENSIS RESPECTIVE CUSTODI
ET CANONICO, NEC NON ECCLESIÆ IMPER. S. BARTHOLOMÆI FRANCO-
FURTI PRÆPOSITO &c. DOMINO MEO PERQUAM GRATIOSO.

[illegible]

PRÆSIDE R.P. CRAFFTONE MICHAEL è Soc. JESU AA. LL. & Philosophiæ Magistro, ejusdemque Professore ordinario.

MOGUNTIÆ, Ex Typographia CHRISTOPHORI KÜCHLERI

FIGURE 2. A 1664 sheet of theses from the University of Mainz. Seventeenth-century thesis sheets are relatively rare. This one is particularly interesting, as it depicts a student (right), a Jesuit professor (left), and in the background the Jesuit college (now the Institute for European History). Underneath are theses from the three-year curriculum defended in a disputation by the student, Hugo Adolph Heidelberger. Courtesy of the Staatsbibliothek Bamberg; GM 80/M1664, 1.2.

FIGURE 3. A more typically baroque example of the genre from the University of Bamberg, 1750. The allegorical figures of Faith, Hope, and Love carry the Eucharist in an oversized monstrance. The lower text gives the names of the year's sixty-four graduating bachelors and seventy-four masters, along with the topics of two graduation addresses, one dealing with cabbala and the other with physiognomy. Courtesy of the Staatsbibliothek Bamberg; GM 57/15.

or necessity of experimental philosophy became a frequent topic in the eighteenth century.[36] Eventually experiments were performed at such events.

The Professors

The careers of Jesuit Physics professors followed a standard pattern. The early career of Friedrich Spee von Langenfeld (1591–1635) can serve as a representative example.[37] Spee is perhaps the most famous seventeenth-century German Jesuit due to his *Trutz-Nachtigall,* a collection of mystic poetry, and, primarily, his *Cautio Criminalis,* a scathing condemnation of witch trials. The son of lesser Rhenish nobility, Spee was sent to school at the University of Cologne, where he completed his humanistic studies, although probably not at the Jesuit Tricoronatum. After entering the Society in 1610 and completing the required two-year novitiate at Fulda, he studied philosophy at the Jesuit college at the University of Würzburg. Spee then taught humanities classes for several years at the Jesuit colleges in the Rhenish cities of Speyer and Worms. Like many young Jesuits, Spee petitioned to be sent to the missions in the Indies, but General Vitelleschi refused his request. From 1618 to 1623 Spee studied at the Jesuit theological faculty at the University of Mainz, where he was ordained in 1622. Spee then taught the three-year philosophy course at the Jesuit university in Paderborn. Throughout his training and teaching, Spee, like most Jesuits, had many pastoral duties: teaching catechism, hearing confessions, preaching at parishes, and, according to his *Cautio Criminalis,* serving as confessor to accused—and almost inevitably condemned—witches. Spee carried out mission work, being sent to recatholicize a Protestant region near Hildesheim in 1628–29, where he was severely wounded in an ambush. He briefly taught moral theology at Paderborn but was removed from his chair by the rector, who, it appears, disagreed with Spee's interpretation of the Jesuit vow of poverty. He was transferred to Cologne, where his repeated publication of the *Cautio Criminalis* without the Society's imprimatur brought him into conflict with his superiors. Nevertheless, he continued to teach moral theology and later Scripture in Trier. His life was prematurely ended by the plague he contracted while caring for war wounded—another of the Jesuits' tasks.[38]

Spee's career was quite typical: humanistic education followed by entry into the Society, then philosophical studies. This led to a stint of teaching in the humanities classes followed by the completion of theological training and ordination and then a rotation through the triennium as professor of philosophy. Career paths could then diverge substantially: undertaking missions both within Europe and overseas, teaching in a faculty of theology, acting as confessor to a prominent personage, performing pastoral duties, or a combination of several options. This pattern is well charted.[39]

Several generalizations that retained their validity well into the eighteenth century can be made about Jesuit philosophy professors in Germany. First, they had survived a long process of formation aimed at ensuring that each member of the Society was directed to those tasks to which he was best suited. The *Ratio studiorum* specified that capabilities and talents of those selected to teach in the higher faculties were to be well above average.[40] Second, philosophy professors had to have completed their theological education before being permitted to teach philosophy. The importance of this requirement, which aimed at ensuring that philosophy stood in harmony with theology, was recognized early on in the Society's teaching activities and was confirmed in the 1599 *Ratio studiorum.*[41]

Third, no Jesuit philosophy professors in Germany were "professionals" in the sense that they devoted the bulk of their careers to teaching philosophy. Professors generally would have had one year of education in Physics as they went through the triennium as students and, although they may have been given some time to revise the material, were usually teaching it for their first and only time.[42] They often had very little exposure to contemporary developments in natural philosophy, other than what they had encountered in scholastic compendia.[43] As they were required to have completed their training in theology, they were familiar with the connections between scholastic theology and philosophy. The responsibility for ensuring that such inexperienced faculty were competent and orthodox rested upon senior faculty, in particular the prefects of studies, whose duties were outlined in the *Ratio studiorum.* As with all aspects of the *Ratio,* the manner in which the provisions governing philosophy were enforced depended to a large extent upon the personality of the prefect and on the constantly changing circumstances in which the Society and its members found themselves.

Fourth, these senior Jesuits were the authors of textbooks and compendia, while the junior professors who actually taught philosophy composed dissertations and theses for disputations. Fifth, as education became the primary mission of the Society in Europe, virtually all senior Jesuits, including the general and his staff in Rome as well as the delegates from the provinces to the general congregations, had had similar early careers. The vast majority of them had at some stage taught Physics.[44] It would not, therefore, be an exaggeration to say that the Society was a society of natural philosophers.

The nationality of Jesuit Physics professors in Germany also follows clear patterns. It was the original Jesuits' intent that the Society rise above national boundaries—its members were to serve in any land where they were needed, regardless of their origin. For its first half-century, the Society was plagued by chronic personnel shortages. This greatly retarded the rate of foundation of colleges, which, rapid as it was, would have been several times faster if the Society had had the human resources to meet all the requests it received for educational institutions. The shortage of qualified professors was particularly acute in Germany in the early period of the Society's activity there, and, through a combination of the Society's conscious policy and sheer necessity, many foreigners taught at the colleges in Germany, in particular Spaniards, Italians, and Belgians. In the 1560s and 1570s, young Italians were sent straight from the Collegio Romano to teach in Germany.

The colleges' successes in Germany overcame these shortages, although periodic crises, such as the effort to recatholicize parts of the empire during the Thirty Years' War, meant that professors had to be assigned to Germany from other parts of Europe. As part of the process of recatholicizing Bohemia after the Battle of White Mountain, Rodrigo de Arriaga was sent in 1624 from Spain to the Jesuit college in Prague, where he taught with distinction.[45]

By 1600, however, virtually all Jesuit professors in Germany were Germans and most had been educated in Germany. While the German Jesuits contributed several theologians of note, in particular in controversial theology and casuistry, they produced no philosophers of the stature of Suárez. Several German mathematicians, such as Christoph Clavius, Christoph Scheiner, and Athanasius Kircher, achieved fame beyond the Society and empire, but there was no German natural philosopher of comparable renown.[46]

PART 2

The Seventeenth Century

CHAPTER 4

The Curriculum in the Seventeenth Century

The Structure of the Curriculum

In their first decades in Germany, the Jesuits often had to integrate their colleges into the preexisting institution of the university, which in many cases was openly antagonistic toward their presence. Consequently, the Jesuits made great efforts to adapt their educational enterprise to the local conditions they inherited. This resulted in considerable variation in their teaching from college to college and from Jesuit practices in other countries. Throughout the 1580s and 1590s, however, the Society sought to standardize its educational practices as much as possible throughout the world. This culminated in 1599 with the final version of the *Ratio studiorum,* the concrete manifestation of the Society's pedagogical philosophy. Once their institutional position was secure, the German Jesuits sought to implement the provisions of the *Ratio* and eliminate as far as was feasible local variations. On the whole they were successful in this endeavor.[1]

In contrast to university curricula in Germany and Italy, with their prevailing lack of structure, Jesuit schooling was very methodical. Having learned from his own experience and that of others, Ignatius was determined to prevent students from aimlessly pursuing their own interests and consequently frittering their time away on a range of classes for which they were unprepared.[2] Accordingly, the first Jesuits adopted much from the structured model of the *modus Parisiensis* from the University of Paris, where they had met as students and masters.[3] The Jesuits' version of the *modus Parisiensis,* their *Ratio studiorum,* established a thorough, methodical curriculum. There was no choice of subjects; all students followed the same progression. The order of classes was firmly established and their content highly standardized. No student could proceed to the next class until passing an exam.[4] The philosophy triennium followed the five-year humanities curriculum that provided students with the Latin and Greek necessary to study philosophy. Students were not to be admitted to philosophy unless they could first demonstrate sufficient ability in letters. The material taught in lectures was reinforced in daily revision and repetition sessions as well as in frequent disputations.

Skill at disputation was actively fostered. The *Ratio* required that "[f]rom the very beginning of Logic [i.e., the first year of philosophical study] . . . the young men be so trained that nothing would make them more ashamed than to fail in the form of disputation. And let the instructor demand nothing more severely from them than the laws and method of disputing." Shorter disputations were held weekly, and longer disputations, often accompanied by printed lists of theses, were held monthly.[5] Disputation was a fundamental part of Jesuit education until the suppression of the Society. As in all early modern universities, the goal was not to discover new knowledge but to enable students to master and reproduce a defined body of knowledge as effectively as possible.

The *Ratio studiorum* specified that the philosophical curriculum was to be divided into three years, called Logic, Physics, and Metaphysics, and was to be structured around the works of Aristotle.[6] Two noteworthy features of the philosophical curriculum illustrate the Jesuits' concern with effective pedagogy. First, the whole curriculum was generally taught by one professor, who stayed with his class as it moved through the triennium.[7] This provided continuity and allowed the three-year course to be treated as a whole, so that the professor could draw connections between different parts of the curriculum. Second, material was to be dealt with in its "natural place" regardless of where Aristotle treated it. For example, metaphysical issues that arose in the Logic or Physics years were to be postponed until the third year unless they were questions about the nature of God, in which case they were better treated in theology; as Suárez had stated, certain subjects could not be treated until a firm foundation had first been laid.[8]

Although the *Ratio studiorum* specified that Logic was to be taught by explicating the textbook of either Toledo or Fonseca, Porphyry's *Isagoge* remained a favorite textbook for the course.[9] The latter part of the first year was to prepare the students for the second year, which dealt with natural philosophy:

> [I]n order that the whole second year may be devoted to matters of physics, at the end of the first year a full disputation on *scientia* should be prepared; into which should be put mainly a prolegomena of physics, such as the divisions of science, abstractions, speculative, practical, subalternate, also the different methods of proceeding in physics and mathematics, about which Aristotle [writes] in the second book of the *Physics*. Finally something should be said about definition in book two of *De Anima*.[10]

Physics, the second year—its students were *Physici*—was to deal with the eight books of Aristotle's *Physics*, the books of *De caelo*, and the first book of *De generatione*.[11] The *Ratio* gave only very brief instructions on which parts were to be treated. Those parts of book 8 of the *Physics* that dealt with the number of intelligences and the infinity of the Prime Mover were to be dealt with in the third year as part of Metaphysics. The second, third, and fourth books of *De caelo* were to be treated very briefly except for those parts on the elements and on the substance and influence of the heavens. The bulk of *De caelo* was to be treated by the professor of mathematics.

The final year, Metaphysics, dealt with the second book of Aristotle's *De generatione*, his *De anima*, and his *Metaphysics*.[12] Professors were advised when treating *De anima* not to be led into anatomical questions, which were the province of the medical faculty, yet numerous texts did cover anatomical and physiological questions.[13] In *Metaphysics*, questions about God and the intelligences that depended on the revealed truths of faith were to be skipped over, as they belonged to the province of theology.

After completing Metaphysics, the student who successfully passed his oral examination could progress to one of the higher faculties. In accordance with their rules, the Jesuits themselves did not teach medicine or law. In preexisting universities, such as Ingolstadt and Mainz, these disciplines continued to be taught by non-Jesuits. Some universities, such as Dillingen and Bamberg, that began as two-faculty Jesuit institutions, later developed faculties of law and medicine staffed by non-Jesuits.

Genres of Physics Texts

Because it is a normative document, the *Ratio studiorum* is static. It does not reflect changes in teaching after 1599. Further, the very vagueness of the *Ratio* left much room for professors to choose how much attention to devote to particular questions and how to treat those questions. Consequently, the *Ratio* tells us very little about what the Jesuits actually taught in the Physics year around 1630. However, several genres of sources enable us to reconstruct what was taught: Aristotle commentaries, textbooks, lecture notes, and dissertations.

The commentary genre flourished in the Society in the second half of the sixteenth century and first decades of the seventeenth century, with Fonseca, Toledo, Ruvio, and the Conimbricenses being among the most prominent authors.[14] These commentaries, while contributing to the sixteenth-century revival of scholasticism, had moved away from classical medieval forms of the commentary, which followed the text of Aristotle line by line. Early modern commentators—and the Jesuits were no exception in this—successively abandoned such close dependence on the text of Aristotle, a process culminating in Suárez's *Disputationes metaphysicae,* which freed itself completely from the traditional Aristotelian order of the material and, according to Martin Grabmann, marked the beginning of the autonomous philosophical handbook. It was a bridge between the commentary and the next genre, the compendium or *cursus.*[15]

Professors did use commentaries in preparing their lectures, but increasingly throughout the seventeenth century they came to rely on the philosophical compendium or *cursus philosophicus.*[16] The compendia began to bloom in the early seventeenth century and flourished until the end of the Society and beyond. Whereas commentaries generally focused on one of Aristotle's texts, compendia covered the whole of the triennium. These were often very large works, consisting of one or even two folio volumes. The most influential example in use in Germany in the seventeenth century was Rodrigo de Arriaga's *Cursus philosophicus.*[17] First published in 1632, it underwent numerous editions and reprints, appearing posthumously in 1669. Textbooks both reflected the teaching of philosophy in schools and were sources for it; Arriaga's *Cursus* was frequently cited in lectures, dissertations, and other textbooks, and Arriaga, like his fellow authors, claimed that his *Cursus* presented philosophy in the same way as it was taught in the three-year curriculum in schools.[18]

Another example of the genre that was well known in Germany was the 1649 *Philosophia universa* of the English Jesuit Thomas Compton Carleton (1591–1666), who taught at the Society's English "college-in-exile" at Liège.[19] In 1657 Melchior Cornaeus published the first compendium by a German Jesuit, the *Curriculum philosophiae peripateticae.* This work was in quarto format rather than folio and dealt with questions more briefly than the others, but as its author announced, it was meant not for masters and doctors but for students and those who wanted to refresh their memories of what they had studied at an earlier age. In contrast to the vast majority of compendia, Cornaeus's book included a section on mathematics.[20] Professors used numerous other examples in this genre in lectures and disputations in Germany.[21]

The compendium employed a vestigial version of the medieval *quaestio*, a dispute carried on with an imaginary respondent.[22] Cornaeus conducted the dispute in particularly lively fashion. The work was divided into questions (e.g., "What is place and whereness?"), which were further divided into doubts (*Dubitationes*). He posed a doubt ("How does Aristotle define place?"), supplied an answer (*Resp.* for *Respondeo*—I answer), and then raised an objection on behalf of his opponent (*Dices*—you will say). Sometimes the validity of the objection was granted (*Concedo*); other times the opponent stood his ground (*Instas*—you insist), and Cornaeus had to provide another answer to overcome the objection. Arriaga's work was also divided into disputations, but the debate was not carried on as vividly as in Cornaeus's work. The headings of sections were statements rather than questions and in place of Cornaeus's parry-and-thrust presented the occasional "you say" or "you will ask" (*Rogabis*). Nevertheless, the structure of the text consisted of a cycle of assertion, counterargument, and refutation.

Another genre, similar in many ways to the compendium, consisted of manuscript copies of lectures. Jesuit professors did not lecture directly from compendia but used them to prepare their own lectures. Many survive in libraries throughout Europe.[23] Most in Germany seem to be students' recordings of the lectures rather than the professor's own lecture notes; the *Constitutions* stated that: "[t]hose [students] who are more advanced in the humanities and the other faculties should carry with them paper to jot down what they hear or anything which strikes them as noteworthy. Later on they should rewrite in the paper notebooks, with better arrangement and order, what they desire to keep for the future."[24] The thoroughness of these manuscripts indicates that the professors dictated their material so that students could record it word for word. The *Ratio* also required afternoon repetition sessions in which students could repair any gaps in their notes and knowledge.[25] Therefore it is safe to assume that students' notes accurately reproduced the lectures. In structure and format the lectures were virtually identical to the *cursus* genre, following the same order of material and employing the *quaestio* format.

An important academic genre was the printed disputations that accompanied verbal disputations held at the colleges, as the *Ratio studiorum* required.[26] These survive in great numbers in different formats.[27] They could be as short as lists of one-sentence theses taken from the entire philosophical curriculum (*Theses ex universa philosophia*). Often they appeared as tiny parts of elaborate engravings that accompanied public academic events (figures 2 and 3).

There were more substantial versions. From the late sixteenth to the later seventeenth centuries they usually presented a summary of the course's treatment of a substantial section of the curriculum—the first book of the *Physics* for example—in sixteen or so quarto pages. In general they dealt with the section of the course that had just been taught. Thus, by the end of the year, the disputations had covered the whole of the Physics curriculum and could be bound together to provide a treatment of the entire curriculum. Others consisted of paragraph-length theses selected from the whole of the philosophical curriculum. Most were quite conventional in their presentation, but others were a little unusual. A 1648 philosophy disputation was divided into fifty paragraphs covering the life of the Jesuit missionary St. Francis Xavier. Each stage of Xavier's life presented the author with an opportunity to discuss philosophical questions. Paragraph 13, for example, describing Xavier's voyage over the seemingly endless oceans, allowed the author to consider local motion and the continuum; his death opened an examination of the corruption of composite bodies.[28] There were also several examples of disputations organized alphabetically that thereby functioned as short dictionaries of scholastic philosophy.[29]

By the later seventeenth century, another form of "disputation" became increasingly common: that on a specific problem, such as Georg Saur's refutation of Cartesian philosophy or Anton Nebel's series of four dissertations on the barometer.[30] These treatises were increasingly referred to as dissertations. Their length varied, from around twelve to over two hundred pages. Coverage of the broader curriculum was still included but was reduced to a number of one-sentence theses attached as an appendix to the dissertation. Presumably it was these theses that were actually defended in the public disputation. Although publication was funded by the student, both the theses defended by the student and the disputation or dissertation they accompanied were written by the professor who was listed as the *praeses* (supervisor). Students did not have the expertise to write the treatise, and the fact that identical lists of theses were often defended by different students indicates that they did not write the appended theses either.[31]

These disputations and dissertations should not be confused with modern dissertations, which are meant to be a new contribution to knowledge.[32] In 1670 Wolfgang Leinberer, S.J., of the University of Ingolstadt responded to the twofold charge that too many books were published that contained things unknown to the ancients, while on the other hand too many books merely "reboiled the same old cabbage ad nauseum."[33] He described the purpose of disputations in the preface to one defended by his student. By preparing for disputations, the students "stimulate their diligence, cultivate their minds, promote their success," and arouse in the minds of their listeners "the desire to attend the college and investigate the truth." One should not expect novelties in disputations, Leinberer warned, since "the Truth by its nature is immutable; if Philosophers investigate the truth, it is necessary that they will frequently print the same assertions and theses, unless perhaps in order to say something new, they speak falsely or prefer to paint the unknown truth with the colors of adulterated dye, which by its brilliance seizes the minds of its admirers." Nor was it right for students to fashion new words for explicating ancient opinions; they should instead adhere to those that had been proven in "the marketplaces of the good arts."[34]

The Curriculum and Aristotle

To understand the transformations that occurred over the seventeenth and eighteenth centuries, it is first necessary to outline the content of the Jesuit Physics curriculum taught in Germany. This brief treatment does not provide a detailed analysis of all aspects of the Jesuit natural philosophical curriculum but merely a framework in which our discussion can be situated.[35] This summary draws upon compendia, lecture notes, and dissertations to present an "ideal type" of the curriculum in the early to mid-seventeenth century; individual texts generally differed in some way from this archetype.

The order of the Physics curriculum followed by the Jesuits in the early seventeenth century was provided by Aristotle. Physics immediately followed Logic, in which many of the categories and concepts essential to Physics were treated. The course was first divided into *physica generalis* and *physica particularis.* The first part followed the order of Aristotle's *Physics* and dealt with the characteristics common to all natural bodies. The second followed Aristotle's books of natural philosophy—*De caelo, Meteorologia, De generatione et corruptione* (also referred to as *De ortu et interitu*), and *De anima*—which dealt with particular bodies, namely heavenly bodies, atmospheric bodies, the elements, and living bodies respectively.

While Aristotle's influence on the course was omnipresent, the course was by no means a line-by-line commentary on Aristotle. The Society's normative documents placed considerable emphasis on the study of Aristotle, and the 1599 version of the *Ratio studiorum* specified that philosophy professors had to explicate the text of Aristotle.[36] Nevertheless, the Jesuits' students did not actually read much Aristotle in the classroom. The Society's visitor to the Austrian province reported in 1583 that in Vienna the Logic students "read hardly a line of Aristotle."[37]

One could study Aristotle in the early modern period without actually reading him. As Charles Schmitt has emphasized, the sixteenth century saw a blossoming of different genres of "Aristotle," which differed in the amount of Aristotle they provided and their faithfulness to the original text.[38] Although the Coimbran commentators provided at the start of each question the passage of Aristotle to be commented upon, the *cursus* or compendium omitted it. The only passages of the Aristotelian text cited directly were definitions or brief quotations integrated into the discussion, much like the opinions of any other authorities. Aristotle's opinion was generally the first and set the question to be discussed, but the textbook was no longer a direct commentary upon Aristotle.

The Jesuits' interpretation of Aristotle was filtered through the long, often competing commentary traditions. Some of these commentators were considered more acceptable than others. Arab commentators in general and Averroes in particular were not regarded favorably.[39] Nevertheless, this still left a huge number of authorities. Despite popular stereotype, there was no monolithic consensus among Jesuit professors on every question of natural philosophy. In one sense peripatetic physics was the evaluation of different textual sources for and against any particular opinion. Jesuits adhered to different authorities on virtually all aspects of philosophy. Reinhard Ziegler, rector of the Mainz college, wrote to General Acquaviva in 1613, "[T]hose of us who write differ greatly from each other in their opinions."[40] As the Prague Jesuit Kaspar Knittel showed in his 1682 *Pleasant and Useful Aristotle,* Jesuits stood on both sides of virtually every argument among peripatetic schoolmen and freely disagreed with their fellows.[41] Debate on numerous issues occurred not only between but also within generations of Jesuit philosophers.

Unlike the older orders, the Society of Jesus was not committed to any particular school within scholasticism and was careful to remain so. One of the reasons why in 1585 the professors of the Collegio Romano questioned the value of having one canonical textbook for the entire Society was that by fixing the Jesuits' teaching, "We will irritate many rivals just like hornets, who want to write against 'the doctrine of the Jesuits,' as they call it."[42] Jesuit philosophy was always marked by its eclecticism. Jesuit philosophers disagreed with St. Thomas on many issues, as was permitted them by the *Ratio studiorum;* Suárez himself disagreed with Thomas frequently.[43] Jesuit natural philosophers in Germany held views derived from the Thomist, Scotist, and even Ockhamist traditions.

The kinds of authorities the Jesuits used changed radically over the seventeenth century. In contrast to the Jesuit textbook authors of the late sixteenth century, who displayed great familiarity with the works of the medieval high scholastics, the textbook authors of the seventeenth century generally relied on secondhand sources when citing the medievals.[44] While it is clear that textbook authors such as Arriaga were familiar with the writings of the more prominent doctors such as Aquinas and Scotus, they relied primarily on the previous generation of Jesuit philosophers such as Suárez and Ruvio rather than directly on the medievals. The young professors who had to produce lectures and theses for disputation in turn tended to rely on compendia and even less on high and late scholastic commentaries. The process of declining familiarity with the scholastic authorities continued and accelerated throughout the seventeenth century, so that by the eighteenth century German Jesuit authors and professors drew exclusively on modern sources, both Jesuit and non-Jesuit.

The Curriculum

The course always opened with a definition of what kind of science physics was and what its aim and object was. Following Aristotle, scholastic philosophy divided the sciences into the practical and the speculative. The practical, such as ethics and politics, taught precepts for action. The speculative engaged in the contemplation of the truth alone. Following a tradition passed on from Aristotle through Boethius and Thomas Aquinas, Jesuit philosophers divided the three branches of speculative science into metaphysics, physics, and mathematics.[45] Whereas metaphysics considered "being" removed from matter and mathematics considered quantity removed from body, physics was concerned with natural bodies.[46] Physics, as a speculative science, was not concerned with the manipulation of nature. Technological concerns, being a form of art, had no place in the Physics curriculum.

The aim of natural philosophy was attainment of *scientia,* which had been defined, following Aristotle, in the Logic course as knowledge of an effect through its causes. This was to be attained through demonstration that proceeded from indubitable propositions through syllogism. Since the attainment of true *scientia*—that is, a demonstration of the true cause of an effect through the elimination of all other possible causes—was very rare, the Logic curriculum also defined other standards of knowledge, having lesser degrees of certainty, that were employed in physics. Nevertheless, physics aimed at attaining eternally true conclusions about its object, which was defined in various ways but by most definitions was natural bodies—that is, bodies subject to change.[47] Supernatural bodies were excluded from physics. Thus physics essentially aimed at providing the causes of the composition and changes of natural bodies.

Peripatetic physics was marked by binary oppositions or dualisms: act and potency, matter and form, substance and accidents, nature and art. The first of these, act and potency (*actus* and *potentia*), was perhaps the most significant. All natural bodies were in a state of both being (act) and becoming (potency). Although these two concepts were in general not systematically discussed in the Physics curriculum, falling rather into the province of Metaphysics, they permeated Physics, as shown by the course's treatment of principles, to which it turned next. This discussion dealt with the subject matter of book 1 of Aristotle's *Physics*. Principles were divided into extrinsic, generally termed causes, and intrinsic, the principles per se. All bodies were composed of the union of the two fundamental intrinsic principles of prime matter and substantial form. Peripatetic physics rested upon this foundation of hylemorphism (from *hyle*, matter, and *morphe*, form). In addition to these two principles, any change required a further principle, privation, as the starting point at which the form was absent. In the process of generation prime matter functioned as the subject, substantial form as the *terminus ad quem*, and privation as the *terminus a quo* of any natural process. Form was act, and matter was potency, although the degree to which matter was pure potential and lacking act, and form the opposite, was subject to endless dispute among the Jesuits.

Prime matter was the first subject from which all material bodies came and into which they decayed. Whether it could exist independently of form was hotly debated by philosophers. Thomas specifically stated that matter could not exist without form, whereas Suárez said that it could. In this Suárez was supported by other Jesuits; Arriaga wrote that "prime matter must be granted its own existence, as Scotus says . . . and this [opinion] is so common among all the teachers of our Society that none may deviate from it."[48] They agreed, however, that it had the appetite, aptitude, or potential to receive any form and in this sense could be described as pure potency.

The second principle of natural body was substantial form. This was the nobler and more important part of the composition, as it gave it its essence and individuality or "whatness" (*quidditas*). A 1629 Ingolstadt dissertation summed up the role of prime matter and form in the constant process of becoming as follows:

> Matter is in itself ungenerable and incorruptible, and clearly producible through [divine] creation alone and destructible through annihilation alone. Nevertheless it is the root and foundation of all generation and corruption since it is in itself indifferent to any substantial forms whatsoever and makes itself subservient to any agents indiscriminately, so that they may introduce the dispositions toward any form whatsoever. With sufficient dispositions established, a new substantial form is generated and the old is corrupted. And so according to the diverse application of diverse and contrary agents the generation and corruption of forms are continued perpetually.[49]

Substantial forms were of a completely different order of being from accidental forms. Accidental forms were not capable of combining with prime matter to constitute a complete being, as were substantial forms. Rather, accidental forms served to "ornament and perfect" the substance in which they inhered.[50] If the substantial form made a body what it was—an apple, for example—the accidental forms provided its size, color, flavor, and other perceptible attributes.

Since nothing could come from nothing, only from contraries, a third principle was needed to account for coming to be, namely privation. Things did not come to be through opposition but from nonbeing. Thus privation was the nonbeing of a particular form. However, in the new composite being, privation played no further role except as the nonbeing of a future substantial form, and it received considerably less treatment than the two other principles. Over the seventeenth century the course's treatment of the role of privation in natural change dwindled, and by the eighteenth it was completely neglected.

After dealing with intrinsic principles, the course turned to extrinsic principles, more commonly referred to as causes, the subject matter of book 2 of the *Physics.* There were four causes: formal, material, efficient, and final. The classic example was that of a marble statue: the material cause was the marble, the formal cause the shape given to the marble by the sculptor, the efficient cause was the sculptor himself, and the final cause the purpose for which the statue was being made, such as the worship of a god. As the origin and purpose of all creation, God was the first efficient cause and final cause. God was then ultimately the cause of all effects, but there was great debate over the extent to which God was directly the cause of all natural changes and how natural, created agents and causality could be reconciled with God's omnipotence. Jesuit texts generally held that God immediately "concurred" with the action of all secondary, created causes.[51] Numerous aspects of causality were disputed in the course: the relationship of cause and effect, the necessity of a cause for any effect, whether single effects could have multiple causes and vice versa, and so on. Since there were numerous examples in nature of the normal workings of causality being inhibited, with the result that "monsters" occurred, there was also sufficient room in the course to deal with a wide range of phenomena that fell into this category, such as monstrous births and portents.

The Physics course was primarily concerned with the natural action of substantial forms upon prime matter to produce the endless sequence of ever-changing bodies. The Jesuits generally adopted some form of Aristotle's definition of nature as the principle and cause of motion and rest.[52] There was, however, another order of causes—the artificial. Although the various arts were excluded from natural philosophy, natural magic, an area on the cusp between natural and artificial change in that it used an understanding and application of the principles of nature to manipulate nature to practical ends, was often dealt with in considerable depth. While some Jesuit texts virtually ignored the subject, many throughout the seventeenth century contained lengthy discussions of magic, classifying and assessing which of its forms were licit. A 1623 Würzburg dissertation was not unusual in having sections on natural magic, alchemy, minerals as alchemical ingredients, the creation of gold, alchemical sublimation, black magic, and remedies.[53] A 1694 dissertation from the same university devoted nearly four hundred pages to natural magical subjects.[54]

This distinction between natural and artificial change was crucial. Since the scope of physics was restricted to natural motions, artificial bodies and motions, such as machines and their operations, were excluded from it. The study of mechanical devices was classified under not natural philosophy but mathematics.[55] This disciplinary division accounted for the difficulty in integrating experimental physics into the peripatetic curriculum in the later seventeenth and eighteenth centuries. While such experiments may have dealt with natural phenomena, which fell under the province of physics, they often used machines or instruments—that is, artificial means. Thus, in the mid–seventeenth century when the Jesuits encountered new experiments such as those using an air pump, there was no obvious place to put them in the Physics curriculum. Rather, they were scattered ad hoc throughout the curriculum.

The course then proceeded to the consideration of the properties (*proprietates*) of bodies, namely motion, quantity, the continuum, place, the infinite, and time. Books 3, 5, and 6 of the *Physics* were commonly treated together, as they all dealt with motion. Following Aristotle, Jesuit philosophers defined motion as "the act of a being in potential insofar as it is in potential."[56] It was, then, any change in which a natural body's potential became act. The scholastic term *motus* referred to natural change in general, not just local motion. There were four kinds of change: change of substance or *Quid,* termed *generation;* change of quantity or *Quantum,* termed *augmentation;* change of quality or *Quales,* termed *alteration;* and change of place or *Ubi,* termed *latio* or *local motion.*[57] The first was a change in substantial form; the last three were merely accidental changes.

The remaining properties of natural bodies were then covered in the order of the fourth book of Aristotle's *Physics.* This included much of the most difficult material of the course, which was often acknowledged to be so by authors. The discussion of the continuum was particularly fraught with difficulties, as it dealt with the problem of indivisibles. The question of quantity was also highly complex but of great importance. The treatment of place included the problem of the vacuum, which was by definition a place that lacked body but was capable of receiving it. The Jesuits agreed with

Aristotle upon the impossibility of a vacuum in nature, but the question of how bodies would behave in it should God use His divine powers to create a vacuum was open to much discussion. General Physics usually concluded here. Since books 7 and 8 of the *Physics* considered the Prime Mover, which the scholastic tradition identified with God, their subject matter was better considered in metaphysics or theology.[58]

The second half of the Physics curriculum, Particular Physics (*physica particularis*), was structured around Aristotle's other works of natural philosophy. Here the course moved from natural bodies in general to specific natural bodies. The first section of Particular Physics followed the structure of Aristotle's *De caelo* (On the heavens) and dealt with the heavenly bodies. It began with a discussion of the origin of the cosmos and considered whether it could have existed eternally and whether it could yet exist eternally. It then considered the physical composition of the heavens—their substance, qualities, and influence on the sublunar world. The *Ratio studiorum* merely followed the traditional division of the disciplines in decreeing that all other matters pertaining to the heavens, including their motions, were to be treated in the mathematics class.[59]

After celestial bodies, which were treated first because of their nobility deriving from their immutability, the course descended to mutable bodies, beginning with atmospheric phenomena. In most texts, this section was relatively short and followed the structure of Aristotle's *Meteorologia.* It included comets and meteors, which were regarded as exhalations from the earth.[60]

The course then proceeded to a discussion of terrestrial bodies following Aristotle's *De generatione et corruptione,* which covered types of change other than motion (generation, augmentation, and alteration) as well as the four elements (fire, air, water, and earth) that in mutual combination formed the natural bodies that underwent such changes. In such processes of change two further sets of binaries figured prominently as the instruments of change: the four qualities consisting of the opposite pairs of hot and cold and dry and wet.

Although not qualities in the strict sense, another binary pair figured in this section of the curriculum: heavy and light (*gravitas* and *levitas*). This last pair was of particular significance. According to Aristotle all natural bodies were either heavy or light—that is, all had an inherent tendency to move toward or away from the center of the earth. Earth and water were naturally heavy, fire was naturally light. The classification of air was more difficult but was generally held to be heavy. Bodies tended to move naturally toward their natural place: that is, heavy bodies above the earth tried to actuate their potential to move downwards by falling until their natural motion was interrupted by an intervening body. Thus the scholastic mechanics of motion and fall were neatly integrated into the structure of act and potential that underlay all forms of natural change.

Coverage of natural philosophy continued into the third year with treatment of *De anima,* which lies outside the scope of this discussion. Because their rules did not permit Jesuits to teach medicine, and because the various fields of natural history were generally taught within the medical faculty, there was no treatment per se of natural history in the Jesuit philosophy curriculum. There was, however, room in the course for some treatment of natural historical material. For example, in their treatment of the element of earth, some professors discussed the characteristics and membership of the mineral kingdom at some length. In the treatment of *De anima,* which distinguished between vegetative, animate, and intellectual souls, texts could discuss the creatures that had these souls, listing the kinds of plants and animals and their various characteristics. There was, however, no attempt to cover Aristotle's books of natural history in any detail comparable to the coverage of his books of natural philosophy.

The Jesuit Physics curriculum provided an ambitious coverage of the natural world. Equipped with the tools provided by logic, it identified the principles of change in natural bodies in general. Once it had set forth the principles that governed all of creation, it proceeded to particular natural bodies, descending from the more perfect to the less. In this project the Jesuits fell squarely within the scholastic university tradition.

The Jesuits certainly had their critics. Kaspar Manz, professor of law at Ingolstadt, published a thorough critique of the Jesuit philosophy curriculum in 1648. He claimed that it lacked clarity, utility, and pleasure. It did not provide adequate preparation for law and probably not even for theology. Making students read it destroyed their health. If one asked philosophy students to talk about nature, they just "sit there like mute fishes." Manz's discourse was typical of attacks on speculative philosophy. But he presented no viable alternative beyond a return to a kind of *prisca sapientia* based on the study of particulars, which he claimed was the source of Solomon's wisdom. Manz himself might just as well have sat there like a mute fish; the Jesuits determined the content, structure, and method of the philosophy curriculum and would do so until 1773.[61]

God in the Curriculum

Scholastic natural philosophy was permeated with questions relating to the supernatural, and the Jesuits' natural philosophy was no exception. While philosophy and theology were distinct sciences, their subject matter often overlapped. Even though physics was the science of natural bodies and their changes in general, concerns about the supernatural figured throughout the Physics curriculum. For example, as all natural change had a cause, the study of causes was an essential part of the curriculum. Yet because God was the first cause and the ultimate final cause, the course's treatment of causes required a discussion of the relationship between God as the first cause and created secondary causes in order to determine how creatures could themselves act as causes. Often theology determined that particular philosophical doctrines had to be taught as true and others as false. As Catholic priests, Jesuit professors were subject to the relevant decrees of the popes and Church councils on such matters. Because the Fifth Lateran Council of 1521, for example, had decreed that the rational soul was the substantial form of a human being, the Physics course had to teach that substantial forms existed; otherwise they would appear to be denying the existence of the soul.[62]

Jesuit professors did not rely solely on Church authority, for they provided numerous purely philosophical arguments in favor of the existence of substantial forms. But they also argued that the theological need for substantial forms put the question of their existence beyond doubt. The Fifth Lateran Council had also decreed that the world had been created *ex nihilo*. Jesuit Physics texts often explicitly referred to the rulings, and thus the course taught this opinion in its treatment of *De caelo*, although it carefully considered rival opinions, refuting them on both philosophical and theological grounds.[63] In addition, scriptural evidence could be deployed in favor of a philosophical position, one prominent example of course being the immobility of the earth.[64]

The course also had to take into account that just because something was naturally impossible did not mean that it was impossible for God—the Jesuits, following a long scholastic tradition, took great care not to limit God's omnipotence beyond what was a logical contradiction.[65] Thus many questions received a twofold treatment in the course. While bodies were naturally impenetrable, for example, the philosopher also had to consider whether this impenetrability could be supernaturally overcome and what would happen if it were. It was clearly within God's powers to do so: the virgin birth and Christ's appearance through closed doors after his resurrection among the apostles at Emmaeus (John 20:26) showed that supernatural intervention could suspend the natural impenetrability of bodies to allow them to pass through each other. Such episodes raised fundamental questions about the natural body and its attributes.

The Propaedeutic Function of Philosophy

Just as it had been at the medieval university, the Jesuits saw philosophy as an essential part of training future theologians, who had to complete the philosophical triennium before moving to theology. Teaching philosophy was also part of training Jesuits who had completed their theological education but had not yet begun teaching in the faculty of theology.

The conceptual vocabulary of eucharistic theology illustrates the need for Catholic theologians to have undertaken an extensive study of philosophy. The view of the Eucharist confirmed at Trent and adopted by the Society of Jesus was essentially that of Saint Thomas Aquinas as presented in his *Summa theologiae.* The extent to which Thomas conducted this discussion with concepts drawn from scholastic philosophy, in particular from physics and metaphysics, is revealed by the following passage from the *Summa:*

> Now it is manifest that every *agent* acts to the extent that is in *act.* But every created agent is limited in its *actuality* since it is limited in *genus* and *species.* Hence, the *action* of every created agent is sustained by a limited act. However, what limits each thing in its actual *being* is its *form.* Hence, no natural or created agent can act except to change a form. For this reason every *conversion* that takes place according to the *laws of nature* is formal. But God is infinite act, therefore his action extends to the whole *nature* of the being. Therefore, it cannot bring about only a formal conversion, so that different forms follow after one another in the same *subject;* rather, it brings about a conversion of the whole being, so that the complete *substance* of this is changed into the complete substance of that. (italics added)[66]

All the italicized terms had technical meanings that the student first had to master in philosophy.[67]

Late-sixteenth- and seventeenth-century Jesuit works of scholastic theology bore a relation to Thomas's *Summa theologiae* similar to the one Jesuit Physics texts bore to the Aristotelian corpus. They followed the same order of the material without being line-by-line commentaries and maintained a respectful, if sometimes critical, posture toward the source. Accordingly, they also used much the same terminology as Thomas did. They might not have always used the terms in precisely the same way as Thomas or with precisely the same content, but it was impossible for them to discuss crucial elements of scholastic theology such as transubstantiation without such terms. Thus all Jesuit theological works, from authoritative textbooks by the most esteemed theologians such as Bellarmine, Suárez, and de Lugo, through to theological disputations published at colleges in the provinces, shared such terminology. One can then understand the Jesuits' firm conviction that it was impossible to study theology without first obtaining a firm foundation in philosophy. Without it, the student would lack the conceptual tools to understand the writings of the medieval doctors of the Church and of the modern Jesuit authors who explicated and developed upon them.

In addition to providing the fundamental conceptual vocabulary shared by scholastic philosophy and theology, the philosophy course treated more specific questions that were of importance for scholastic theology's treatment of the Eucharist. The philosophy course did not explicitly link many of these questions to the Eucharist, but they make much more sense to modern readers who understand that such questions were preparing the ground for eucharistic theology.

The subject matter of book 4 of the *Physics,* which examined how bodies were in a place, provides an example. Discussions of this issue were usually quite detailed. They posed questions such as whether a body could naturally be in more than one place and whether more than one body could be in the same place, whether this could occur supernaturally, and, if so, through what kinds of supernatural virtue. They considered the difference between occupying a place "definitively"—that is, with the whole in every single part of the place, the way our soul was in us or an angel was in a place—and "circumscriptively"—that is, with each part of the whole corresponding to a different place within the whole, the way a natural body was in a place.[68]

These arcane discussions may now seem like scholastic hair-splitting. Cornaeus considered whether a man who was placed by divine power in two places simultaneously would have all his accidents in both places. The discussion became particularly bizarre when Cornaeus pondered whether Peter would die in Mainz if he were simultaneously in Rome and died there or whether he would see in Mainz what he was seeing in Rome.[69]

These discussions become more comprehensible when one sees them developing the foundation for an explanation of how Christ and the accidents of bread were in the Eucharist. The Jesuit theologian and cardinal Juan de Lugo's treatment of the Eucharist in his influential theology textbook *Disputationes scholasticae et morales* delved deeply into this issue of place and drew heavily on Physics texts.[70] Martin Becanus, a German theologian, pondered at considerable length "[w]hether the same body can simultaneously be present in diverse places" in his treatment of the Eucharist in his *Theologia scholastica.*[71] It was identical to a treatment of the question of place one would encounter in a Physics textbook. The distinctions Becanus developed in this *quaestio,* such as the difference between "circumscriptive" and "definitive" place, were then applied to the issue of how Christ was locally in the Eucharist.

Theological concerns not only influenced the questions covered by philosophical instruction but also limited the possible answers. Thomas's account of the Eucharist required that the student be well versed in the physics of place but also limited the acceptable answers to certain questions to just one. Jesuits, both philosophers and theologians, stated that a body could be in more than one place through supernatural intervention. Cornaeus stated in his philosophical compendium that this "is proven through the faith. In the Venerable Eucharist Our Savior is de facto wholly, simultaneously, indivisibly, and definitively in all churches, altars, and ciboria in which the properly consecrated host is preserved, indeed in individual hosts and even in every single part of the host. Nor does anyone other than the unfaithful deny this."[72] A body had to be able to exist supernaturally in more than one place because this was undeniably the case with the Eucharist. To deny that would place one in the ranks of the "unfaithful" along with the Protestant "heretics."

The New Physics

In many of the disputes in the Physics curriculum both sides could draw on long traditions supported by numerous scholastic doctors. But with the developments occurring in natural philosophy in the seventeenth century, the Jesuits were increasingly confronted by a vast range of opinions on natural phenomena and methods for conducting natural philosophy that had no precedents in the scholastic tradition. In this sense they were novel and consequently posed major problems for professors, who were supposed to be wary of novelties. Nevertheless, these novelties eventually transformed the entire Jesuit Physics curriculum. The most contested areas included mechanical doctrines of matter, motion, and mechanics, theories of the composition and motion of heavens, and the question of method in natural philosophy.

Many of these novelties were in fact revivals of ancient doctrines that had been refuted by Aristotle. The best example of this is atomism, which was experiencing new life in the seventeenth century. This doctrine undermined the hylemorphic foundation of peripatetic physics and consequently all the doctrines constructed upon it. Already by 1630 Jesuit philosophers were responding in their texts to the problems posed by atomism. The Jesuit response to Descartes over the coming decades in many ways drew upon an ongoing debate with a materialist view of body that preceded Descartes.

Another site where Jesuits philosophers had to confront novelties was in their treatment of book 4 of the *Physics*, where they had to deal with novel views of place, space, and the void, including those postulating that a vacuum was possible in nature.[73] Many of these opinions were not merely speculative but based upon experiments, such as Evangelista Torricelli's mercury tube.[74] Linked to these developments was a challenge to the Aristotelian doctrine of heaviness and lightness: if bodies were not absolutely heavy or light but only relatively so, and if lighter bodies rose only because they were pushed up by heavy bodies trying to descend, the entire Aristotelian doctrine of natural place and natural motion, a linchpin of scholastic physics, was thrown into question. The doctrine of absolute heaviness and lightness and other peripatetic views of free fall were further being challenged by the new Galilean mechanics.

Perhaps no field was subject to as much questioning as that of the heavens; the astronomical developments of the late sixteenth and early seventeenth centuries posed numerous difficulties for the Jesuits' treatment of the heavens. Tycho Brahe's determination that comets and novas were in fact celestial phenomena challenged both the peripatetic doctrine of comets' meteorological nature as well as that of the solidity of the celestial spheres. Galileo's observation of features on the moon and his and the Jesuit Christoph Scheiner's discovery of sunspots posed problems for the doctrine of the immutability of the heavens. The very distinction between perfect, immutable celestial matter and the terrestrial world of change and decay was no longer secure.

All these debates, and the many others being carried out in the Physics course, raised a more fundamental and underlying issue, that of the epistemological and methodological foundations of natural philosophy. Scholastic natural philosophy was essentially a speculative discipline. As Peter Dear has pointed out, observation did play an important role in the development of the propositions from which syllogistic demonstrations proceeded; this was, however, repeated observation of everyday phenomena which was termed experience.[75] Observation in the sense of experiment, of particular, individual, or constructed events, was foreign to scholastic physics, as it was not based on the everyday repeated observations that produced experience. In contrast, Jesuit mathematics had a long tradition of employing particular observations, primarily in the fields of optics and astronomy. Also absent from Jesuit Physics was mechanics, in the sense of the study of machines. As machines were artificial and did not therefore reveal natural phenomena, they could not be treated in Physics and were consigned to mathematics.

Thus for experiment to enter the Physics curriculum it was necessary that observation of particular, individual events be used as the basis of physical demonstrations. The disciplinary boundaries between mathematics and physics would have to be renegotiated so that machines and the phenomena they revealed could be treated as physical problems and not solely mathematical.

CHAPTER 5

The Physics of the Eucharist

THE SACRAMENT OF THE EUCHARIST WAS ONE OF THE MOST BITTERLY contested battlefields of the Confessional Age. Nowhere were the intellectual and cultural struggles surrounding it more intense than in Germany. Exploiting their mastery of the medium of print, Protestant theologians assaulted the Catholic Church's teachings on the Eucharist, yet fought just as bitterly among themselves over the true nature and meaning of the sacrament. In response the Council of Trent sought to define the Church's teaching on the Eucharist. Catholic controversial theologians penned untold masses of treatises attacking the myriad Protestant teachings and justifying their own. Disputes over the Eucharist remained among the main doctrinal stumbling blocks in the doomed attempts at reconciliation between the confessional parties in the sixteenth century. Efforts to revive popular Catholic piety often focused on this sacrament with the foundation of confraternities dedicated to veneration of and frequent participation in the Eucharist. Another renewed form of popular piety, the Corpus Christi procession, literally became a battleground as Protestants clashed physically with participants in this most visible demonstration of Catholic devotion and solidarity. In addition to founding colleges, the first Jesuits in Germany, in particular Peter Canisius, the most influential of them, engaged in all these forms of response and renewal, delivering sermons extolling the benefits of frequent participation in the Eucharist, founding pious confraternities, writing polemical tracts against the Reformers, and debating Protestant theologians at imperial diets.

The natural philosophy curriculum at the Jesuit colleges in Germany also played a vital role in the Society's defense of the Catholic understanding of the Eucharist. The Society of Jesus regarded philosophy as institutionally and epistemically subordinate to theology but also as an extremely important and useful pursuit in itself. Thus philosophical opinions had to conform to the truths of the Catholic faith as established by theology, but philosophy also made an important contribution to the defense of the faith as a propaedeutic training for the study of theology. This chapter will move from generally considering the ideal role of philosophy and its relationship with theology to examining a particular case. Since theological issues were found throughout Jesuit philosophy, it is impossible to discuss all the ways in which they influenced the philosophical curriculum. Rather, an examination of the physics of the Eucharist reveals how the subordinate yet vital role of philosophy was manifested in concrete ways in the Physics curriculum.

Two related yet distinct questions with crucial consequences for the Eucharist became standard elements of Jesuit Physics texts. The first was whether the quantity of any particular natural body was identical to its substance. The second was whether the external appearances of the bread and wine could be accounted for only by the scholastic category of absolute accidents. In both these cases the Jesuits felt that the real presence of Christ in the sacrament of the Eucharist, one of the central tenets of the Catholic faith, was ultimately at stake. These two studies demonstrate theology's power not only to determine the subject matter of the Physics curriculum but also to restrict the range of possible opinions on certain crucial issues. One of the most visible results of this interaction was that Jesuits demonized Descartes, rejecting central aspects of his philosophy, particularly his view of natural bodies, which they regarded as incapable of accounting for the real presence in the Eucharist.

The Jesuits and the Eucharist

From its earliest days the Society of Jesus placed particular emphasis on the Eucharist, which, along with the sacrament of penance, played an unusually significant role in their pastoral activities. Starting with Ignatius, the Jesuits were advocates of frequent communion—that is, monthly, weekly, or even more often—at a time when most Catholics took communion only once a year. Ignatius's *Spiritual Exercises,* a crucial element in the spiritual formation of every Jesuit and also in the Society's pastoral mission to the community, listed several rules that Catholics should observe "in order to have the proper attitude of mind in the Church Militant." The second of these rules was "[t]o praise sacramental confession and the reception of the Most Holy Sacrament once a month, and better still every week, with the requisite and proper dispositions."[1]

In Germany Peter Canisius recommended frequent communion, "the more often, the better," much to the concern of older orders.[2] While the Jesuits did not begin the movement encouraging frequent communion, they soon became its leading proponents: "Our Society uses this sacrament often, unlike virtually any other order," admitted Canisius in 1586.[3] They believed they were re-creating the primitive Church and, more importantly, offering to the faithful the spiritual health provided by the Eucharist. The Council of Trent stated that the sacrament was "spiritual food for the soul" and "an antidote through which the faithful may be freed from everyday guilt and preserved from mortal sin."[4] Canisius proclaimed in his catechism the benefits bestowed by the Eucharist:

> *[T]his is the holy banquet wherein Christ is received, the memory of his Passion is solemnified, the minde is replenished with grace, & a pledge of future glory is given us:* As the Church mooved by the feeling and experience of these fruites, doth notablely sing. . . . This is the Bread that descended from heaven, and giveth life to the worlde, & upholdeth & strengthenth our mindes in spiritual life. . . . This bread is a medicine causing immortallitie, a preservative, never to die, but to live in God through Iesus Christ.[5]

For the Eucharist to have this efficacy, however, it was crucial that Christ be truly and substantially present in the Eucharist. But the real presence was one of the most important loci of the Reformation assault on Catholic dogma and was subjected to extremely bitter attacks. Much of the debate on the real presence hung on what *real* really meant. Zwingli's followers claimed that it was absurd to believe that Christ was substantially present; his presence was merely symbolic. The Calvinists claimed his presence was real but spiritual.[6] Catholics insisted that their doctrine rested upon a literal reading of the gospel: "Hoc est corpus meum," Christ proclaimed at the Last Supper (Matthew 26:26; Mark 14:22; Luke 22:19). This point was declared to the faithful in the Corpus Christi procession, which stopped four times to listen to the Gospels, and to the Reformers in countless polemical texts, thereby turning their doctrine of the primacy of Scripture against them. The Council of Trent, which met to "fix" Catholic doctrine in response to the Protestants' attacks on it, anathematized those who denied the real presence and claimed the sacrament was merely a sign.[7]

But the question of how Christ was truly present in the bread and wine was difficult. It could be in part explained by reason, but much had to be taken on faith. Some were fortunate enough to have achieved an understanding of the mystery of the real presence through mystical experience. During his retreat at Manresa in 1522, when Ignatius underwent a fundamental conversion experience, he received an illumination on the great mysteries of the Catholic faith, including the Eucharist. As his "autobiography" recounts, "[W]hile he was hearing mass . . . at the elevation of the Body of the Lord, he saw with interior eyes something like white rays coming from above. Although he cannot explain this very well after so long a time, nevertheless what he clearly saw with his understanding was how Jesus Christ our Lord was there in that most holy sacrament."[8]

Less fortunate mortals had to rely on the Church and its theologians for an explanation of how Christ was present. Trent not only reaffirmed the doctrine of the real presence but also enshrined one particular account of how bread and wine became the body and blood of Christ, namely transubstantiation. It considered all other accounts of how the transformation occurred to be incompatible with the real presence. Thus if someone held that Christ was substantially present but stated that the presence was the result of a process other than transubstantiation—as Luther did in his doctrine of consubstantiation—he was still committing a heresy. The Jesuits were committed to the Tridentine account of the real presence and transubstantiation. Furthermore, as fundamental truths of the Catholic faith, the real presence and the doctrine of transubstantiation took epistemological precedence over any philosophical doctrine. This hierarchy had consequences for the Physics curriculum.

Quantity and Substance

Two important questions in the Physics curriculum had crucial ramifications for the doctrine of transubstantiation and consequently for the real presence: the relationship between substance and quantity and the identity of the eucharistic species with Aristotelian absolute accidents. In the first case, the Jesuits inherited a debate that had been in progress since the fourteenth century but that they continued to shape. In the second case, the central problems received new force and urgency during the seventeenth century in response to fundamental claims made by the increasingly popular schools of the mechanical philosophy.

To grasp the significance of both of these questions it is necessary to have some understanding of Thomas Aquinas's exposition of transubstantiation in the *Summa theologiae,* where he considered the question of how Christ exists in the sacrament. Thomas was attempting to synthesize two centuries of discussion prompted by the claims of the heretic Berengar of Tours (ca. 1000–1088) that the substance of Christ could not be present in the sacrament. Berengar's argument was simple: what looked, felt, and tasted like bread was bread, not Christ. Thomas abhorred this view. On the other hand he sought to avoid the "excessively sensual" responses to Berengar that seemed to imply that the faithful were chewing Christ's raw and bloody flesh.

After affirming the real presence, Thomas conceded that it seemed absurd that Christ's body could be contained in the tiny dimensions of a host. Nevertheless, he declared that "according to the Catholic faith it is absolutely necessary to profess that the whole of Christ is in this sacrament."[9] But it was necessary to distinguish between the two ways the body of Christ was present in the sacrament. First, it was present through the "sacramental sign": through the priest's words of consecration, the substance (both substantial form and prime matter) of the bread and wine was changed into the substance of the body and blood of Christ. Second, it was present through "natural concomitance": since the body of Christ in the sacrament was the same as the body of Christ in heaven, it was joined to the attributes of the body of Christ in heaven, such as his quantity, soul, and divinity. These heavenly attributes, however, were not actually present in the Eucharist; only the substance of Christ's body was the sacrament (which was why the faithful did not eat chunks of bloody flesh). Therefore, the quantity and dimension of his body were not actually in the host. Thomas wrote that

> Christ's body is in this sacrament by means of substance and not by means of quantity. But the actual totality of a substance is as truly contained by small as by large dimensions; for example, the whole nature of air is equally in a large as in a small amount of air, and the whole nature of man is equally in a large or a small man. Therefore, the whole substance of the body and the blood of Christ is contained in this sacrament after the consecration in the way that the substance of bread and wine was there before it.[10]

In his reasoning Thomas made use of a clear distinction between substance and accidents to avoid equating substance and quantity. For him, if substance was the same as quantity, then the sacrament would have to have Christ's dimensions as a consequence of having his substance. Therefore, any attempt to identify quantity or extension with substance would automatically run into the problem of explaining how then the substance of Christ could exist in the sacrament without his extension.[11]

After discussing the exact nature of Christ's presence in the sacrament, Thomas turned to the question of the accidents that remained in the sacrament. His treatment contained two articles that were crucial for the later debates both on the question of substance and quantity and on the question of absolute accidents. The first article considered whether the accidents remaining in the sacrament had a subject in which to inhere. For them not to have a subject would be deceptive and against the natural order of things, would contradict the very definition of an accident,[12] and would render the accidents without that which individuated them—that is, what made them different from other accidents.[13]

To this Thomas replied that the accidents could not inhere in the substance or substantial form of the bread and wine because, as he had already shown in an earlier question, it did not remain. After eliminating all the other possible subjects (such as the air around the sacrament), Thomas declared that the accidents did not inhere in any substance. But this could not occur naturally and, in fact, had to be a miracle brought about by God acting through primary causes. He explained this by stating that "since an effect depends more on a first cause than a secondary cause, God, who is the first cause of substance and accident, is able to conserve an accident in being when its substance has been removed through his infinite power, through which He conserves the accident in being just as its own [secondary] cause did."[14] As the primary cause, God could replace any secondary cause, in this case a substance conserving an accident, with his own power.

Thomas's second article stated that although the accidents did not inhere in any substance they did inhere in something, namely the "dimensive quantity" of the bread and wine. Quantity was the first accident, and all other accidents adhered to their subject through the mediation of quantity. It was not in the nature of an accident to be the subject of another accident, but when through divine power an accident could exist in itself without a substance as its subject, it could also function as the subject of another accident. In this way quantity of the bread and wine was the subject of the other accidents, such as whiteness and taste.[15] There is no doubt, then, that for Thomas quantity was not substance (or either of substance's two components, substantial form or prime matter) but an accident, albeit one that could be supernaturally endowed with unusual powers in the sacrament of the Eucharist.

Even though Trent adopted an essentially Thomistic eucharistic theology in the mid–sixteenth century, Thomas's was by no means the only tradition within eucharistic theology. In the thirteenth and early fourteenth centuries Thomas's theology had not yet taken on the canonical status it was to achieve after Trent under the impulse of the scholastic revival of the later sixteenth century. While Thomas held that transubstantiation was the only doctrinally correct way to explain the real presence, in the later thirteenth and early fourteenth centuries, many theologians did not consider transubstantiation to be *de fide.* Even as transubstantiation asserted itself, many theologians insisted that it was by no means the most reasonable explanation, and they granted their assent to it only under compulsion of the Church's authority.[16] Nevertheless, by granting transubstantiation their assent, theologians were able to continue debating many elements of the doctrine.

On the question of substance and quantity, the English Franciscan William of Ockham (ca. 1285–ca. 1349) became the main authority for the rival Nominalist tradition. In response to Thomas's view that quantity was an accident, Ockham argued that quantity was not different from substance and that any distinction between the two could not be proven. Ockham said that quantity was a connotative term, not an absolute one, signifying only that a substance or a quality had "part outside of part." Substance was not quantity per se; rather, substance was quantified, or a quantum, if it had "part extended outside of part." Furthermore, qualities such as color were also quanta as they had "part outside of part." Ockham explicitly said that the body of Christ in the sacrament was not a "quantum," as it clearly was not extended.[17]

Although Nominalists claimed that this view of quantity was doctrinally correct, it was constantly subjected to harsh criticism as being incompatible with transubstantiation and consequently the real presence. For example, in 1325, at the urging of Pope John XXII, who had beatified Thomas Aquinas, John Lutterell drew up a list of fifty-six errors in Ockham's "pestiferous teachings." The twelfth of these was "[t]hat both the quantity which is continuous and the quantity which is discrete are the very substance itself," which led to the error that "what is seen with the bodily eye is the body of Christ in the sacrament."[18] To say that substance and quantity were the same implied that Christ must be quantitatively in the sacrament, something that Thomas had emphatically denied. It followed that if Christ's quantity were not there, then his substance must also not be there. For the Thomists, Ockham's view led ineluctably to a denial of the real presence.

When Trent ruled that the substance of the bread and wine ceased to exist in the sacrament, those who followed Thomas's eucharistic theology read this as confirmation of the distinction between substance and quantity, for the quantity of the bread and wine certainly remained after their substance was gone. Nevertheless, the Nominalist view persisted and had its supporters across the centuries. Not only did an ontologically distinct entity of quantity offend against Ockham's razor by postulating a superfluous entity, but the separation of quantity, which bestowed extension and impenetrability, from matter was for many intuitively implausible. How was it meaningful to speak of a body that could be separated from its quantity and consequently not be extended? Could nonquantified body really be body?

When the Jesuits began their teaching in the mid–sixteenth century, they inherited a long tradition of refutations of the Ockhamist or Nominalist view of quantity based on its incompatibility with the Thomistic account of transubstantiation. While it would be wrong to say the Jesuits were strict Thomists, Thomas held great authority in the Society. The *Ratio studiorum* instructed Jesuit professors to "follow the teaching of St. Thomas in scholastic theology, and consider him as their special teacher; they shall center all their efforts in him so that their pupils may esteem him as highly as possible."[19] However, on the whole, Jesuit support for the Thomist position on the question of substance and quantity was at best lukewarm. The appeal of the Nominalist account, which had allowed the doctrine to survive across the centuries in the face of persistent attacks, worked upon the Jesuits as well.

In his *Disputationes metaphysicae* Francisco Suárez accurately reported the most telling Nominalist arguments—for example, that there was no observable effect requiring a distinction between substance and quantity. Nevertheless, Suárez rejected Ockham's view of the identity of quantity and substance, stating that it contradicted the common opinion of the theologians, philosophers, and Aristotle. However, he was forced to make a significant admission. The opinion that distinguished quantity from substance, he wrote, "must be held completely, for although it cannot be sufficiently demonstrated through natural reason, nevertheless from the principles of theology, particularly because of the mystery of the Eucharist, we are convinced it is true."[20] God could separate quantity from the substance of bread and convert the latter into his own body and blood only if quantity was naturally (*ex natura rei*) distinct from substance.[21] Only through an appeal to Catholic dogma could Suárez settle a hotly contested philosophical debate in favor of an opinion that even its defenders admitted could not be rationally demonstrated.

In the German assistancy the Nominalist view of quantity was supported even more enthusiastically. Rodrigo de Arriaga in Prague wrote favorably of the Nominalist view of quantity and substance in his *Cursus philosophicus*. He claimed that "ex Philosophia" there was no reason to postulate quantity distinct from prime matter and that this opinion had sufficient supporters both ancient and modern who provided very strong arguments. The function of quantity was to render a thing impenetrable. Arriaga admitted that it was impossible to determine whether something was impenetrable by having something added to it or by its essence, but then he drew the typically Ockhamist conclusion that when we do not have positive arguments for proving a plurality of entities (i.e., quantity being something additional to the thing itself), we must deny that they exist. Arriaga conceded that his opinion was not certain because of the argument derived from the Eucharist that quantity was distinct from prime matter, but, in contrast to Suárez, he maintained that the latter argument was not convincing and was merely probable.[22]

These debates were also rehearsed in academic disputations, but there Suárez's and Arriaga's openness to Nominalism was generally rejected in favor of a safe Thomism. An unusually detailed treatment of the issue occurs in a 1594 disputation from the University of Würzburg entitled *On Quantity, Place and Time.*[23] Its authors wheeled out the standard arguments against the Nominalist position. They stated in Thomistic fashion that quantity was an accident and disagreed with those who assigned to substance some inherent extension. The Nominalist position that the essence of prime matter was quantity was therefore absurd. The disputation argued that the Nominalists' attempts to reconcile their view with the mystery of the Eucharist were ultimately incapable of accounting for Christ's real presence in the Eucharist without having to resort to creating different types of quantity—that is, Ockham failed his own test.

In fact, a rejection of the Nominalist view was a standard part of Jesuit academic Physics texts. Nevertheless, all texts that dealt with quantity were forced to resort to an appeal to the real presence of Christ in the Eucharist in order to deny the identity of quantity and substance. A disputation held under the supervision of Kaspar Lechner at Ingolstadt in 1616 was typical in asserting that "[q]uantity is really distinguished from substance. This is revealed in the venerable sacrament where the quantity of the bread remains once the substance of the bread has been destroyed."[24]

However, if the frequency with which the revisers general had to address the issue is any indication, classroom lectures may have been as open to Nominalism as Suárez and Arriaga, and the provinces were full of professors espousing the Nominalist position. In response to queries from the provinces, the revisers general wrote judgments rejecting this opinion fourteen times between 1612 and 1649.[25] For example, on 14 July 1612 Revisers General Camerota and Lorini wrote:

> Regarding the proposition of a certain [unnamed] Master of Philosophy who said that he had taught that quantity is not an accidental thing distinct from material substance and the other material accidents and that he had been instructed by the provincial either to retract it or to propose publicly no statements concerning

> quantity: it does not appear to us that the author should be forced to retract it. Indeed, that opinion is held by almost all Nominalist authors. However, the provincial correctly ruled that the opinion may not be defended in theses, because without doubt it is not in harmony with either Aristotle, Saint Thomas, or many other theologians and provides insufficient support for the principles of the faith.[26]

The "principles of the faith" no doubt referred to transubstantiation. The revisers permitted treating the matter in class but prohibited a defense of it in public.

But the revisers' position hardened, and in 1615 they wrote in response to the same proposition that it could be "neither taught nor defended."[27] In 1624 they complained in frustration to General Vitelleschi that some members were not adhering to the required solidity of doctrine in that they were teaching and publishing forbidden opinions, including that the heavens were fluid, that quantified things were composed of indivisibles, and that "quantity is not an accident distinct from the quantified thing."[28] Because of this repeated need to rule against the opinion, the Ordinatio of 1651 included on its list of forbidden philosophical opinions (derived largely from the revisers' records) Proposition 23: "It is doubtful whether quantity is distinguished from matter. Similarly whether quantity and impenetrability are distinguished from substantial forms."[29]

The Ordinatio did not immediately dull enthusiasm for the opinion. Melchior Cornaeus's support for the Ordinatio was anemic at best. In his 1657 compendium Cornaeus treated substance and quantity, swapping "You say—I respond" arguments in his typically lively style. But it was clear that the "I" identified with the Nominalist opinion and that the "you" could raise only inadequate objections. According to Cornaeus the Nominalists identified quantity with prime matter itself, and its function was to render a thing impenetrable. A form could not provide impenetrability, as a form was a "spiritus," which was by its nature penetrable.

But in a fashion that recalls his grudging adherence to the Society's ruling on the absolute nature of heaviness and lightness, Cornaeus made an abrupt about-face, conceding that it was probable that quantity was an accident distinct from prime matter but making clear that he was submitting to authority: "For the authority of those people to whom I am rightly subject orders that this side be taught in our schools; I yield to them and approve of their answer." He added, "[W]e must assert concerning quantity that it is an accident distinct from substance and truly separable, and actually separate in the Eucharist."[30] Once again only the weight of the Eucharist compelled Cornaeus to declared his allegiance to the Realist position. But even then he denied that quantity was the immediate subject of the other accidents, as the Thomists claimed, asserting instead that God himself maintained them in the Eucharist miraculously.[31] The view of prime matter that Cornaeus espoused was becoming strikingly "material." Prime matter had its own existence and its own impenetrability. Far from being the traditional scholastic prime matter lacking any characteristics beyond the ability to receive form, it was coming to resemble the mechanical philosophers' view of body; Cornaeus's prime matter was becoming extended stuff.

So even on a natural philosophical question with direct consequences for the Eucharist, there was considerable dispute within the Society. The revisers general in Rome were increasingly frustrated as they repeatedly ruled against Nominalism yet seemed unable to compel consent. Even when consent was given, the Society's hierarchy could not make professors stop presenting the forbidden opinion in a more convincing fashion than the approved one.

Absolute Accidents

Although the question of absolute accidents and the Eucharist was related to the issue just treated in its concern with providing a philosophical account of an aspect of the doctrine of transubstantiation, it was in many respects distinct. The issue at stake was whether the species or external appearances of the Eucharist could be explained only in terms of scholastic accidents (often termed *absolute* or *real,* as they existed absolutely and not merely in the perceiver). For the Jesuits, any other account of the species led to a denial of the real presence. In response to the Protestant reformers who had vociferously denied the validity of transubstantiation or even the real presence, Trent reaffirmed both the real presence and the doctrine of transubstantiation. However, its chapter on transubstantiation merely used the term *species* to refer to the external appearances of the sacrament and made no mention of accidents.[32]

The fundamental question at the crux of all the debates on absolute accidents and the Eucharist was what the fathers at Trent meant by *species.* When the fathers at Trent used the term, did they, having been raised in the scholastic tradition, automatically assume that *species* meant absolute accidents and use the terms synonymously, or did they use a general term such as *species* because they deliberately sought to avoid tying theological orthodoxy to any particular philosophical system?

Modern theology can easily avoid these poles and solve the problem by admitting the historical contingency of Trent's vocabulary. Edmund Schillebeeckx, for example, has argued that the fathers at Trent were not consciously trying to enshrine or avoid enshrining any particular philosophical school as orthodoxy. All the fathers thought of the dogma in Aristotelian terms and quite un-self-consciously used the philosophical vernacular available to them, which happened to be Thomistic-Aristotelian. For them, *species* and *accidents* meant exactly the same thing. "Putting a distance between themselves and this way of thinking would, for them, have been equivalent to a refusal to think at all meaningfully about this mystery of the faith." But since Trent's vocabulary was historically contingent, Schillebeeckx argued, the vocabulary used to express Trent's theological truths could change while the truths remained.[33]

This historical contingency was of course not apparent to participants in debates at the time. The options were more stark. The majority of theologians were just as embedded in scholastic thought as the fathers at Trent and regarded species as being equivalent to absolute accidents. This was not contingent but truth for all time. In contrast, philosophers and theologians who rejected the scholastic categories of substance and accident—and while there may have been few in the mid–sixteenth century, there were more and more in the seventeenth century—argued that Trent had deliberately avoided using the term *accidents,* as it had not wanted to make any philosophical system into dogma. Therefore, other accounts of species were acceptable.

Both sides thought historical investigation could determine what Trent intended. One of the most striking features of these debates is the extent to which philosophical texts on the subject of absolute accidents are punctuated by lengthy excursions into the history of dogma. Opponents of absolute accidents, for example, argued that conciliar and papal rulings had consistently avoided the term *accidents.* The main concern of canonical rulings was to confirm the real presence of Christ in the sacrament in response to heretics such as Berengar of Tours, John Wyclif, and Jan Hus, who asserted that the presence was only symbolic. Virtually no medieval canonical statements on eucharistic matters mentioned the species at all, let alone offered a philosophical definition of what they were. For example, Berengar's forced recantations at Rome in 1059 and 1079 merely affirmed the real presence.[34] Compared to the issue of the real presence, what the species were was unimportant, so why make absolute accidents obligatory?

Certainly some conciliar statements on the doctrine of transubstantiation necessarily referred to the external appearances of the sacrament in order to emphasize their distinction from the substance. Usually councils used the term *species.* But to account for the instances that the councils used *accidents,* opponents of absolute accidents dismissed this usage by arguing these rulings were concerned primarily with confirming the real presence and had not explicitly sought to provide a philosophical account of what the species were; the councils had not sought to make absolute accidents orthodoxy.[35] While the persistence of the species of bread and wine without their substance of bread and wine was a matter of faith, Trent certainly had not decreed that the equivalence of the eucharistic species and absolute accidents was as well.[36] Similarly one could argue that while Thomas had stated that to deny transubstantiation was heretical, he did not similarly judge the use of philosophical terms other than *absolute accidents* for the eucharistic species.

In contrast, those who insisted that by *species* Trent meant accidents and that nothing else was acceptable could respond that Thomas clearly used term *accidents* to refer to the species remaining in the sacrament. Moreover Thomas used several terms—*accidents, species,* and *qualities*—almost interchangeably when discussing the Eucharist. And these philosophers and theologians could list pages of authorities whose usage was similar to Thomas's. Furthermore, when the councils said accidents, they meant accidents. And if accidents were necessary to account for transubstantiation, and transubstantiation was necessary to uphold the real presence, then denying absolute accidents undermined explanations of the real presence. This was what was at stake.

For the Jesuits there was no question that by *species* Trent meant accidents. The first generations of Jesuits were raised on a diet of scholastic philosophy and Thomistic theology. Their intellectual world was virtually that of the fathers at Trent. In fact two of the first Jesuits, Diego Lainez and Alfonso Salmerón, had been at Trent and had participated in debates on the Eucharist. The category of absolute accident was the only acceptable interpretation of *species.* Furthermore, Jesuit theologians stated that this was a matter of faith—that is, something Catholics had to believe. Suárez wrote that it was a "certain matter in the teaching of the faith." In Germany the theologian Adam Tanner also stated in his widely read 1627 textbook, *Theologia scholastica,* that it was a matter of faith.[37] This was something Trent had not explicitly stated.

Certainly Jesuit theologians granted that many aspects of transubstantiation were open to debate—for example, whether the substance of the bread and wine was transformed into the substance of Christ or annihilated and replaced by his substance. But they regarded it as certain that the species that remained were absolute accidents. This meant that any philosophical system that did not employ accidents, or that could not be reconciled with accidents in some way, would in the eyes of the Jesuits face considerable, if not insurmountable, difficulty in explaining the orthodox account of transubstantiation and consequently the real presence in philosophical terms.

The Philosophical Repercussions

This theological doctrine of transubstantiation held considerable significance for natural philosophy. Claudio Costantini and Pietro Redondi have drawn attention to how the issue of absolute accidents linked theology and physics in a way that had important consequences for Jesuit physics.[38] Both have shown that from the 1620s onward the Jesuits became increasingly disconcerted by materialist physics such as atomism. Costantini and Redondi have identified the great importance of eucharistic accidents in making such physical systems questionable on theological grounds. Although atomism was problematic on many levels in both philosophy and theology and had always been linked by its opponents to atheism, one major source of the Jesuits' unease was their conviction that it could not account for accidents. Since there was no room for an ontological entity such as accidents in a world that consisted solely of extended particles of matter, the Jesuit argued that atomism lacked the entities needed to support the orthodox account of transubstantiation and, ultimately, the real presence.

One way the issue of absolute accidents came to prominence in the 1630s and 1640s was through various vacuum and pneumatic experiments used by some natural philosophers to suggest the corporeality of light. Galileo had already asserted the corporeality of light in *The Assayer.* Whether this was the reason for his 1633 condemnation, as Redondi asserts, is dubious. But there is no doubt that the problem of the substance of light increasingly occupied Jesuit natural philosophers throughout the 1640s. On the basis of a series of experiments begun with Gasparo Berti's (ca. 1600–1643) water column experiment and followed by Evangelista Torricelli's (1608–47) mercury tube experiment, several natural philosophers came to assert that a vacuum was possible in nature.[39] Inspired by Berti's earlier experiment, around 1643 Torricelli filled a three-foot-long glass tube to the top with mercury and then inverted the tube, placing its opening in a bowl of mercury. The mercury flowed out of the tube until it stood at a height of around twenty-nine inches.

The question of what was in the top of the tube immediately arose. If the space in the tube above the mercury was empty yet light could still pass through it, significant problems concerning the nature of light were raised for scholastic philosophers. If there was a vacuum in the tube, then by definition there could be no substance there, as a vacuum contained nothing and substance was clearly something. But scholastic philosophy held that light was an accident. Consequently there would be an accident, namely light, existing without a substance as its subject in the void in the tube.

For those philosophers who did not feel that absolute accidents were a necessary element of the physics of the Eucharist, the solution to this was obvious. The French Minim Emmanuel Maignan (1601–76) grasped the implications that the vacuum held for the nature of light in his 1653 account of Berti's experiment. One did not have to admit "that an accident, that is to say light, would be without a subject or else would be generated from nothing inside that empty flask." Rather, the experiments showed that light to be "a real body," a substance.[40] Faced with the choice of denying the vacuum or reclassifying light as a substance, Maignan chose the latter.

For the Jesuits Maignan's solution could not hold. They insisted that light was an accident; it was neither matter, substance, nor substantial form. To suggest that light was a real body questioned the necessity of accidents, and without accidents the orthodox account of the Eucharist could not be upheld. The categories that formed the very foundations of Catholic eucharistic theology were being undermined. This was a very weighty and dangerous issue; in the claims of the "vacuists" one heard the shouts of the heterodox, wrote the Italian Jesuit Paolo Casati.[41] In contrast to Maignan, the Jesuits refused to reconsider the nature of light and instead claimed that there could be no vacuum in the mercury tube. In response to the texts published supporting the existence of the vacuum, the Jesuits responded with texts arguing for its impossibility. Along with Casati, the Jesuits Honoré Fabri, Nicolò Cabeo, and Nicolò Zucchi all published texts opposing any possibility of the vacuum and indicating the dangers posed to the Eucharist by such corpuscular physics.[42]

Athanasius Kircher had been present at Berti's water column experiment and was familiar with Torricelli's mercury tube experiment. In his 1650 *Musurgia universalis,* Kircher wrote that some people had claimed that the space at the top of the Torricellian tube was truly empty, as no body could have entered to take the place of the fallen mercury. Such people, "like insolent and annoying braggers crowing their triumph before victory, have babbled many things that are clearly not only repugnant to the principles of nature but also dangerous to the orthodox faith; they boast that they can demonstrate with this most subtle experiment that a located thing can naturally subsist without a location, and accidents without subjects."[43] In the debates of the day, "accidents without subjects" could have been read only as a reference to eucharistic accidents.

Despite the Italian and French Jesuits' perception of the seriousness of the question, there is very little evidence to show that by the middle of the century a concern with such issues had entered into the teaching of Jesuit professors in Germany. The German Jesuits had little awareness of these experiments and consequently of their implications. So in marked contrast to their ubiquitous treatment of the much older debate around the identity of substance and quantity, German Physics texts did not yet touch upon the theological significance of absolute accidents. Neither Arriaga's nor Cornaeus's textbooks, for example, addressed the problems that mechanical theories of matter posed for absolute accidents and the Eucharist. This situation was to change after the middle of the century as the Jesuits mobilized against the implications of Descartes's doctrine of matter.

The Cartesian Threat

It may be ironic that René Descartes (1596–1650), who was educated by the Jesuits and counted several Jesuits among his correspondents, became the target of their bitter attacks, including those in the German provinces. But Descartes transgressed against the Eucharist in a way the Jesuits could not tolerate.[44]

In the first edition of his *Meditations* in 1641, Descartes stated that bodies consisted of extension and that this was clear and distinct and guaranteed by the existence of God. All perception was the result of bodies in motion acting upon other bodies. Descartes first circulated the work in manuscript form, giving several scholars the opportunity to consider it before publication. Antoine Arnauld (1612–94) asked Descartes to explain how his view of body could be reconciled with the Catholic faith.[45]

Initially Descartes evaded the issue, writing to Arnauld that the Council of Trent had itself said that Christ was in the Eucharist in a way that "we can hardly express in words."[46] Descartes drew on the very vagueness of Trent's wording to grant himself space on the issue. But Arnauld pursued Descartes on the issue and made his point in a letter to Mersenne worth citing at length due to its admirable clarity in its exposition of the issues at stake:

> [T]he chief ground of offence to theologians that I anticipate is that, according to M. Descartes' doctrines, the teachings of the Church relative to the sacred mysteries of the Eucharist cannot remain unaffected and intact.
>
> For it is an article of our faith that the substance of the bread passes out of the bread of the Eucharist, and that only its accidents remain. Now these are extension, figure, colour, odour, savour and the other sensible qualities.
>
> But M. Descartes recognizes no sense-qualities, but only certain motions of the minute bodies that surround us, by means of which we perceive the different impressions to which we afterwards give the names of colour, savour and odour. Hence there remain figure, extension and mobility. But M. Descartes denies that those powers can be comprehended apart from the substance in which they inhere and that hence they cannot exist apart from it; and this is repeated in his reply to his theological critic.[47]

Arnauld shared the Jesuit view that the eucharistic species were accidents, yet the Cartesian view of body, which explained all perception through the motion of particles, made no provision for absolute accidents.

In 1641 Descartes published Arnauld's first letter in a set of objections in the appendix to his *Meditations.* This time he accepted the challenge to vindicate his matter theory and responded to Arnauld's queries about its implications for the Eucharist. Our senses, he wrote, are stimulated by the "superficies which forms the boundary of the dimensions of the perceived body." This boundary is composed not only of the particles themselves but also of the spaces between, filled with air and water. This superficies is not part of the substance or quantity of any body but is merely "that limit which is conceived to lie between the single particles of a body and the bodies that surround it, a boundary which has absolutely none but a modal reality."[48]

After transubstantiation the new substance is contained precisely within the same limit as the old and continues to act upon our senses in precisely the same way. After paraphrasing the canons of Trent to the effect that the species of the bread remains unaltered, he wrote, "Here I do not see what can be meant by the appearance of the bread, except that superficies which intervenes between its single particles and the bodies surrounding them."[49] Anticipating the objection that Trent really intended *species* to mean real accidents, Descartes stated that to the best of his knowledge the Church had never taught that the species were real accidents miraculously existing by themselves.[50]

At this point Descartes added to the second edition of the *Meditations* a passage he had omitted from the first at Marin Mersenne's recommendation. In the first edition, Descartes claimed he had never denied the existence of real accidents; in the second, he was not so coy and launched an attack on the doctrine of real accidents.[51] Descartes suggested that those who first stated that external species were accidents had done so unthinkingly, for it was an absurd explanation. "The human mind," he wrote, "is unable to think that the accidents of bread are realities and yet exist apart from the substance of the bread, without thinking of them after the fashion of a substance."[52] Thus his opponents were maintaining a contradiction: they said that the substance of the bread was changed, yet they wanted that part of substance called a real accident to remain. "Why should we not choose that opinion which offers the least opportunity to turn aside from the truth of the faith?" he asked, boasting that "I have here shown with sufficient clearness that the doctrine that assumes the existence of real accidents does not harmonize with theological reasoning" and promising to show in his forthcoming *Principles of Philosophy* that it was wholly in conflict with theological thought.

After criticizing the theory of accidents for unnecessarily increasing the number of miracles required to account for transubstantiation, Descartes concluded with the remarkably optimistic statement, "[I]f I here may speak the truth freely and without offence, I avow that I venture to hope that a time will some day come when the doctrine that postulates the existence of real accidents will be banished by theologians as being foreign to rational thought, incomprehensible, and causing uncertainty; and mine will be accepted as being certain and indubitable."[53] Descartes contended that it was actually his adversaries who were truly impious; in suggesting that Descartes's doctrine was contrary to Scripture and the faith, they attempted to employ the authority of the Church to "overthrow the truth."[54]

Thomas Compton Carleton, S.J.

By committing his view so forcefully to writing, Descartes exposed himself to severe criticism. His antagonistic and critical tone was a great provocation, virtually an insult, to theologians who adhered to the Thomistic interpretation that employed absolute accidents. A philosopher, one untrained in theology at that, was making pronouncements on matters of the faith and, even more galling, had leveled the charge that the theologians were not only wrong but abusing their authority to harm the faith. Descartes's charges were certainly taken badly by Thomas Compton Carleton, an English Jesuit who was a professor of theology at the Society's English college-in-exile in Liège. Carleton responded to Descartes in his 1649 textbook, *Philosophia universa.* It had the classic form of the textbook, with three sections paralleling the curriculum: Logic, Physics, and Metaphysics. It was a conservative work in matters of natural philosophy: it included no section on mathematics, took an extremely cautious line on subjects such as the novelties in the heavens, and criticized other Jesuits such as Arriaga.

Carleton announced his polemical intent in his preface, denouncing novel, "clanging" dogmas that attracted the ignorant, who, "seduced by pimps, are lured into the quicksand of error as if by the song of the Sirens." Carleton wanted to prevent "this evil from crawling further" by showing that these opinions were "figments."[55] Carleton here drew upon some traditional tropes: the learned and the ignorant, the silent and the noisy, true philosophy and empty novelty. It was time for the silent majority to stand up and stop the further spread of tainted philosophy.

One source of "evil" was identified in the section dealing with substantial forms. Carleton referred to "a certain recent author who seems to cast all substantial forms out of the universe (beyond that in man and perhaps other living things) and wants fire, earth, water, etc., to be nothing other than prime matter with diverse heterogeneous motions by which their parts are moved and agitated, and in this diversity of motion he seems to attribute the difference between inanimate things at least."[56] A marginal note identified the author as Rene Descartes and the text as part 4 of his *Principles of Philosophy* (published in 1644). Carleton here argued that it was philosophically absurd to deny substantial forms but did not yet draw any links to the Eucharist. Not until the next section, on accidental forms, did he reveal the real danger in Descartes's philosophy. He was amazed, he claimed, to see that a Catholic author could profess something so clearly contrary to the faith, namely that there were no real and physical accidents.[57] Carleton admitted that Descartes was a difficult adversary in that he did not follow the scholastic method handed down in the schools—thereby revealing the discomfort felt by those trained in traditional school philosophy when forced to confront a very different style of philosophical argumentation. Nevertheless, he was capable of a forceful reply.

Carleton identified nine points in Descartes's response to Arnauld in the appendix to the *Meditations* where Descartes had attacked the orthodox view of transubstantiation and the Eucharist. These could be gathered into three groups. The first was Descartes's assertion that equating the species with the accidents of bread and wine was incorrect, did not accord with reason, and was theologically unsafe. The second was Descartes's claim that such an equation had never been the official view of the Church and that theologians were harming the Church by maintaining it. Finally, and perhaps most galling, was Descartes's insistence that his own view was superior.[58]

The central problem facing Carleton was to show that the species referred to by Trent were in fact accidents and could not be otherwise. Once that had been established, it was straightforward enough to use the arguments of Thomas, the Jesuit opticians, and Arnauld to show that Descartes's view of matter was incompatible with absolute accidents and consequently contradictory to an article of the faith. It is worth considering Carleton's method in some detail, as his response to Descartes formed a model for later Jesuit theologians and philosophers over the succeeding century.

Descartes's suggestion that the Church had never taught that the species were Aristotelian accidents did have considerable justification. There was little evidence from the Church councils, other than an ambiguous reference at Constance in 1415, to equate the species with accidents.[59] But one could draw upon other sources to define and explicate the Catholic tradition, and Carleton made an appeal to these authorities to refute Descartes. He argued that it was the consensus of the doctors of the Church that the species of the bread remaining in the Eucharist were true accidents, and he named a vast range of medieval and recent scholastics, including many contemporary, primarily Jesuit, theologians. However, when Carleton proclaimed that "all the theologians speak with one voice," he implicitly referred to those he regarded as orthodox, ignoring those on the other side of the dispute, not to mention those theologians who had been formally condemned as heretics, such as Hus and Wyclif.

Having thereby marshaled the unanimous voice of the theologians behind him, Carleton asserted that it had always been the opinion of the Church that the species were accidents, citing numerous councils, including Florence, Lateran, and Trent, to the effect that the species of the bread and wine remained in the sacrament. Carleton claimed that although these councils had not mentioned the term *accidents,* theologians were unanimous in their opinion that the species referred to in these councils were actually accidents. In addition the Council of Constance in 1415 and the regional councils of Rome in 1413 and Cologne in 1536 specifically stated that the species were accidents. Despite this, Descartes had the gall to state that the Church had never taught that the species were actually accidents. Carleton brushed aside any hint of doubt on the matter.

Carleton then lampooned Descartes's temerity in presuming to pronounce in theological matters. Descartes had said that the human mind could not conceive of accidents remaining without a subject. Of course it can't, replied Carleton, it's a miracle. Augustine had said there were many things that reason could not reach; did Descartes really want to abandon all the divine mysteries merely because he could not comprehend them? Trust in one's own reason rather than in God speaking through the Church was the source of all dissent from the Church, Carleton claimed. Descartes clearly did not know the first thing about theology, yet he condemned accidents in the Eucharist as "unsafe for the faith"—how could one so unlearned say what was safe and what not? After demonstrating that the species were indeed accidents, Carleton argued, like Arnauld, that the Cartesian view of body was incompatible with the orthodox account of transubstantiation and consequently was incorrect and dangerous philosophy.[60]

Thus a philosophical textbook contained a long dispute on a theological issue carried out by appeals to the authority of theologians and of the Church. Once the correctness of the theological proposition had been demonstrated on the basis of this authority, it was used to guarantee the correctness of a physical proposition and eliminate from the realm of the acceptable all other interpretations. In addition to his appeal to authority, Carleton included a philosophical defense of the existence of real accidents, but it followed the theological treatment; the issue had in effect already been decided.

The Society's attitude toward Descartes, particularly on those issues relating to matter and its forms, was not conditioned solely by Carleton. Clearly *post quid* does not mean *propter quid,* although Carleton was frequently cited by later Jesuits on this subject. However, Carleton's 1649 counterattack was symptomatic of increasing Jesuit awareness of the implications not only of Cartesian physics but also of all matter theories that endangered the real presence. The Ordinatio of 1651 prohibited several propositions with direct ramifications for the physics of the Eucharist, although the issue of absolute accidents was not mentioned. Proposition 23 banned equating quantity and matter. Propositions 18 and 19 banned atomism, but since the Ordinatio did not provide any commentary, it is not clear whether they were banned on the traditional grounds that atoms implied equating quantity and substance or whether they were banned because of the more recent issue of absolute accidents, or for another reason entirely. The Ordinatio made no direct mention of Descartes.[61]

Descartes and the German Jesuits

Certainly before the middle of the century, concern with the whole question of accidents and the Eucharist did not figure in a significant way in the Physics curriculum in Germany; the question of the identity of quantity and substance received much more attention. But by the 1650s, a basic familiarity with the central tenets of Cartesian philosophy had reached the Society's philosophy professors in Germany.

Typical of the treatment of Descartes in midcentury was Adam Weber's alphabetically arranged three hundred assertions of peripatetic philosophy of 1653. Weber wrote that "René Descartes, a new author more desirous of novelties than the truth, tries to eliminate all accidents from the universe. But because he relies on arguments that are mere trifles that can be refuted without any Herculean labor, we do not judge him worthy of philosophical anger."[62] Weber did not mention the issue of the eucharistic species, and his dismissive attitude does not seem to reflect an awareness of the gravity of the matter. It is significant, however, that this, the sole mention of Descartes in the text, refers to accidents—the philosophical question that had become indissolubly linked with eucharistic theology and physics.

Weber's disdainful dismissal of the Cartesian threat was not borne out by later developments. The Society became increasingly alarmed at the spread of Cartesian philosophy and its implications. In 1687 the provincial congregation of the Upper Rhenish province appealed to General Thyrsus Gonzalez (1624–1705, general from 1687), asking that he command that the provisions of the 1651 Ordinatio be upheld, particularly to prevent the introduction of novel Cartesian opinions that opposed Aristotle.[63]

The Society found it necessary to develop explicit refutations of Descartes and defenses of absolute accidents. These were then deployed in influential philosophical textbooks. Giovanni Battista Tolomei (1653–1726), a professor at the Collegio Romano and cardinal from 1712, published an important textbook at the end of the century that was among the first Jesuit attempts to integrate experimental natural philosophy into a peripatetic framework in a way that could be taught in schools. It was widely cited by

German Jesuits in the first half of the eighteenth century. Tolomei stated that "the truth that there are absolute material accidents is supported by divine revelation and the mystery of the Eucharist." In support he relied on the ruling of Trent, which, he claimed, stated that "the absolute material accidents of bread remain because the species are themselves accidents."[64] Tolomei did not even offer any historical justification for this equation of accidents and species as Carleton had done fifty years earlier; for him it was obvious.

Through such works, the Jesuit colleges in Germany, which at this stage were still heavily dependent on textbooks from the Society's "core," became increasingly familiar with the issues at stake. By the end of the century an affirmation of absolute accidents on theological grounds had become a standard part of philosophical theses. Drawing on Carleton's treatment of the issue, Professor Friderich Geiger (1655–1734) of the University of Würzburg responded in 1694 to the question of whether it was certain that "from the mystery of the Eucharist there are accidents really distinct from substance, as the peripatetics claim." He cited the councils of Lateran and Trent to the effect that the existence of accidents was "de fide" and claimed that the species mentioned by the councils were the same thing as "accidentia peripatetica." The views of moderns who explained all sensible accidents through the extremely rapid motion of insensible particles and maintained that God excited the species of the bread and wine directly in the eyes of the viewer were unacceptable. In contrast, the peripatetic definition of accidents as something inhering in, but ontologically separate from, substance was confirmed through the mysteries of the faith.[65]

Attacks on Cartesian physics did not occur only as short sections of general treatises. Lengthy dissertations dedicated solely to the task were produced at German Jesuit colleges. In 1705 Professor Georg Saur (1669–1735) at Würzburg set out to demolish the entire edifice built by "Cartesius" (whose name, he noted, was an anagram of *Sectarius*, or gelded). Echoing Carleton's rhetorical stance of one responding to great provocation, Saur wrote that it was time to refute those who attacked peripatetic philosophy as being "full of contradictions, inane and pernicious." Saur made explicit the priority of Catholic dogma over philosophical opinions: the point of gathering together Descartes's teachings and refuting them with peripatetic doctrines was so that "peripatetic students might learn that their philosophy was based on more solid foundations and was in greater conformity with the dogmas of the Catholic Church, which was the unshakable foundation of truth."[66] Saur divided Descartes's natural philosophy into six areas, each of which he then refuted.[67] One of the areas concerned natural body. Here, as his fellow Jesuits had done earlier, Saur appealed to the theological authority supporting the doctrine of absolute accidents, stating that "[a]bsolute accidents must not be denied by the

Catholic philosopher, who does not want to and must not prefer his own philosophical opinions to the mysteries of the faith explicated by the Holy Fathers."[68]

In 1708 Saur followed up his 1705 treatise with his *Historical Relation of the Judgments and Censures against Antiperipatetic Philosophers,* which listed over two hundred pages of refutations and censures of Cartesian views.[69] His overriding purpose was to demonstrate that the peripatetics had centuries of authority on their side in their assertion that eucharistic species were accidents and that attacks on absolute accidents endangered the faith. The first 150 pages traced the Church's condemnations of antiperipatetic positions on absolute accidents.

Saur retraced much of the ground Carleton had covered but went even further. For Saur, the equation of substance and quantity was effectively the same thing as a denial of absolute accidents. This does not seem to be completely philosophically justifiable. Granted, Thomas Aquinas had clearly stated that quantity was an accident, but even if one regarded quantity as substance, this would not necessarily mean one would have to deny the existence of all other accidents. Yet by conflating the two issues Saur could state that all the condemnations of opinions that equated quantity and substance also condemned the denial of absolute accidents. For example, he stated that Bishop Etienne Tempier's 1277 condemnation of the view that quantity exists in itself was motivated by a desire to protect absolute accidents.[70] Saur was thereby able to deploy against Descartes armies of authorities who had never even mentioned absolute accidents.

Furthermore, Saur followed the strategy taken by other Jesuits such as Carleton and argued that it was the unanimous opinion of all the scholastic doctors that the eucharistic species were in fact the sensible accidents of the bread and wine. All the medieval doctors used the term *species* and *accident* interchangeably, he claimed, citing numerous authorities such as Thomas Aquinas, Albert the Great, Bonaventura, Scotus, Durandus, Ockham, and Peter Lombard. Not only the high scholastics but also the earlier ones were drawn into the batteries of authorities Saur deployed: Lanfranc, Guitmund, and Alger, arguing against Berengar, all believed that the eucharistic species were accidents; Rabanus Maurus in the ninth century thought the same; and even the Church fathers Basil, Augustine, Damascene, Gregory, and Ambrose had taught the separability and permanence of accidents. Thus, Saur claimed, "from the first centuries the peripatetic doctrine comes to us through a continuous succession. Can the antiperipatetics support their doctrine with a similar authority?"[71] Saur not only commanded a vast array of older authorities but also cited almost fifty post-Tridentine authors, many of them Jesuit. Not surprisingly, they all confirmed his view.

Saur's historical survey reveals the extent to which, for the Jesuits, real presence, transubstantiation, species, and absolute accidents had become indissolubly linked. This is a dubious association. The Church fathers did assert the real presence of Christ in the Eucharist, but they certainly had not taught that the species were in fact peripatetic accidents—they did not have an articulated theory of transubstantiation in which such accidents could be situated in a meaningful sense.[72] Such a theory was developed only in response to Berengar's heresy, in part to demonstrate that it was heresy, and was not codified until Thomas's magisterial work in the thirteenth century. But for Saur, any mention of species by a council or doctor of the Church was an affirmation of accidents, and any denial of the existence of absolute accidents led ineluctably to a denial of the real presence.

Saur also did not miss any opportunity to compare the words of condemned heretics with those of Descartes and other moderns. He reported that Wyclif had argued that there were no absolute accidents because whatever could exist without a subject was substance and the human mind could not conceive otherwise. Had not Descartes said the same thing? "Compare the texts of Wyclif, Descartes, and other contemporaries, and you will read there the same sense, indeed, even the same words."[73] Saur concluded his historical survey with the pronouncement that "since the opinion of Descartes and the antiperipatetics does not agree with these propositions, as I have demonstrated, they cannot be taught in Catholic academies, nor by men of the orthodox faith, nor can they be supported by sound doctrine."[74]

By the end of the seventeenth century, a denunciation of the long-dead Descartes had become a standard, virtually ritualized, element of all Jesuit Physics texts in Germany, from textbooks to disputations. Nevertheless, just as in the case of Copernicus, Jesuits professors discussed Descartes's views on body at length before denouncing them as false and dangerous to the mysteries of the faith.

CHAPTER 6

The Tension between Mathematics and Physics

ONE OF THE MOST IMPORTANT ASPECTS OF THE TRANSFORMATION OF natural philosophy over the seventeenth and eighteenth centuries was the mathematization of physics. No work announced this new way of studying nature more clearly than Isaac Newton's masterpiece *Mathematical Principles of Natural Philosophy.* This change also occurred in Jesuit natural philosophy, although it was a process that lasted until long after the appearance of Newton's work in 1687. Two aspects of that process are crucial for this story. First, subjects traditionally considered to be part of mathematics, such as mechanics, came to be integrated into the Physics curriculum. Second, particular observations came to be seen as capable of providing the evidential basis for authoritative statements about nature and causal relationships in nature. To understand how and why these transformations occurred, we must first understand the crucial distinctions between the two disciplines that existed at the beginning of the seventeenth century and then show how these distinctions were challenged and refashioned over the course of the century.

General Background

By the sixteenth century mathematics was a vast field that comprised both theoretical and applied disciplines. Mathematics extended far beyond the "pure" mathematical disciplines such as arithmetic and geometry to mixed mathematics, including astronomy, optics, and music, and even to various practical disciplines or *artes mechanicae* such as civil and military engineering, architecture, drainage, and gunnery. The seventeenth-century encyclopedic texts that sought to present mathematics in its entirety reveal its massive scope.[1] A particularly impressive example of the genre is provided by the *Cursus mathematicus* (1661), by the Jesuit mathematician Kaspar Schott.[2] Its twenty-three books began with a summary of all the branches of mathematics. It then treated the branches fundamental to all mathematics: arithmetic and elementary geometry derived from the first six books of Euclid's *Elements.* These led to introductory and practical trigonometry. After exercises in practical geometry the reader progressed to chapters on the sphere or elementary astronomy, theoretical astronomy, practical astronomy, astrology, chronography (measuring time), geography, hydrography, horography (the construction of clocks), mechanics, statics, hydrostatics, hydrotechnics (hydraulic machines), optics, catoptrics (reflection and mirrors), dioptrics (refraction and lenses), military architecture, offensive and defensive siege tactics, the construction of camps, harmonics or music, and finally a synopsis of new mathematical disciplines. Schott was by no means unusual in assigning such a large number of applied or practical arts to mathematics.

Despite its broad scope, in many regards mathematics did not enjoy the same prestige as other branches of philosophy such as physics and metaphysics. Indeed there was considerable question whether it was truly philosophy at all. The debates about mathematics' very status as a branch of philosophy and consequently as a true science extended through the Middle Ages back to antiquity. First, practical arts such as mechanics were tainted as base because they manipulated nature for gain, as opposed to philosophy, which sought only to apprehend the causes of things through pure speculation. A locus classicus in antiquity of this denigration of mechanics was in Plutarch's (ca. 50–ca. 125 A.D.) biography of Archimedes (ca. 285–212 B.C.).

> [H]e yet would not deign to leave behind him any commentary or writing on [military engines]; but, repudiating as sordid and ignoble the whole trade of engineering, and every sort of art that lends itself to mere use and profit, he placed his whole affection and ambition in those purer speculations where there can be

> no reference to the vulgar needs of life; studies, the superiority of which to all others is unquestioned, and in which the only doubt can be whether the beauty and grandeur of the subjects examined, or the precision and cogency of the methods and means of proof, most deserve our admiration.[3]

It was precisely the practicality of applied mathematics, which could be turned toward utilitarian ends and the pursuit of material gain, that tainted the discipline.

Second, for the ancients, certain branches of mathematics such as mechanics and its affiliated disciplines of statics, hydrostatics, pneumatics, and the study of machines were not considered to be part of natural philosophy precisely because they did not study natural motions. While physics described the natural motion of bodies, the devices studied in mechanics moved bodies against their nature. Mechanics moved bodies upwards, against their natural inclination to fall down.[4] The *Mechanical Questions,* a text widely read in the Renaissance and attributed (incorrectly) to Aristotle, stated that art worked against nature; mechanics was the mathematical description of the artificial motion of natural bodies.[5] To separate a part of nature from the organic whole and create phenomena that did not occur in nature would be art and therefore not part of physics. Thus the effects created by machines stood outside the realm of what physics could study. In the Jesuit curriculum of the late sixteenth and early seventeenth centuries, even simple machines such as the wedge, lever, pulley, and screw were all dealt with in mathematical texts and classes, not as a part of the Physics course, because they moved bodies against their natures.

A more modern attack on the philosophical status of mathematics was launched by Alessandro Piccolomini's 1547 *Commentarium de certitudine mathematicarum disciplinarum,* which stated that mathematics failed to meet Aristotle's criterion for a science as it did not and could not provide any causal account of the objects that it studied. While Euclid's geometrical proofs could be held up as the surest and consequently highest form of demonstration, such demonstrations said nothing about the actual existence of geometrical bodies or their causes. For example, one could prove that a triangle contained two right angles, but this did not show that there were such actual things as triangles, nor did it or could it give a causal account of triangles. Piccolomini's attack prompted numerous responses. Jesuit professors were to be found on both sides of the debate sparked by Piccolomini.[6]

Not only was the epistemological status of the branches of mixed mathematics questioned, but they did not enjoy the same prestige as natural philosophy. Many considered mixed mathematics to hold a subordinate or "subalternate" position to physics. Thomas Aquinas explained that "one science is contained under another as subalternated to it . . . [when] in a higher science there is given the reason for what a lower science knows only as a fact.[7] That is, a subalternated science could examine and describe phenomena, but only the higher science could provide the causes of the phenomena. A typical example of this relationship was that of astronomy and physics. Astronomy as a branch of mixed mathematics used geometry and arithmetic to observe and predict the position of the stars, but only natural philosophy could provide an account of the stars' nature—for example, why they moved and what they consisted of.

The classic Renaissance formulation of mathematics' subordinate epistemological and disciplinary position can be found in Andreas Osiander's anonymous prefatory letter in Copernicus's *De revolutionibus,* in which he reassured the reader that Copernicus's system was only a mathematical device, not a physical reality. Robert Westman has commented, "[T]he upshot of Osiander's skillfully argued *Letter* is striking: in believing that he has demonstrated the astronomer's *inability* to draw conclusions in natural philosophy, he denies him the *right* to do so. An epistemological pronouncement becomes a defense of traditional disciplinary boundary lines."[8] Thus a range of factors could be used to exclude mathematics from being truly philosophical.

Practitioners of mechanics and mathematics, however, did not necessarily accept such derogatory views of their disciplines and could draw on ancient traditions that held more favorable opinions of mathematics. In contrast to Archimedes, Hero of Alexandria (ca. 10–70 A.D.) reveled in applied mechanics. The huge number of early modern editions of his works along with the descriptions of his devices in numerous early modern works of mechanics testifies to a growing fascination with applied mathematics from the Renaissance onwards.[9] Jesuit mathematicians such as Schott and Kircher knew Hero's work well.

Furthermore, mathematical practitioners increasingly claimed for themselves the ability to speak on matters traditionally reserved for natural philosophy and physics—that is, the structure and nature of the world. Osiander felt obliged to insert a preface to *De revolutionibus* to deflect Copernicus's implicit claim to the right to speak in matters of natural philosophy even though he was writing mathematics. Although most academic astronomers continued to observe the boundaries reaffirmed by Osiander, preferring not to take the risk of adopting the new role of what Kepler called the "philosophizing astronomer," the sixteenth century nevertheless saw the development of a genre of work promoting mathematics on many grounds. In 1570 John Dee proclaimed the certitude and practicality of the mathematical arts. Philipp Melanchthon valued astronomy for its role in helping man understand Creation. Tycho Brahe also stressed the themes of the certitude and utility of mathematics. Kepler further developed the

Copernican tradition of the mathematical astronomer making new claims in natural philosophy.[10] Of these, only Melanchthon was active in an academic setting.

Another constituency outside the academy played an important role in these developments. During the Renaissance, exponents of practical mechanics such as engineers, architects, artists, and assorted humanists increasingly resented the lowly status of their discipline and began to insist on the scientific and truly philosophical standing of mechanics. Its practitioners, often engineers in princely service, responded to the scorn that schoolmen heaped upon their discipline. Not only did they publish numerous works on various branches of applied mathematics, but they claimed for it an epistemological and social status that would have puzzled academic philosophers. The engineer Agostino Ramelli (1531–ca. 1600), for example, proclaimed that after long debate the philosophers had finally granted mathematics the highest place among the liberal arts. While the philosophers were unable to agree on anything in the natural world as it was so "dark and confused," mathematicians could state things "as sure and irrefutable as if the *Oraculum Apollinis* had said it itself."[11] Not only did the mathematical arts provide certainty, but they had been useful to humanity since the time of Adam. Mathematics helped merchants, surveyors, astronomers, doctors (for how could they heal without astrology?), and generals. It was a double attack against a philosophy that, Ramelli charged, lacked either utility or certainty.

Practical mathematicians and mechanics subverted not only the traditional hierarchy of the arts and sciences but also their classification. Ramelli, challenging the traditional scholarly conception of the liberal arts, wrote: "I have sufficiently recognized that without doubt the single solid and sure pillar of the mechanical as well as the other liberal arts consists of the true understanding of the esteemed mathematics."[12] Thus Ramelli actually counted mechanics as one of the liberal arts, a classification quite abhorrent to academic disciplinary divisions. This view was also to be found in Germany. Josef Furttenbach, Augsburg's prolific seventeenth-century author of books on applied mathematics, wrote, "Whoever amuses himself in the liberal arts can calculate particularly in civil exercises, such as the place of a house or other site, very skillfully and without particular trouble."[13] Placing practical mathematics among the liberal arts undermined their very status as liberal—that is, the preoccupation of those who did not need to engage in banausic activities.

Mechanics was gradually claiming a place as a science as opposed to a practical art. Important for this process in the sixteenth century was the series of commentaries on *Mechanical Questions* ascribed erroneously to Aristotle. Now engineers and others could appeal to the authority of Aristotle, *the* Philosopher, to provide explanatory accounts of the phenomena they described, something mechanics needed in order to be considered a science. For example, Ramelli's explanation of the functioning of a ship's mast and rudder by comparing it to the action of a lever was taken directly from the *Mechanical Questions*, problems 5 and 6.[14] But the commentaries did more than make Aristotle a mechanic. W. R. Laird argues that commentators on the *Mechanical Questions* negotiated a place for mechanics as a science between practical mechanics and natural philosophy.[15] While interest in the pseudo-Aristotle's dynamic approach to mechanics had waned by the end of the sixteenth century and was replaced by the dominance of Archimedean statics, a process furthered by the work of Simon Stevin (1548–1620) and eventually Galileo (1564–1642), the tradition of commentaries on the *Mechanical Questions* had raised mechanics from a largely applied, craft tradition and given it a foot in the academic door. For example, by the early seventeenth century, Jesuit mathematicians at the Collegio Romano such as Giuseppe Biancani believed that mechanics could be a mathematical *scientia* and deliver *demonstrationes potissimae* of greater certainty than the demonstrations of most other disciplines. Such Jesuit mathematicians saw the *Physics* and the *Quaestiones mechanicae* as being essentially continuous.[16]

Jesuit Mathematics

The *Constitutions* enjoined the Society to teach mathematics as part of its curriculum of higher studies but specified only that it be taught "in the measure appropriate to secure the end which is being sought."[17] And the ends sought and the methods used to secure them were as broad as the scope of mathematics itself. Both the pure and practical parts of mathematics could be deployed in the service of the Society's apostolic mission. From the colleges in Europe to the furthest missions, the Jesuits found mathematical disciplines useful in a myriad of ways. Their skill in military engineering was used in China, where they participated in sieges and cast cannons for the emperor.[18] Also in China Jesuit mathematicians such as Matteo Ricci and Adam Schall rose to high rank in the civil service as directors of the imperial observatories. Such uses of practical mathematical expertise were extremely valuable in providing the Jesuits with coveted access to dignitaries both in the missions in China and in the Society's European provinces. In Italy Christoph Clavius won great prestige and the gratitude of Pope Gregory for his contributions to calendrical reform. Clavius was not the only Jesuit with expertise at determining the time and date; as horological skill was essential for missionaries in remote parts of the world where there were no adequate clocks or calendars, the construction and use of sundials was included in the Jesuit colleges' mathematical curriculum.[19]

Despite mathematics' undeniable utility, there was still considerable debate within the Society over its epistemological and institutional place. In a continuation of the debate over Piccolomini's *Commentarium de certitudine mathematicarum disciplinarum,* there was a vigorous dispute at the Collegio Romano over the status of mathematics. Clavius's debate with Benito Pereira, a professor of philosophy there, reveals the tensions between the two disciplines and the resistance encountered by the supporters of mathematics. Pereira, restating Piccolomini, argued that mathematics could not have the status of a true science, as it did not provide demonstrations of causes. While a true demonstration considered essence or affections as they flowed from essence, mathematics considered only common and accidental predicates. That is, mathematics studied quantity abstracted from body or motion and thus could not have access to true natures.[20]

The Society of Jesus' teaching enterprise developed at a time when the status of mathematics and its relationship with physics were undergoing thorough reevaluation. Both university-trained astronomers and engineers were claiming that mathematics was the noblest science for two reasons; it provided both certainty and utility in its multifarious applications. Generally practitioners of practical mathematics did not bother to make their case on the basis of scholastic categories or to respond to the alleged superiority of causal knowledge over all other forms of knowledge but argued solely on the level of certitude and utility.

In the late sixteenth and early seventeenth centuries Jesuit mathematicians, including their most influential representative, Clavius, appealed to mathematics' utility and certitude in their struggle to raise the disciplinary standing of mathematics. But they also sought to refute Piccolomini's charges that mathematics was not a true science as it could not provide causal explanations. Biancani specifically rebutted this claim in his *Aristotelis loca mathematica.*[21] After citing many ancient and medieval authorities, including Aristotle himself, who had believed that mathematics was a science because its definitions were causal, Biancani reasoned that mathematics could fulfill Aristotle's criteria for true scientificity in the *Posterior Analytics* and could provide *demonstrationes potissimae*—that is, true causal demonstrations.

Clavius's own response to attacks on the status of mathematics shared much with the mechanical practitioners', who sought to raise the status of their discipline by asserting its utility. But in contrast to the mechanics and engineers, who in general worked outside the academy at princely courts or in armies, Clavius was attempting to establish a secure place for mathematics within the Jesuit colleges. When the *Ratio studiorum* was being developed in the latter sixteenth century, Clavius, professor of mathematics at the Collegio Romano from 1565 to 1612 and teacher of several generations of

Jesuit mathematicians, attempted to institutionalize a level of mathematical instruction in the curriculum that was advanced for the times.[22]

The Society certainly valued the utilitarian benefits that Clavius said were delivered by mathematics; this ensured that mathematics was granted a place in the Jesuit curriculum. Yet in the face of Jesuit philosophers' opposition or apathy, the result of Clavius's campaign was something of a compromise. The final 1599 version of the *Ratio studiorum* adopted much but by no means all of Clavius's program: mathematical instruction was to be given to all students in the second year of the philosophical triennium for one hour a day, and provision was to be made to offer advanced study in mathematics for selected students. Clavius undertook other measures to flesh out the *Constitutions'* general recommendation in favor of mathematical instruction and to support a pedagogical program in mathematical subjects. He trained several generations of mathematicians who served in the Society's colleges around the world, and throughout his career he methodically produced the textbooks and reference works these mathematics teachers needed.

However, the integration of mathematics into the Society's pedagogical program did not by any means follow a smooth path. The tensions between Jesuit mathematicians and philosophers can be seen in numerous questions along the porous border between mathematics and natural philosophy. For example, when mathematicians treated natural philosophical questions, a task they increasingly took upon themselves in the seventeenth century, they often did not adhere to the strictures of peripatetic philosophy. Consequently, they encountered problems with the *censurae* of the Roman revisers.

Athanasius Kircher was told by the revisers general to omit those passages from his *Iter exstaticum* where they felt he argued against the peripatetic doctrine of prime matter and substantial forms. They claimed he went so far as to insult peripatetic philosophy.[23] Reviser Brunellus was even harsher in his evaluation of the *Iter exstaticum,* stating that it contained many things "inconsistent with the common sense of the schools."[24]

When acting as consultants to the revisers, providing "expert testimony" on manuscripts, mathematicians and natural philosophers could hold very different opinions on the same work. The mathematician Kaspar Schott evaluated Paulo Casati's mechanical works very favorably in his *censura,* stating, "I have read the two dissertations by Father Paulo Casati on the earth moved by machines and on its weight and size and I think that the things they consider are most worthy, for the author proceeds very solidly in those matters, and completely from the principles of statics."[25] But Reviser Gabriel Beati (1607–73) was worried that Casati ascribed weight to the air as if it were his own doctrine and as if he were not merely reporting the opinion of others.

Beati even recommended that Casati omit any mention of the weight of air—a remarkably cautious recommendation, considering that Aristotle himself had stated that air had weight and there was nothing specifically prohibiting this opinion in the *Ordinatio* of 1651.[26]

Mathematics in the German Provinces

Despite Clavius's success in finding a place for mathematics in the *Ratio studiorum*, in the German provinces the structure it decreed remained an ideal rather than reality. While mathematics was taught at a high level at select colleges such as the Collegio Romano, this was not the case in Germany. Two problems confronted efforts to promote mathematics in the colleges in the German provinces: a persistent lack of mathematicians qualified to teach and the continuing controversy over the institutional and epistemological status of mathematics. The first of these was largely the result of the second: mathematics was not regarded highly enough to warrant devoting personnel and resources to it. Many professors viewed mathematics with disdain and did not support Clavius's program.

In 1613 Ferdinand Alber (1548–1617), provincial of Upper Germany, had to respond to a complaint that the Jesuit philosophy professors who made up the arts faculty at Ingolstadt had not been proposing mathematical questions in examinations. Also, the other professors had been treating the professor of mathematics, not as a full member of the faculty, but only as an extraordinary professor. In response Alber urged the faculty not to neglect instruction in mathematics.[27] But Christoph Scheiner (1575–1650), professor of mathematics at Ingolstadt and one of the most prominent astronomers of the age, still complained that mathematics was neglected at Ingolstadt and looked down upon by the other members of the faculty of philosophy. He warned that if things continued in this way "soon we will produce masters of arts without arts."[28] Furthermore, there is little evidence that there were mathematical schools imparting advanced instruction to talented students that could be compared to Clavius's in Rome or Gregory of Saint Vincent's in Belgium.

Despite these problems, the German provinces boasted several prominent mathematicians in the first third of the seventeenth century, although many of these mathematicians had been trained in Rome by Clavius and his successors. The Upper German province was particularly well endowed in this regard, featuring mathematicians of the caliber of Scheiner, Christoph Grienberger, and Johann Baptista Cysat (1588–1657). Their surviving correspondence and publications attest that the first third of the seventeenth century was a particularly lively period for mathematics in the German colleges.[29]

The Swedish invasion and occupation of the 1630s, however, had a catastrophic effect on the development of mathematics at the Jesuit colleges in Germany. All instruction ceased at several universities and was not restored after the Swedish occupation ended in 1636. Mathematical instruction was reinstated only with considerable difficulty. In Mainz, for example, the philosophical triennium, including Physics, was taught again without interruption from 1641, but mathematical instruction was intermittent throughout the 1640s, and the college lacked a mathematics professor on occasion throughout the rest of the seventeenth century.[30]

Despite its strong mathematical tradition, the Upper German province also faced difficulties filling its mathematics chairs in the middle third of the century. The chair of mathematics at Dillingen was empty for thirteen of the twenty-five years following 1631/32. From 1634/35 to 1663/64 the chair of mathematics was occupied for only two years at Freiburg im Breisgau, even though it was a university. The college at Ingolstadt lacked a mathematician from 1640/41 to 1645/46.[31] Evidently the Society's leadership considered it a better use of scarce personnel to ensure that there were no interruptions to the philosophical triennium than to provide mathematical instruction. The colleges in the Austrian province seemed to have had more success in filling their mathematics chairs. At Graz, for example, the chair was occupied virtually continuously throughout the seventeenth century, although most professors held the chair for only one year and only one professor held it for as long as three.

Nevertheless, the situation with mathematics was better in the German provinces than in some others. In 1664 the Polish Jesuit Adam Kochanski (1631–1700) wrote from Germany to Athanasius Kircher in Rome, asking him to intervene with the general on his behalf so that he could stay in a German province, where he would be better able to practice mathematics than in his homeland and would thereby bring greater honor to God and the Society of Jesus.[32]

Another significant contrast between Jesuit mathematicians and their philosopher colleagues can be seen in their career paths. Mathematicians initially shared the standard career path, first teaching the triennium. Most then taught mathematics for only a year or two. But some, after completing their education, were able to dedicate themselves to mathematics, often remaining in the same chair for many years.[33] In this they were quite different from philosophy professors, who cycled once, or occasionally twice, through the triennium and then departed to other duties. For example, Kaspar Schott was professor of mathematics for ten years at Würzburg from 1656 until his death in 1666. He had previously taught mathematics in Sicily and before his return to

Germany had spent two years as assistant to the mathematician Athanasius Kircher in his museum at the Collegio Romano. Because of this long-term commitment to mathematics and related subjects, mathematicians were in general much more familiar with contemporary developments in natural philosophy than their inexperienced colleagues who taught Physics in the triennium.

The mathematics professor had only one hour a day for one year to impart some knowledge of his subject to his students. Therefore, the mathematical curriculum as taught in the classroom could touch upon only a small fragment of the substantial riches treated by a work such as Schott's *Cursus mathematicus.* The lectures of Otto Cattenius (1583–1635) at the University of Mainz in 1610/11 give us a detailed knowledge of the mathematical curriculum as it was actually taught in the classroom. Cattenius was a young mathematics professor who had been trained in the Jesuit college at the University of Mainz by Johannes Reinhard Ziegler (1569–1636). Ziegler had supervised the publication of Clavius's complete works in Mainz and had also corresponded with Kepler.[34]

Cattenius's course commences with a brief treatment of the first book of Euclid. The second section is extremely brief and deals with roots. It assumes some knowledge of arithmetic, which the students must have learned in the lower classes. The third section begins with lessons in practical geometry, including the construction of geometrical instruments. This is followed by exercises on measuring distances, heights, areas, and volumes. The final and by far the lengthiest section is a treatment of the sphere that deliberately departs from the order of Sacrobosco's *Sphere.* The first part covers the circles that make up the sphere of the world. This is followed by a geographical appendix. The second part covers the motions and structure of the heavens and the size, number, and distance of the stars. Another appendix is then included, this one on the theory and construction of sundials. The third section on the sphere covers the vital skill of computing the ecclesiastical calendar. The fourth and final part of the treatment of the sphere treats optics, including mirrors and lenses.

One factor that immediately distinguishes Cattenius's mathematical course from the courses taught by his colleagues in Physics is the practical and applied nature of his subject matter. While many fields of mathematics are omitted from his course, it nevertheless has a clear structure. The earlier lectures in Euclid, practical geometry, and the circles of the sphere provided students with the knowledge and skills necessary to comprehend Cattenius's lessons on constructing sundials and computing the ecclesiastical calendar. Throughout the course there was constant instruction on how to actually do things, such as construct and use a particular instrument or measure the position of stars. Thus the course built on the earlier material to transmit the skills necessary to perform tasks essential to the Jesuits' apostolic mission.

In this regard the mathematical course was quite different from Physics, which proceeded speculatively. An important part of the transformation of natural philosophy over the course of the seventeenth and eighteenth centuries was the shift from an essentially speculative way of proceeding to the practical, applied approach using instruments or at least devices that characterized mathematics. Thus, with the advent of experimental natural philosophy, Jesuit mathematicians, with their emphasis on hands-on, practical training, were much more open to using experimental devices than were philosophers.

What Can Mathematicians Talk About?

Another very significant feature of the changing relationship between mathematics and natural philosophy is clear from Cattenius's lectures. In his treatment of the sphere, Cattenius did not limit himself to those elements necessary to teach his students enough astronomy to create sundials and calendars; he also talked about the structure and substance of the universe. The 1599 *Ratio studiorum* stated that *De caelo* was to be covered by the professor of mathematics, except for the elements and the substance and influences of the heavens, which would be taught in Physics.[35]

But the traditional distinctions between these two disciplines were being undermined. In particular mathematicians increasingly took for granted their ability to make claims about the physical nature of the world. Jesuit mathematicians shared in this view. By the time Cattenius delivered his lectures in 1611, they did not avoid discussing the substance of the heavens.

Mathematicians and philosophers proceeded in different ways. Peter Dear has shown that in natural philosophy the basis for deductively argued syllogisms about nature was inductively reached experiences. For Aristotelian philosophers, an experience was a self-evident statement about how things occurred in nature based on repeated, everyday exposure to such occurrences. An experience was something that an author could assume his readers were familiar with, as they had also had the same experience many times. The reader acknowledged the validity of experiences as the premises of syllogistic reasoning about nature because they were evident. Therefore, reports of singular, historical events were inherently problematic in that they were not everyday occurrences the reader was familiar with and could grant his assent to.[36]

This presented difficulties to the mathematician-astronomer who wished to use astronomical observations, particularly those gained by the use of instruments such as the telescope, which were not evident and immediately accessible to the reader, to argue for a particular statement about the structure of the heavens. Similarly, the experimenter who wished to use phenomena displayed only by a particular machine in a particular place at a particular time to make a general statement about nature faced the problem that the phenomena displayed by his apparatus were not evident to the reader. The fundamental problem, as Dear puts it, was "how can a *universal* knowledge-claim about the natural world be justified on the basis of *singular* items of individual experience."[37] Jesuit mathematicians and, to a lesser extent, philosophers contributed in significant ways to the process of developing the epistemological status of the singular observation to the point where it could be regarded as evident.

Although Biancani had argued that mathematics could provide a causal demonstration, most mathematicians did not take this approach to defending their discipline. "The usual, and most effective, approach," Dear states, "was to carry on as if the mathematical discipline in question were obviously and unproblematically a science."[38] In general this was the strategy adopted by Jesuit mathematicians in Germany; they simply assumed they could talk about the nature of things.

The Fluidity of the Heavens

The debate over the solidity or fluidity of the heavens reveals not just the tensions between Jesuit mathematicians and natural philosophers in the German provinces but also the growing acceptance by natural philosophers of the epistemological reliability of individual observations.[39] Dear notes that Jesuit mixed mathematicians acknowledged that certain observations or events were accessible only to a small number of people—those skilled with instruments and who were there at the time. So to make these things evident, they relied on emphasizing the expertise of the practitioner and on the repetition of observations: either the event was repeatable, or, in the case of astronomical observations, the combined observations of numerous astronomers over many years granted it reliability.

In Germany Jesuit natural philosophers slowly came to accept both of these strategies. The repeated observations of numerous astronomers granted their observations credibility and reliability as the basis for claims about the natural world. Like most mathematicians, the natural philosophers did not explicitly state this, but it was an assumption underlying their arguments in those sections of their texts dealing with the issue of the fluidity of the heavens.

Aristotle's view in *De caelo* was that the planets were carried around by a number of solid spheres.[40] However he did not state whether *solid* meant hard and impenetrable or merely corporeal, which would mean they could actually be fluid. Even the term *crystalline* could be taken by later commentators to support either interpretation. According to Edward Grant, early scholastic commentators took *solid* to imply corporeal yet fluid, although under the influence of the rediscovery of Aristotle's cosmological texts many later scholastics came to read *solid* as hard and impenetrable. However, no supporters of either position considered the matter significant enough to directly address it with a *quaestio* in a commentary. Only after Tycho Brahe published his theory of the heavens in which the sphere of Mars cut the sphere of the sun were scholastic commentaries forced to treat in detail the specific question of whether heavens were fluid or composed of hard spheres. While the bulk of scholastic opinion initially favored hard spheres in this debate, the theory of fluid heavens also had its supporters who could draw on a long tradition.[41]

Jesuits stood on both sides of the questions. Robert Bellarmine had himself taught as a young Physics professor that the heavens were fluid and that the planets swam in them just like fish in the ocean. That Bellarmine had taught this was well known to later Jesuits such as Scheiner.[42] But as the Society's referees of philosophical orthodoxy increasingly denied the legitimacy of multiple readings of numerous questions in the early seventeenth century, this opinion too was consigned to the ranks of the unacceptably heterodox. In his 1614 *censura* of Blancanus's *Loca mathematica,* Reviser General Camerota upheld the distinction between mathematics and physics and ruled that the author was permitted to use materials from Tycho's observations and the telescope but could not assert the truth of the liquidity and corruptibility of the heavens. Consequently, the author had to say that the matter was not yet solved.[43]

General Acquaviva himself also rejected Bellarmine's opinion on 11 October 1614, saying (incorrectly, it must be said) that the opinion that the stars traveled through the heavens like fish in the sea was contrary not only to Aristotle but to the entire peripatetic school and thus should be avoided by the Society and its members. Shortly thereafter he once again warned against the forging of novelties and urged that on the matter of the liquidity of heavens the most common opinion of the philosophers be followed.[44] Acquaviva did not acknowledge that Aristotle's text itself was not at all definitive on the subject or that its vagueness had allowed for considerable space for interpretation through the Middle Ages. Until the middle of the seventeenth century numerous rulings against fluidity made both in Rome and in the provinces referred to Acquaviva's initial precedent insisting that Aristotle's opinion, or at least the officially defined reading of it, be followed.[45] General Muzio Vitelleschi cited Acquaviva in once again prohibiting this opinion from being taught but acknowledged that should it become the more common opinion it would be permitted.[46]

Despite the constant rejection of the fluidity of the heavens by the Society's hierarchy after 1614, a survey of opinions on the question held by Jesuits in the German provinces shows that professors were divided on the matter. More important for this story is the basis on which they disagreed with each other. Although it would be an exaggeration to say that Jesuits took sides on this issue solely according to their disciplinary identities, broad patterns initially distinguished mathematicians and philosophers. Mathematicians felt confident in stating on the basis of astronomical observations that the heavens were fluid. Such observations, including Tycho Brahe's, revealed two phenomena difficult to reconcile with hard, impenetrable spheres. First, certain planets were at times above the sun and at others below. Second, Tycho's observations of cometary parallax suggested that comets were celestial phenomena and not in the sublunary realm. If there were in fact hard spheres carrying the planets around in their orbits, the comets would crash into them. Therefore, there could be no such spheres. In contrast, philosophers were much slower to grant the physical consequences that mathematicians claimed followed from their observations and used various means to account for the observed phenomena—ranging from the astronomers' incompetence to divine intervention—while maintaining the officially sanctioned celestial physics.

A couple of years before Acquaviva's 1614 ruling, Otto Cattenius taught the fluidity of the heavens in Mainz. In his course of mathematical lectures Cattenius stated that if the spheres were solid they would overlap and intersect with each other, "for there is mutual penetration and cutting of the orbs. The reason for this conclusion is that the same planets have been observed to exist sometimes higher and sometimes lower than the sun. This could not happen unless the orbs of the sun and planets mutually intersected and penetrated each other."[47] Comets raised a similar problem. Many comets, including the one that appeared in 1577, "observe paths other than any celestial orb hitherto accounts or could account for." From this Cattenius concluded that "the whole of heaven is completely clear and liquid and not stuffed with hard, real orbs."[48] For the mathematician Cattenius and, most probably, his mentor Ziegler, the crucial factor was that observations of celestial phenomena could not be reconciled with the cosmology of hard orbs, which necessitated another physical account of the nature of the heavens—an account they were not reticent in providing.

Ziegler and Cattenius were by no means the only Jesuit mathematicians to reject an interpretation of celestial physics on the basis of astronomical observations. Even after Acquaviva and the revisers' rulings against the fluidity of the heavens, the revisers were forced to admit in 1624 that the provinces were not adhering to their rulings on the matter.[49] Dissent was common in the German provinces; in his text on the 1618 comet the Ingolstadt mathematics professor Johann Baptista Cysat affirmed the accuracy of Tycho's observations and the truth of his claims about the superlunary position of the comet:

> Wherefore, sure about the certainty of the Tychonic observations, certain about the clarity of his demonstrations, we assert undoubtedly with Tycho that *those seven comets observed by Tycho exist above the moon,* and moreover we indicate as probable that all other comets in the proper sense are not aerial, but celestial, since they all had the same evidence of heavenly seat and motion. We recently declared this opinion, just as we also confirm it by our experimental [i.e., observational] science.[50]

For Cysat, Tycho's conclusions were correct because they were based on astronomical observations.

This faith in the ability of observations to serve as the foundation of conclusions about the heavens was also shared by other Jesuit mathematicians in Germany. Cysat at Ingolstadt, Johannes Lantz (1564–1638) at Munich, and Hieronymus König (or Kinig, 1582–1645) at Dillingen carried out a lengthy correspondence about the 1618 comet with each other and with Rome.[51] The mathematicians did not necessarily agree with each other's conclusions. For example, König stated in a letter to Cysat on 22 March 1622 that the comet was sublunar. König claimed that he could demonstrate on the basis of astronomical observations that the path of the comet was actually beneath the moon. The Roman Jesuits had claimed that the comet was above the moon, but after receiving a letter containing their observations, König concluded that the comet was removed from the earth by only one-tenth of the distance between the earth and the moon, which put it in the upper reaches of the realm of air.[52] By comparing König's interpretation of the Roman observations of the 1618 comet to the Roman Jesuits' published account of these observations, Ziggelaar has shown that König misread the Romans' observations. This accounted for the considerable discrepancy between his conclusion about the location of the comet and that of the other mathematicians. But despite this discrepancy in their conclusions, all these Jesuit mathematicians were agreed on the validity of using observations of the heavens to draw conclusions about their nature.

What was of importance for the Physics course was that Jesuits philosophers also came, albeit gradually, to acknowledge the ability of mathematics to make statements about the nature of the world. A 1634 Physics dissertation presented under the chairmanship of Georg Gobat (1600–1679) at the Jesuit college in Fribourg in Switzerland in the Upper German province asserted that "the opinion of those people who think that the heaven of the planets is now also liquid is not improbable." Gobat gave three reasons for this. First, the majority of the Church fathers taught that the planets swam in the ether "just like fish in water." Second, "because the text of Holy Scripture can be sufficiently well explained [through this opinion]." And third, "because Aristotle himself in the *Metaphysics,* book 12, chapter 8, teaches that astronomers should be consulted in astronomical matters and elsewhere he ordains who should be consulted in his art; and the more recent astronomers are more suitable than ancient. . . . Instructed by instruments [organis] they testify that they have observed in the heavens phenomena whose explanation can be either hardly or not at all reconciled with a motionless solidity of the heavens."[53] That is, the claims of the mathematicians to speak on physical matters could actually be granted by appealing to the natural philosopher's chief authority, Aristotle himself. Gobat also conceded that the solidity of the heavens was a question that fell within the domain of competence of the astronomer. Interestingly, in the passage of the *Metaphysics* that Gobat refers to, Aristotle merely states that he was quoting mathematicians on the subject of the *movements* of the heavens.[54] Thus Gobat's claim that the study of heavenly motions made with instruments could support reliable statements about the *physical structure* of the world goes beyond what Aristotle actually said.

Gobat, despite his generosity toward the mathematicians, believed that the autonomy of the astronomer had limits. He made it clear that the findings of the astronomers on this matter were acceptable only so far as they were compatible with the Church fathers and Scripture—the astronomers were not free to draw whatever conclusions they wished. This was, after all, only one year after the trial and condemnation of Galileo.

The following year at Fribourg another Physics dissertation, chaired by Peter Udr, S.J. (1599–1640), also upheld in no uncertain terms the ability of mathematicians to make physical statements. This was perhaps not surprising, as Udr had himself taught mathematics earlier in his career at Freiburg in 1623/24. He wrote:

> How certain it is that many comets existed in the heavens, and traversed their course in the celestial region, has been demonstrated so clearly and effectively by Tycho through the doctrine of parallaxes concerning the seven comets appearing

> in his age, that Christopher Rothmann, the most expert mathematician of the Landgrave of Hesse, said after much discussion with Tycho, "If any appear, teaching the contrary, they ought to be considered bungling and stupid, and be exposed to laughter rather than response."[55]

These were harsh words, considering that many Jesuits did teach the contrary opinion. Udr did not abandon the peripatetic opinion on all matters of celestial physics; the spots on the moon, for example, were caused, not by "mountains, forests, or valleys," but "by rarer parts of the moon, less capable of imbibing and retaining light."[56]

But other Jesuit philosophers were not as generous to the mathematicians. Johannes Feirabent, professor of mathematics at the University of Mainz from 1617 until its occupation by the Swedes in 1632, echoed the complaints made earlier by Scheiner and revealed the divisions between himself and his philosopher colleagues in a letter to Cysat on 3 February 1620:

> I am [obscure—"happy"?] that Your Reverence is of the same opinion as I on the fluidity of the heavens, which I greatly desire we might establish and confirm with arguments and refutations of objections collected into one book, especially since my peripatetic adversaries amongst my colleagues cannot be convinced that they should have greater faith in the observations of the moderns than in those of the ancients, Ptolemy, etc. On the contrary, there are quite a few who say that with the rejected, useless furniture of eccentrics, epicycles, deferents, etc., they can save our phenomena and the danger of the penetration of bodies.[57]

What caused such skepticism on the part of the philosophers? Discrepancies among the mathematicians' conclusions, such as those produced by König and the Roman professors, could only have lent credence to the conviction of many philosophers that astronomical observations were inherently unreliable. Thomas Compton Carleton, for example, took a skeptical view of the epistemological strength of astronomical observations. As late as 1649 in his *Philosophia universa,* which took a conservative view of novelties in general, he argued that not all celestial observations were to be trusted. Drawing on the doyen of Jesuit mathematicians, he cited Clavius's assessment of Tycho. Tycho's *experientias,* Clavius had said, were "*suspectas,*" conveniently glossing over the complexities of Clavius's opinion of Tycho's work.[58] Consequently, Carleton argued, it was better to follow Scripture, even when its meaning was not completely clear, than to trust the mathematicians.

While Gobat had claimed that Scripture and the Church fathers stated that the heavens were fluid, Carleton presented another reading. Carleton granted that astronomy was a most noble science but argued that

> nevertheless, where any passage of Scripture is cited that, even if it is not completely clear, indicates sufficiently the contrary of that which the astronomers claim to have ascertained through their own observations [*experimentiis*] (such as the solidity of the heavens . . .), especially if in that matter the authority and explication of the Fathers agree, I consider that the certainty of those observations can validly be doubted. Nor are those observations of very great value, so they can and should be rejected, and an uncertain opinion that is in greater keeping with Scripture should be held.[59]

Although Carleton made no reference to Bellarmine's letter to Foscarini, a *locus classicus* for the correct relationship between Scripture and astronomy, the similarity in sentiment is striking. Concerning the phenomena of new stars and comets appearing above the sphere of the moon, Carleton stated, "If those experiences are true, then those new and amazing phenomena ought to be ascribed to miracles, which God performs on occasion for ends known to himself."[60] In short, around the middle of the seventeenth century a prominent Jesuit philosopher could still argue that it was preferable to rely on miracles as explanatory devices for unusual celestial phenomena than to accept the physical implication of astronomical observations.

Moreover, Carleton was by no means alone in the Society in arguing that Scripture and miracles should be accorded greater weight in astronomical matters than the observations of astronomers. A 1642 Ingolstadt disputation on *De caelo* under the direction of Conrad Calmelet (1605–58) argued that it was the authority of Scripture and the Church fathers that determined whether the heavens were solid or liquid:[61] "First of all, Elihu, a friend of the great Job, said that they spread out most solidly, just like bronze.[62] And moreover Sacred Scripture [confirms this] in many places. Likewise the Holy Fathers in many places, so that I may pass over in silence innumerable scriptural interpreters, theologians and philosophers, although strong and most distinguished patrons of the contrary opinion are not lacking."[63] Calmelet granted de facto that the comets could appear in the celestial realm above the sphere of the moon. This did not mean that the heavens were fluid. Citing Arriaga, Calmelet claimed that comets were miraculous and that "they are suspended from whichever part of heaven God wishes: they pass through, stand still and pass away, when God wishes; and God removes impenetrability either from the heavens or, as is more probable, from the comet so that they might pass through."[64] If the heavens were solid, he continued, how was it that certain stars were sometimes seen below the sun and sometimes above? This implied that "many observations, particularly those by Tycho of Mars, are called into doubt. For the diverse motions and circuits of many stars are comfortably saved through epicycles and eccentric orbs. Nor for this reason does any confusion of celestial orbs exist, as some people fear."[65] Unlike the astronomer Feierabent, who wanted to toss out the "rejected, useless furniture of eccentrics, epicycles, deferents," Calmelet questioned the reliability of the astronomers' observations rather than the traditional machinery of the spheres. Traditional astronomy combined with God's active intervention in the world was capable of accounting for all the novelties in the heavens while maintaining the doctrine of the solidity of the heavens. The claims of the astronomers, based on observations, could not stand against the authority of Scripture and the Church fathers.

Edward Grant has argued that scholastic authors gradually abandoned solid orbs for fluid heavens in a process that was largely complete by the 1670s. Despite the views of Carleton and Calmelet, it appears that Grant's thesis is generally correct in the case of the German Jesuits. In contrast to Carleton's conservatism, other textbooks accepted the fluidity of the heavens well before the middle of the century. The first edition of Arriaga's philosophy textbook, which was extremely influential in Germany, stated already in 1632 that because of the observations of mathematicians and astronomers the idea of fluid and corruptible heavens had replaced previous theories. That is, astronomers' observations were reliable.[66] Melchior Cornaeus's 1657 *Curriculum philosophiae* also supported the fluidity thesis. Significantly, the Ordinatio of 1651 did not refer to the issue of the solidity or fluidity of the heavens, implying that even if the Society did not officially recognize the fluid heavens as the "most common" opinion, it had enough supporters to be at least tolerated.

Mathematics and Other Physical Questions

Mathematicians were asserting their right to speak on physical matters in many other fields beside celestial physics. Dear has noted that this can be seen in the titles of numerous dissertations that sought to address physical questions mathematically, using the term *physico-mathematica*.[67] The preeminent example of a mathematician who conducted natural philosophy was Athanasius Kircher. In his museum in the Collegio Romano, Kircher was to a large extent freed from the disciplinary constraints that encumbered professors in the colleges. But as early as 1631, when he was still professor of mathematics in Würzburg, Kircher wrote a dissertation titled *Ars magnesia, hoc est, disquisitio bipartita-empeirica seu experimentalis, physico-mathematica de natura, viribus, et prodigiosis effectibus magnetis,* in which he discoursed on the properties of magnets and their nature.

As mathematicians raised the institutional and epistemological status of their discipline and attempted to use traditionally mathematical ways of proceeding to answer physical problems, mathematics increasingly entered and altered the Physics curriculum. Gradually the new-found prestige of the various branches of mathematics and its proven ability to solve problems clearly made it more acceptable for Physics professors to adopt elements of mathematics. This was a long and by no means uniform process. The almost complete absence of any mathematics from Thomas Compton Carleton's compendium as late as 1649 shows that not all Jesuit philosophers had acknowledged that mathematics had any role in natural philosophy. But there are numerous examples of great openness toward mathematics. In Prague Rodrigo de Arriaga conducted "experiments" with falling bodies in the 1630s in which he did not narrate specific events but was, according to Dear, "framing claims in already universalized statements of experience."[68] However, in the final edition of his philosophical textbook published posthumously in 1669, Arriaga introduced historical descriptions of specific trials. That is, individual observations of discrete historical events had become a satisfactory source of evident experiences upon which one could argue about a physical problem—in this case, natural motion.

In 1645 the Dillingen professor Christophor Haunold (1610–89) sought to combine mathematics with physics, as part of the title of his disputation, "physical theorems mixed with mathematical," indicated. According to Haunold, Plato had said that those who were unskilled in geometry could not enter the Academy

> since he understood that in many arguments of philosophical disputations excellent work could not be done without an understanding of the mathematical disciplines: for many things occur in physical speculations that one unskilled in mathematics cannot penetrate, nor can he explain or treat them according to their dignity. Because of this, we wish to add to these theses a treatment of these sciences so connected among themselves by nature, so that it may be clear how much philosophy is helped by mathematical knowledge.[69]

Haunold also knew Blancanus's *Loca mathematica Aristotelis,* the classic example of a Jesuit mathematical text advocating the relevance of mathematics for physics.[70]

But what did Haunold mean by *mathematics,* given that there is no geometry or algebra in the work? Primarily he meant mechanical, non-natural phenomena. Consider an example from Haunold's work. Physics commentaries had traditionally used devices as "experiences" to argue against the possibility of the vacuum or how rarefaction and condensation occurred. These devices included everyday items, such as watering cans, water pumps, and cupping jars, which were commonplace enough to fall under the category of evident "experience." In the seventeenth century, new devices were pressed into service as experiences in much the same way that these older objects had been used since antiquity. Thus in 1645 Haunold tested Arriaga and Oviedo's explanation of rarefaction and condensation through the intromission and expulsion of corpuscles by conducting an "experientia" using a pump. This device consisted of a piston in a cylinder of the kind used by jesters to inflate bladders—an early modern version of the bicycle pump. With the hole at the end of the pump blocked, it was increasingly difficult to push the piston into the cylinder. With the hole covered it was also increasingly difficult to extract the piston, and if it was released it was sucked back into the cylinder. Haunold concluded that "in the first case the air is condensed with the tube and in the second it is rarefied, and this nobody denies. Now I will prove that this cannot occur through the expulsion or intromission of corpuscles."[71]

Haunold's proof once again consisted of an experience drawn from everyday observations. When a musket was fired, the ball was expelled through the rapid expansion of the gunpowder. But how could the powder expand through the intromission of corpuscles if there was no way for them to get into the powder? This led into a discussion of the intensely difficult and disputed question of the composition of the continuum and of the problems with both Aristotle's view that it was composed of a continuous, infinitely divisible matter and the view that it was composed of discrete particles of matter. This was a problem so difficult that "by the confession of virtually all moderns one could not escape from it," and it was virtually impossible to find a solution that would satisfy everyone.[72] Haunold proposed a theory that combined elements of Aristotle's view with Zeno's theory of indivisible minims.[73]

Haunold was not really conducting experiments in the sense that he was constructing an apparatus in order to create a particular phenomenon, nor was he testing a deductively generated theory. His use of the pump and the musket lay in the tradition of "thought experiments" based on everyday experience, which were a prominent feature of scholastic physics. However, the experiences he drew upon were not quite evident, although the musket was probably within the everyday experience of many people of Haunold's day. But in contrast to most experiences in which nature displayed its normal behavior, these two devices did create "spaces" that were unnatural. The musket in particular created a phenomenon that was distinct from any truly naturally occurring phenomenon. This was a typical example of the way mathematics began to enter the Physics curriculum. Machines, devices that created the artificial, the traditional realm of the mechanic, were being used as the basis for physical speculations.

A more explicit attempt to integrate mathematics into the Physics curriculum occurred just after the middle of the seventeenth century in Melchior Cornaeus's 1657 *Curriculum philosophiae peripateticae,* whose full title declared that it was "embellished with many figures and curiosities drawn from mathematics and restored to physics."[74] The textbook contained a lengthy section on mathematics—quite unusual for a Jesuit philosophy compendium at this time—that covered much of the material normally dealt with in mathematics lectures: geometry, the sphere, and sundials. Apart from including a separate section on mathematics, Cornaeus also adopted a mathematical way of proceeding in examining physical questions.

The question of heaviness and lightness was traditionally a question of physics, as it dealt with the natural motion of bodies, not of statics, as the sub-branch of mathematics that dealt with artificial motion. However, in a fashion reminiscent of Copernicus's statement that he wrote mathematics for mathematicians, Cornaeus claimed that his arguments were drawn from statics and were meant for "staticists" and could not be judged by those unskilled in mathematics.[75] While Aristotle was among his sources favoring the opinion that there was no absolute lightness and that all upward movement was violent (that is, not from an inherent, natural principle of upwards motion) and was caused by the downward motion of denser bodies taking the place of lighter bodies and forcing them upwards, the most influential authority was Archimedes, Galileo's chief influence in mechanics. Also, Cornaeus went beyond relying on everyday experiences to support his argument, citing experiments constructed specifically to examine this particular question.

However, censorship and the injunction to follow Aristotle or the more common opinion were hindrances to philosophers who truly wanted to adopt elements of mathematics, such as experiments, because this often led to novel conclusions. Cornaeus's treatment of the issue of absolute lightness and heaviness in his *Curriculum philosophiae* displeased the Society's hierarchy precisely because he sought to address physical problems by mathematical means and immediately ran afoul of the Ordinatio of 1651. As noted in chapter 2, Cornaeus's conclusion was considered unacceptable by the hierarchy, and he only grudgingly declared his allegiance to the officially sanctioned opinion of two contrary principles of absolute heaviness and lightness. The problem was not that the generals and revisers rejected mathematics per se; rather, it was that the use of mathematical methods led to conclusions at odds with the Society's essentially Aristotelian physics. This was a difficulty that Jesuit mathematicians confronted throughout the first half of the seventeenth century and that philosophers who adopted their methods would also necessarily encounter.

Over the first half of the seventeenth century astronomical observations had become an acceptable basis for claims about the real nature of the world, as long as they conformed to theology and Scripture. But this was not necessarily the case with other fields of mathematics such as statics. What then was to be the place of experiment in the Jesuit Physics curriculum? Could experimental devices provide evident experience about the world if they produced phenomena that were not natural in the traditional sense but created and artificial? For experiment to be accepted as a reliable way of examining nature, it was essential that the action of machines and the creation of events that did not occur naturally could also be regarded as natural by more than just isolated philosophers. Furthermore, what if an experiment produced a result that led to conclusions that undermined fundamental principles of theology and peripatetic philosophy? All of these issues arose when the German Jesuits grappled with one of the most important experimental devices of the seventeenth century, the air pump.

CHAPTER 7

The Peregrinations of the Pump

MATHEMATICS AND PHYSICS AS CONCEIVED AND PRACTICED IN THE Society of Jesus stood in a tense relationship. In the first half of the seventeenth century, the practitioners of these disciplines held markedly different views on what constituted a valid experiential basis for claims about the nature of bodies. Natural philosophers, or physicists, relied primarily on an Aristotelian concept of repeated, everyday experience (*experientia*) as the basis of true statements about the world. Mathematicians, however, relied on observations of particular phenomena. These could be observations of astronomical phenomena in which the observer was assisted by instruments. But they could also be observations of phenomena created with the aid of machines or instruments—that is, experiments. This chapter continues to explore the relationship of experience to experiment and the growing acceptability of the latter among Jesuits mathematicians and natural philosophers by examining the Jesuits' encounter with the paradigmatic instrument of seventeenth-century experimental natural philosophy, Otto von Guericke's air pump.

The meandering path of the air pump from its birthplace in Magdeburg to Würzburg, where a Jesuit priest, Kaspar Schott, published the first account of the device in 1657, and then on to England and back again reveals much about the uses of instruments and machines in mid-seventeenth-century Europe. The pump was seen by many as a curious novelty. There is no doubt that it was quite an extraordinary device, but it was much more than this. According to its inventor, it could create a vacuum—a claim that for the Jesuits was theologically loaded. It also challenged one of the most central tenets of Aristotelian physics, namely that nature abhorred a vacuum. The air pump, then, presented a threat that the Jesuits had to tame.

The Jesuits, including Schott, first did this by treating the pump in the same way they had treated other novel pneumatic devices: they regarded it simply as another of the plethora of *experientiae* that demonstrated the impossibility of the vacuum. However, Schott, a mathematician and amateur of mechanical devices, eventually came to see the pump not as an experience displaying the impossibility of the vacuum but as an experiment that said little about the vacuum but much about the characteristics of atmospheric air. In doing so, Schott presented a classic example of the Jesuit ability to defend the central tenets of Aristotelian natural philosophy while taking full advantage of the benefits offered by the blossoming experimental philosophy.

Kaspar Schott, S.J.

In early 1631 Athanasius Kircher, a young Jesuit professor of mathematics and oriental languages at the University of Würzburg, woke in the middle of the night to a vision of armed horsemen riding into the courtyard of the Jesuit college.[1] The prophecy was soon realized. The victorious march to the Baltic of the Catholic generalissimo Albrecht von Wallenstein had prompted King Gustavus Adolphus of Sweden to intervene in the war in Germany in July 1630 both to protect the German Protestants and to preserve Sweden's position as the Baltic great power. Although he was too late to save the beleaguered Protestant city of Magdeburg from a brutal sack at the hands of the imperial army, Gustavus Adolphus and his army swept south, brushing the overextended Catholic forces aside. Despite Kircher's premonition, Würzburg was ill prepared for the arrival of the Lion of the North. After the rumor circulated that Gustavus Adolphus would slaughter the Jesuits, Prince-Bishop Franz von Hatzfeld ordered them to leave the city. On 11 October 1631 the prince-bishop rode off, saying he was going to get help from Frankfurt, but did not stop until he reached Cologne and never returned. The city, left to its own devices, soon surrendered on 15 October.[2] Not only was the college at Würzburg lost, but the Jesuits were forced to abandon the whole of the "Priests' Alley" (*Pfaffengasse*), the line of colleges stretching along the Main and Rhine Rivers, much to the glee of Protestant pamphleteers.

The events of 1631 link, through their mutual misfortune, the two major human figures of this chapter: Otto von Guericke (1602–86), mayor of Magdeburg and the inventor of the air pump, and the Jesuit mathematician and publicizer of Guericke's invention, Kaspar Schott. In 1631 Schott was a philosophy student and was among the Jesuits who had to flee Würzburg. Schott eventually reached Sicily, where he remained for almost twenty years. He followed the standard career pattern of a Jesuit mathematician and taught at the various colleges on the island.[3]

From Sicily Schott reestablished contact with his old professor, Athanasius Kircher.[4] Kircher had reached Rome where he had established his famous museum of marvels at the Collegio Romano and had begun the prolific authorial output that made him one of the foremost authorities of his age in matters of natural philosophy and magic. In August 1652 Schott was summoned to Rome to be Kircher's "compagno."[5] Whether he was called at Kircher's wish or because his superiors were already preparing him to return to Germany to help rectify the poor state of mathematics in the German provinces is not clear. Whatever the case, in 1655 Schott was sent back to the Upper Rhenish province to teach mathematics.[6] He was sent to Würzburg to be professor of mathematics and ethics at the university he had left over twenty years earlier, and he held this post until his death in 1666.

Letters to Schott from the generals emphasize that he was to write mathematical works. And write he did, despite his problems with the cold German weather.[7] He brought with him from Rome a mass of material that he converted into an authorial output that was phenomenally prolific, even by baroque standards.[8] First came the *Mechanica hydraulico-pneumatica* (487 pages). From 1657 to 1659 his four-volume *Magia universalis naturae et artis* (2,372 pages) appeared. There followed in 1661 his *Cursus mathematicus* (660 pages), a massive compendium of all the branches of mathematics, of which he was particularly proud. It was, he boasted to Kircher, "a work attempted by many but hitherto finished by none."[9] His *Physica curiosa, sive Mirabilia naturae et artis* appeared in 1662 (770 pages; the second edition, in 1667, was augmented to 1389 pages). The *Anatomia physico-hydrostatica fontium ac fluminum* (433 pages) was published in 1663, followed the next year by the *Technica curiosa, sive Mirabilia artis* (1,044 pages). In 1665 his *Schola steganographica* (346 pages) appeared, and in 1666, the year of his death, the anonymous *Joco-seriorum naturae et artis* (363 pages) was published.[10]

It was a massive achievement, not least considering that he also had to fulfill his time-consuming pastoral and teaching duties, he received little assistance, and the university library had been stolen by the Swedes.[11] Still, conditions were good enough in Würzburg that he turned down an offer from Vicar General Oliva in 1661 to return to Rome as Kircher's assistant,[12] although several years later he asked unsuccessfully to be appointed professor of mathematics at the Collegio Romano.[13] In 1665 Schott declined an offer to become rector of the Society's small college at Heiligenstadt. Oliva accepted his response, admitting that a province should not lack mathematicians and that it was easier to find rectors than mathematicians—a telling assessment of the poor state of Jesuit mathematics in Germany at the time.[14] Schott's prolific output continued until his death on 22 October 1666. It is not surprising that his later correspondence shows him having problems with his health and that his necrology stated that he died exhausted by his studies.[15]

Schott consciously took on the role of a disseminator of knowledge, particularly of the mathematical sciences. He recalled that he had noticed in his travels and teaching a great curiosity for such matters not only among boys but also among princes and the learned. While he was Kircher's assistant in Rome there was a constant stream of important visitors to the museum, which he felt redounded to the benefit of the "universal Catholic Church." From that time he had been filled with the desire to pass on whatever he knew that was of interest.[16]

Schott thus shared the Jesuit conception of intellectual production as a form of apostolic activity. One did not have to publish works of controversialist theology to serve the Church; cultural enterprises such as Schott's volumes raised the prestige of the Church in general and the reputation of Jesuit learning and schools in particular.[17] His forte was natural magic, curiosities of nature, and in particular mechanical devices, and it was in this field that he could best serve his Society and Church.

To spread knowledge, Schott first had to gather it, and to this purpose he sought to establish a network of correspondents throughout central Europe. In the *Mechanica hydraulico-pneumatica,* his first book, published only two years after his return to Würzburg, he included an appeal for contributions that he could publish in later works—promising, of course, to acknowledge the contributor.[18] The little that remains of his correspondence shows him asking others for information, passing on reports he had heard, and requesting and sending books. For example, in 1664 he wrote to Baron Johann Christian von Boyneburg asking if he knew of any works containing curiosities suitable for the second edition of the *Physica curiosa.*[19] He also wrote to J. M. Faber, personal physician to the duke of Württemberg, to whom Schott sent lists of all of Athanasius Kircher's publications, advising him on which books the duke's sons should read (including his own *Technica curiosa*) and updating him on Stanislaus Lubienetzski's progress on his book on comets.[20] His later works published the fruits of such correspondence.

Although Schott cultivated the role of a "clearinghouse" of inventions and information in Germany, he saw himself as more than just a disseminator of information.[21] He was a commentator and critic of the material he published. Clearly, he was operating with standards of credulity different from those currently used, but he was quite willing to denounce a claim as false. And if he encountered something particularly noteworthy that warranted careful investigation and evaluation, he was prepared to ask experts for their opinions on it.

The Travels of the Pump

When Schott arrived in Würzburg in 1655 he found there one of the most curious yet important devices of seventeenth-century natural philosophy: Otto von Guericke's air pump. Like Schott, the air pump had reached Würzburg by a roundabout way. Although Guericke, one of the mayors of Magdeburg, began his air pump experiments in his hometown in the later 1640s, the stage on which his pump first received notice was the Imperial Diet in Regensburg. The diet met in 1653–54 to tie up the loose ends left by the Peace of Westphalia, which had finally ended the Thirty Years' War in 1648.

After acquitting himself well as the city's representative throughout the protracted and difficult negotiations in Westphalia, he was sent to the Regensburg Diet. His experience of the tedious negotiations in Westphalia made him well aware that he would have plenty of time on his hands at Regensburg to continue his experiments. Furthermore, such assemblies, which attracted people from many regions and professions, were ideal places to exchange information, a particularly valuable commodity for Guericke, isolated as he was in Magdeburg, a city without a university or reputation for learning.

At Regensburg Guericke met the Capuchin friar Valeriano Magni (1586–1661), from whom, he claimed, he first learned of Torricelli's mercury tube experiments.[22] Magni asserted that he had independently invented the mercury tube while at the court of the king of Poland and published the first book on the new device. This work provoked a bitter dispute over priority with Torricelli's supporters in France, which resulted in Magni's accepting Torricelli's priority. In light of the polemics generated in the priority dispute, Guericke's ignorance of the Torricellian tube experiments is remarkable and reveals the extreme intellectual isolation he endured in Magdeburg.[23]

Guericke was not the only one with time on his hands in Regensburg. As the negotiations dragged on endlessly, the empire's assembled princes, notables, and courtiers generated a high demand for entertainment. There were frequent festivities such as banquets and masques, not to mention constant carousing. Occasionally one of the wealthier princes would arrange something particularly noteworthy. An Italian opera staged by Emperor Ferdinand with fantastical splendor at a cost of forty-six thousand gulden provided a novelty for those present.[24] Such a demand for edification and diversion attracted those who hoped to profit from it. One of those who came was Jacques Royer, a young man with remarkable talents: he could regurgitate liquids in fourteen different colors and project a fountain of water from his mouth "per spatium duorum *Miserere.*" At Regensburg he displayed a fountain of fire in the presence of the emperor.[25]

Others took advantage of the opportunity to sell novelties to the assembled dignitaries. Johann Philipp von Schönborn (1605–73), bishop of Würzburg, archbishop of Mainz, and, by virtue of this latter office, the archchancellor of the empire and most senior of its seven electors, bought what Dr. Johann Weber was offering. A letter from 6 June 1654 reveals that Weber had almost finished the air bed that Johann Philipp had ordered. Weber expected that he would soon solve the problem of the air escaping through small, invisible holes, and he warned that as the pump for the bed was even better than expected, the archbishop should take care that his servants would not overinflate it.[26]

It was in this environment of curiosity, desire for diversion, and showmanship that Guericke performed his air pump experiments. Guericke recalled that

> several enthusiastic followers of these kinds of investigations had heard about my aforementioned experiments: they succeeded in convincing me to demonstrate some of them and I endeavored to do this insofar as I could, considering the limits of my capabilities.
>
> At the conclusion of the diet when it was breaking up, my experiments were brought before the notice of His Imperial Majesty as well as the electors and several princes who were on the point of leaving but wished to see my experiments demonstrated before their departure. I could not refuse their request under the circumstances.
>
> Those experiments were of particular interest to His Eminence the Noble Elector Johann Philipp, archbishop of Mainz and bishop of Würzburg, who of all the spectators present persuaded me to make a similar piece of apparatus for himself. Because of the limitations of time, however, this apparatus could not be reproduced by workmen, and so, at the request of His Eminence, I handed over the experimental instruments I had brought with me to Regensburg for a financial consideration.[27]

Guericke's demonstrations for the "enthusiastic followers" clearly had set people talking about the machine. Despite the modesty imposed on Guericke's description by the conventions of the time, the pump must have attained quite a reputation in Regensburg if it was capable of making the emperor alter his travel plans. But why the instrument was of "particular interest" to Archbishop Johann Philipp von Schönborn is not clear—despite his purchase of Weber's air bed, Johann Philipp may have been motivated by more than just a general interest in pneumatic devices. He was certainly one of the great patrons of the age in Germany and was interested in a wide range of arts and sciences, including alchemy and mechanical arts. He himself wrote devotional poetry. His interests were supported by his chief minister, Johann Christian von Boyneburg, himself a noted patron who, for instance, was responsible for bringing the young Leibniz to Mainz.

There is no direct evidence to show that the elector was aware of the theological implications of Guericke's device. Yet Guericke certainly was not shy in asserting that he had created a vacuum, and Valeriano Magni, one of the most vocal supporters of the existence of the void and bitterest opponents of the Jesuit view of the void and the Eucharist, was present at the diet.[28] Magni and Johann Philipp knew each other well, so it is entirely possible that at the diet they discussed the question of the void, absolute accidents, and the Eucharist.[29] In short, whether Johann Philipp purchased the pump out of a general interest in natural philosophy and mechanics or because he was aware of the theological issues at stake and wished to solve them is unclear. Nevertheless, due to Johann Philipp's interest, the pump's progress from Magdeburg to Regensburg continued on to Würzburg, where the elector entrusted the perplexing device to the Jesuit professors of the university and had the experiments performed in his presence.[30] (See figure 4.)

The *Mechanica Hydraulico-Pneumatica*

Johann Philipp was not noted for being particularly friendly toward the Society. He had been very annoyed by the criticism several members of the Society had leveled at him for his policy of ending the Thirty Years' War quickly, even if the Catholic side had not achieved its goals. He also believed that the Jesuits had intrigued against his election to the position of elector of Mainz.[31] General Goswin Nickel wanted to improve the relationship and wrote to Johannes Kreyling, S.J., at the Diet of Regensburg, "I regret that the Most Eminent Prince of Mainz is imbued with such unfavorable convictions about the Society. . . . Certainly I trust that it shall be your work and effort to

FIGURE 4. Otto von Guericke's air pump is displayed at Prince-Bishop Johann Philipp von Schönborn's residence in Würzburg. The genteel setting and participants stand in dramatic contrast to the burly laborers in Guericke's own illustrations, shown in figure 5. From Kaspar Schott, *Technica curiosa,* 1664. By permission of Houghton Library, Harvard University.

remove completely all his unfavorable opinions and then when he knows the Society more cordially he shall love it more sincerely."[32]

Immediately upon his return, then, Schott encountered a situation in which he could employ all his knowledge to further the Society's cause by winning the goodwill of a prince whose interests closely resembled his own. Thus it is not surprising that Schott presented himself to the elector in Mainz when he returned from Italy in 1655. In keeping with his practice of presenting books and other gifts to the nobility, he asked Kircher to send him an example of his artificial language machine, the "artifici-linguarum," so that he might offer it to the elector.[33] Johann Philipp had found the ideal person to examine his new acquisition.[34]

Schott was fascinated by the pump once he saw it in Würzburg. The device that Johann Philipp had purchased from Guericke was a relatively simple affair (figure 5). It consisted of a round brass receiver approximately a foot in diameter. It had one small opening with a brass stopcock or key that could open and close it. The opening could be connected to the pump itself—basically a giant syringe modeled on those used to fight fires. This brass tube contained a piston and two valves fashioned out of leather flaps. Attached to the piston were leather straps so that several men could pull it. When the piston was extracted, air from the receiver filled the brass tube. The valves ensured that the air did not return to the receiver but was expelled into the atmosphere when the piston was pushed in again. The whole device was placed in a tub of water to ensure a better seal against the air. Operating the machine was a laborious and awkward task and required two or more burly men to heave on the piston. The longer they pumped, the harder it became to extract the piston and the more violently the piston was sucked back into the pump. Evacuating the receiver could take two, three or more hours, depending on its size.[35]

The pump displayed quite remarkable phenomena. If the receiver was weighed before and after pumping, there was a discrepancy of around two ounces. This was in itself strong evidence for the weight and corporeality of the air. If the key sealing the receiver was opened, air was sucked in with great violence. If the opening of the receiver was placed in water and the key opened, the water was sucked in until it almost filled the receiver. The damage inflicted on the lips and fingers of those who foolishly placed them over the opening was further testimony of the violence that was somehow generated by the pump.

Two crucial questions confronted those who encountered the pump: What, if anything, remained in the receiver after the air had been pumped out, and what accounted for the violent suction? To the first question Guericke explicitly answered that he had created a perfect vacuum. To the second, he argued that the suction was due to the

FIGURE 5. The earliest versions of Otto von Guericke's air pumps. The lower one is the one he took to the Imperial Diet at Regensburg in 1654, where it was purchased by Prince-Bishop Johann Philipp von Schönborn. From Guericke, *Experimenta nova* (1672). By permission of Houghton Library, Harvard University.

difference in pressure between the space inside the receiver and atmospheric air. These claims would be difficult for the Jesuits to accept.

By a curious coincidence Schott had virtually finished his *Mechanica hydraulico-pneumatica* when he arrived in Germany. This was not a scholastic philosophy textbook intended for consumption by students at the Jesuit colleges but rather a mathematical work intended for a much broader audience. As an account of a multitude of curious pneumatic and hydraulic machines, many of which were in Kircher's museum, it was the perfect work in which to include an account of the marvelous new air pump.[36] The pump's great novelty earned it the particular status of its own appendix. Schott took advantage of the opportunity to further the Society's cultural-political agenda by dedicating the book to Johann Philipp von Schönborn in a flood of watery metaphors.

Schott can be situated in a number of mathematical traditions. In a surviving manuscript he depicted many variations of machines such as water wheels, gears, and mills designed to transfer power.[37] These rather idealized machines appear to be heavily influenced by Agostino Ramelli, placing Schott firmly in the tradition of practical mechanics, which he knew well. But Schott was just as much a part of the Jesuit academic mathematical tradition, whose practitioners were, by this time, assigning scientific status to mathematics and forcefully asserting their ability to speak philosophically. Schott shared this view of mathematics' disciplinary and epistemological standing.

Schott cited numerous authors on hydraulics and pneumatics, including Hero of Alexander, the ancient master of talking statues and self-opening doors; the Neapolitan nobleman Giovanni Battista della Porta; and moderns such as the Nuremberg patricians Daniel Schwenter and Georg Philipp Harsdörffer, whose entertaining work of mathematics, the *Deliciae physico-mathematicae,* contained chapters on curious pneumatic and hydraulic devices.[38] The weakness of his predecessors, Schott believed, was that "hardly anyone joined theory with praxis, which we do."[39] Schott's self-professed claim to novelty was that he was taking the experience of the practical mathematicians and mechanics and providing it with a theoretical foundation. Thus he divided the *Mechanica hydraulico-pneumatica* into a theoretical section and a practical. The theoretical discussed four principles of mechanical hydraulics and pneumatics: attractive, expulsive, rarefactive, and the natural heaviness of water; the practical provided examples of devices that functioned according to these principles.

While the theoretical section did not adopt the form of scholastic natural philosophical texts, it was heavily influenced by them. Schott's discussion of the first principle of mechanical hydraulics and pneumatics, the attractive force by which nature prevented any potential vacuum from occurring, drew upon a long scholastic tradition of discussion of the vacuum. The experiences that Schott used to show the impossibility of a vacuum and the ability of nature to make bodies move against their natural inclinations were widely current in scholastic textbooks and disputations. They included cupping jars that sucked in the flesh of the patient they were placed upon, bellows that could not be opened when the spout was sealed, water that would not flow out of a glass inverted in a bowl of water, smooth marble blocks that adhered together, and siphons that could actually make water flow upwards.[40]

Schott was writing a work of mathematics, not a work of scholastic physics, so he limited himself to providing *experientia* against the vacuum, referring his readers elsewhere if they wanted speculative philosophical arguments. Similarly, Schott did not delve into the theologically based arguments against the vacuum mentioned. Nevertheless, having been Kircher's assistant in Rome after the controversies over the vacuum and absolute accidents had unfolded in the wake of Berti's and Torricelli's experiments, Schott must have been aware of the theological issues at stake. In short, by the time Schott came to Guericke's experiments, he had a conceptual framework fashioned out of theological, philosophical, and experiential arguments into which Guericke's novelty could be placed. The framework left little room for Guericke's claim that he had created a vacuum.

The Jesuit Assessment

Confronted by the remarkable new device and Guericke's challenging and potentially dangerous explanation, Schott drew upon an array of Jesuit authorities to consider both the device and Guericke's claim to have created a vacuum and published their assessments of the pump in an appendix to the *Mechanica hydraulico-pneumatica.* They provided many of the speculative arguments that Schott had skipped over. First among these authorities was Athanasius Kircher. Schott's old professor had attained great authority in intellectual circles in Europe. He received visitors from all over Europe and correspondence from even further abroad. People with queries concerning natural and magical phenomena turned to him for advice.[41] Faced with Guericke's unsettling claim that the pump produced a vacuum, Schott sought Kircher's assessment. Next was Nicolò Zucchi (1586–1670), a professor at the Collegio Romano who, along with Kircher, had been present at Berti's water column experiment in Rome around 1642. Finally, Schott's colleague at Würzburg, Melchior Cornaeus, wrote an analysis of the experiments that appeared both in his *Curriculum philosophiae peripateticae* and Schott's *Mechanica hydraulico-pneumatica* (figure 6).

FIGURE 6. The frontispiece from Melchior Cornaeus's *Curriculum philosophiae peripateticae* (1657), which, along with Schott's *Mechanica hydraulico-pneumatica,* was the first published work to describe the air pump. More than most Jesuit natural philosophers, Cornaeus sought to integrate mathematics and physics. Courtesy Stadtbibliothek Mainz. Signatur: III l 4°/345h.

What all three Jesuits had in common was their focus on rebutting Guericke's claim that there was a vacuum in the receiver virtually to the point of ignoring his other claims about the role and nature of atmospheric air pressure. Kircher not only rejected the vacuum but saw those who suggested its possibility as dangerous and deluded. In fact, his response to the air pump was virtually the same as his assessment of Berti's water column experiment and the Torricellian tube. Earlier in the *Mechanica hydraulico-pneumatica,* Schott had reprinted Kircher's denunciation in his 1650 *Musurgia universalis* of the babbling braggarts who claimed that "a located thing can naturally subsist without a location, and accidents without subjects."[42]

Reading Schott's description of Guericke's machine did not alter Kircher's opinion of the possibility of the vacuum. Although this time he did not refer directly to the theological issues in his response and instead conducted his argument purely on philosophical lines, he was still amazed at the presumptuousness of those who stated that the experiment showed that there was a vacuum in the receiver—indeed, for Kircher and the rest of the Jesuits, the experiment illustrated much more clearly that there was no vacuum in the receiver. Kircher acknowledged that the piston extracting the air encountered increasingly severe resistance and believed that it reached the point where it could not be withdrawn at all. Like all the Jesuits, Kircher looked *inside* the receiver to account for this resistance, rather than *outside,* as did those who attributed the resistance to atmospheric air. If there truly was no air left in the receiver, then it could not exert any resistance, as it was philosophically absurd that nothing could be the cause of an effect, particularly of such a violent effect. Thus the resistance to the piston had to be attributed to the increasing rarefaction of the air inside and the unwillingness of nature to permit the continuity of bodies to be interrupted.[43]

Zucchi's assessment of the Magdeburg experiment made an argument identical to Kircher's and to his own previous evaluation of those water and mercury column experiments.[44] A vacuum inside the receiver could not cause the resistance, as "the mere negation of body previously contained in the [receiver] cannot exercise resistance against such great effort." Rather, bodies tried to preserve "the unity of continuity in the universe" and resisted being distended. Cornaeus's assessment made the Jesuit opinion unanimous. God had set limits beyond which matter could not be further rarefied. Nature acted to impede a vacuum once this limit was reached with a "retentive power" that no human power could overcome.[45] There could be no vacuum in the receiver. The violent effects were due to nature's determined efforts to ensure that could be no vacuum in the receiver. An explanation drawing on atmospheric air pressure was thus unnecessary.

Guericke's Response

Schott also opened a correspondence with Guericke. One might have expected Schott and Guericke to have borne a certain animosity toward each other. They had both suffered at the hands of the other's confession in the Thirty Years' War. Schott had been driven from his homeland, but this was quite minimal compared to what Guericke had had to endure. In 1632 Magdeburg had been destroyed by the Imperial Catholic Army under Marshall Tilly in the most brutal sack of a city in the whole of the Thirty Years' War. In a bloodbath that outraged even the hardened sensibilities of the time, two-thirds of the city's thirty thousand inhabitants were murdered or consumed in the fire that broke out during the sack and leveled the entire city.[46] For several years Guericke was forced to earn his living as an engineer in the Swedish and later the Saxon armies. Although Guericke was able to rebuild his personal fortune, his city remained a shadow of the earlier trading metropolis on the Elbe.[47] Despite this, the two men were always on cordial and even warm terms. Confessional differences seem not to have played a role in their relationship. Guericke's son stopped in to greet Schott's mentor Kircher in Rome while making his Grand Tour of the Continent.[48] Schott seems to have had no qualms about including Guericke's work in his own books, even though they had substantial differences of opinion. There is no surviving evidence to suggest that the two ever met.

Guericke is rightly regarded as a great experimentalist, but he had also received a traditional scholastic education in philosophy at the Protestant universities of Leipzig and Helmstedt.[49] His *Experimenta nova* shows that he was familiar with scholastic concepts of space and the vacuum. He knew Arriaga's *Cursus philosophicus,* Carleton's *Philosophia universa,* the Coimbran commentaries on Aristotle's *De caelo* and *Physics,* and Riccioli's *Almagestum novum.*[50] With this training, he clearly was able to understand the Jesuit objections to his claims, even if he did not agree at all with them. But Guericke had also been trained at the engineering school in Leiden founded by Simon Stevin and could combine his knowledge of scholastic physics with a practical knowledge of machines, mechanics, and mathematics.

Guericke explicitly claimed that he had established the existence of a vacuum. Although there is a discrepancy between his correspondence with Schott, where he claimed that he had come across the vacuum purely by chance while conducting his experiments, and his published account in the *Experimenta nova,* where he wrote that he specifically designed the experiments to demonstrate the existence of a celestial vacuum, throughout both his correspondence and published work he always claimed that the air pump's receiver contained a real vacuum.[51] Guericke was aware of the weight and elasticity of air—from his earliest correspondence with Schott it is clear that he attributed the great resistance to the action of the pump and the violent suction of the evacuated receiver to atmospheric air pressure—but his stated intention was to end what he considered to be fruitless debates on the existence of the vacuum. This was the claim that the Jesuits latched onto.

Guericke responded to the Jesuits' objections that Schott passed on to him, but he refused to argue in the language of scholasticism. The site of authority was clear for Guericke, and it was not what the Jesuits appealed to. Guericke wrote that neither Aristotle's authority nor that of any other author was valid against a "visual proof," such as what he felt was produced by his air pump.[52] After refusing to play by scholastic rules, he responded to the Jesuits' criticism by arguing that it was based on false theory and unfamiliarity with the operation of the pump. The Jesuits had stated that the resistance to the extraction of the piston increased until eventually it was impossible to withdraw the piston at all and that this meant that the air inside the receiver had reached the limit beyond which it could not be rarefied any further. At that point the natural tendency of matter to maintain its continuity to prevent a vacuum stopped the extraction of the piston. This conclusion was completely incorrect, Guericke countered, for the piston could always be extracted, no matter how long one pumped. Thus there was no limit at which nature would prevent air from being further rarefied.[53]

All effects traditionally ascribed to *horror vacui,* Guericke wrote, were due to air pressure. Guericke even calculated the exact amount of force exerted on the piston by the external air.[54] In fact, Guericke had essentially developed an understanding of the difference between the weight of air and its pressure. The Jesuits' unwillingness to grasp this crucial difference was one of the main stumbling blocks to their acceptance of the atmospheric explanation. Surely a wide column of mercury was heavier than a narrow column, yet they stood at the same height in tubes. Furthermore, if it was the weight of the air pushing on the pool of mercury that supported the column of mercury suspended in the Torricellian tube, how was it that this also occurred inside a building where the roof stopped the bulk of the atmosphere pushing down? Only by differentiating between pressure and weight of the air could a theory that attributed the suspension of the mercury to atmospheric air adequately account for these phenomena.

Schott's Shift: The *Technica Curiosa*

When the *Mechanica hydraulico-pneumatica* was published in 1657, Schott had not been convinced by his correspondence with Guericke either that a vacuum existed in the receiver or that the phenomena displayed by the pump were caused by air pressure. In the *Magia universalis naturae et artis,* also published in 1657, Schott again wrote that the "attractive force of the fear of the vacuum is the principle of hydro-pneumatic machines." Despite faithfully reporting Guericke's opinion once again, he denied that mercury was held suspended by the external air.[55]

By late 1661 Guericke had developed his famous Magdeburger hemispheres: two large metal hemispheres that, once pushed together and the air inside them evacuated, could not be separated by teams of horses. A practicing engineer, he had also begun to explore the ability of the vacuum to perform work. By operating a pump attached to a large cylinder containing a piston that rose as the cylinder was evacuated, even a small boy could lift a hundred-pound weight.[56] Schott, however, was still asking Guericke for an explanation of why he did not ascribe the phenomena to the fear of the vacuum. When Schott asked him why the Magdeburg hemispheres clung together so forcefully when there was no air inside them but not at all when there was air inside, Guericke patiently recounted the same explanation he had been offering since his performance at Regensburg.[57]

But by the time his *Technica curiosa* appeared in 1664, Schott had changed his mind. He still maintained the impossibility of a vacuum but had been completely won over to the view that the cause of the phenomena ascribed to the fear of the vacuum was actually atmospheric air pressure. Much had in fact happened in the seven years since the publication of the *Mechanica hydraulico-pneumatica.* After hearing of Schott's account of Guericke's initial experiment, the Anglo-Irish nobleman Robert Boyle (1627–91) had built an air pump with the help of Robert Hooke (1635–1703). Boyle was interested in exploring the characteristics of a space devoid of air and published an account of his work in his *New Experiments Physico-Mechanical, touching the Spring of the Air, and its Effects; Made for the most Part, in a new Pneumatical Engine* (Oxford, 1660).[58] His pump used a receiver with a lid that allowed access to the receiver so that objects could be placed inside a space that could then be evacuated. While this meant that the receiver leaked, Boyle, unlike Guericke—and herein lies the crucial difference—was not interested in creating an absolute vacuum or demonstrating the possibility of its existence. Boyle stated early in his work that "I here declare once for all, that [by the term *vacuum*] I understand not a space, wherein there is no body at all, but such as is either altogether, or almost totally devoid of air."[59] Boyle refused to be drawn into the polemics over the vacuum, stating that the pump could not show one way or the other whether there was absolutely nothing in the receiver.

Schott was so impressed by Guericke and Boyle's new machines that he did "not hesitate to frankly confess and boldly pronounce, that I have never seen nor heard nor read nor imagined anything so amazing of that kind: nor do I think that the sun has illuminated anything similar, let alone more amazing, since the founding of the world. And the great princes and most learned men to whom I have communicated and explained it are of the same opinion."[60] These developments took pride of place in Schott's new volume, which he did not hesitate to dedicate to Elector Johann Philipp, who had, after all, first brought the pump to Schott's attention and presumably was still interested in new developments concerning the pump.

Whereas Schott and the Jesuits had earlier claimed that the air pump experiment was, in effect, underdetermined—the phenomena produced by the receiver, and the Torricellian tube for that matter, could be accounted for by a range of explanations, in particular by the *horror vacui*—Schott now held that the experimental evidence was unequivocal. Schott had always been fascinated by machines, but he was now filled with wholehearted admiration for experimental natural philosophy, proclaiming that "the chief purpose of the labors undertaken in these four volumes is to excite others toward experimental philosophy. For often already I have indicated by deeds themselves how greatly that kind of philosophizing pleases me, which is not dependent on the subtleties of words and cunning, but which examines thoroughly the concealed heart of Nature itself and which unites 'to know' [*scire*] with 'to be able' [*posse*] in a happy marriage."[61] For Schott, experimental natural philosophy bridged the gap between mathematics and physics. A mechanical device, something that fell in the realm of mathematics, could be part of a philosophical enterprise. The final line, on the marriage of *scire* and *posse,* emphasized this union—the speculative science of physics married to the manipulative art of mechanics to probe the secrets of nature. This was, in effect, the culmination of the program of mathematicians to make their art a science and win for it the epistemological status of philosophy. Despite their disagreements, Schott, the mathematician who specialized in curious devices, and Guericke, the engineer who applied his practical skills to solving natural philosophical questions, shared this preference for experiment, for what the latter had termed a "visual demonstration" of the causes of natural phenomena, over a purely rational physics.[62]

Schott had been particularly impressed by Boyle's witnessing strategy, which placed numerous, "reliable" witnesses at his experiments. Boyle, he wrote, "trusts not only through his own eyes, or through his own judgment, but employs observers everywhere, and faithful witnesses, and the most learned ones at that, whenever he provides experiments." Schott was also impressed by Boyle's repeated performance of the same experiment in different ways, explicitly acknowledging the display of honesty with which Boyle narrated in detail "the machines used, the manner of proceeding, happy or unhappy outcomes, judgments of others different to his own, and then all the circumstances."[63] In short, Schott recognized and was convinced by the central features of the Boylean program of "virtual witnessing," as Shapin and Schaffer termed it three centuries later. By granting the credibility of Boyle's experiments, Schott could be convinced by Boyle's claims.[64]

In an early modern variation of the commentary tradition, Schott described every experiment in Boyle's *New Experiments Touching the Spring of the Air* and added his own comments to them. Schott emphasized Boyle's own admission that he did not use *vacuum* to mean a space totally devoid of any body but simply a space devoid, or nearly so, of air.[65] By not concerning himself with the question of whether there was something other than air in the receiver, Boyle had allowed the question of whether there was a vacuum in the receiver and the consequent question of the possibility of the vacuum to be conveniently shelved, removing the Jesuits' greatest stumbling block to appreciating the possibilities offered by the new experiments.

For Schott, the *experimentum crucis* was Boyle's seventeenth experiment, which combined the two new devices, putting the Torricellian mercury tube inside the air pump. Schott considered this experiment at great length.[66] Boyle put a bowl of mercury inside the receiver and stood an inverted tube of mercury in the bowl. The mercury descended to the usual level of around thirty inches. He then sealed the receiver and began evacuating it. The drop in the level of the mercury in the tube was immediate and continuous. Eventually the mercury in the tube fell almost to the level of the mercury in the bowl in which the tube stood. The clear inverse correlation between the height of the mercury and the amount of air in the receiver was difficult for Boyle and Schott to overlook. This was for Boyle himself a very significant experiment. The experiment also illustrated the crucial distinction between the weight and the pressure, or "spring," as Boyle termed it, of the air.

Such evidence convinced Schott. However, before he acknowledged the causal role of air pressure in the *Technica curiosa,* he argued emphatically against the vacuum. Everyday experience (*experientia quotidiana*), he claimed, showed that nature impedes with every means a vacuum. "Therefore without an evident and irrefutable cause," he warned, "one must not assert that in these experiments there is a vacuum, as nature is always the same."[67] He then recounted all the experiences traditionally deployed against the vacuum.

Thus while he recounted both *experientia* and *experimenta,* the evidence of everyday experience was so overwhelming that particular experiments could not overcome it. The mass of experience showed that there could not be a vacuum in nature, so there could not be a vacuum in the Torricellian tube or the air pump. What filled the top of the Torricellian tube and the air pump, then, was a "subtle, ethereal body" that pervaded all bodies so there could not be any place free of body. Jesuits had used the ether to explain away the vacuum at the top of the tube since the 1640s. Since light, sounds, and magnetic effluvia could pass through glass, there had to be pores through which the ether could enter. However, nowhere did Schott get drawn into the question of absolute accidents and the attendant issue of the Eucharist; he instead limited himself to a philosophical defense of the plenum.

Once Schott had dismissed the possibility of the vacuum, he was able to turn to the question of what kept up the mercury in the Torricellian tube and accounted for the dramatic phenomena displayed by the air pump and the other pneumatic devices invented by Guericke. Water did not rise in the Magdeburg tube (basically a giant water barometer) because of the fear of the vacuum, he wrote, for the simple reason that there was no vacuum there. Rather, "the true and genuine cause why water in the evacuated Magdeburg tube always ascends to a certain and determined altitude is the pressure [*pressio*] of the surrounding air on the water in which the tube is immersed."[68]

At last Schott had acknowledged what Guericke (and Torricelli and others before him) had been arguing for years—that the cause of the phenomena was to be found outside the tube and pump. The explanation for the resistance to the extraction of the air pump's piston, Schott granted, was the difference in pressure between the air inside and outside the receiver.[69] Schott then went through both the modern experiments and traditional experiences that had been used to demonstrate the *horror vacui* but this time used air pressure to explain them.

In an odd twist, Schott now used the fear of the vacuum to justify why air pressure was actually the cause of phenomena formerly ascribed to the fear of the vacuum. Take, for example, the inability of suction pumps to raise water more than about ten meters. The operation of suction pumps had commonly been ascribed to the fear of the vacuum. But, Schott argued, if water pumps functioned through the fear of the vacuum, then they should be able to raise water to any height, as nature's power to prevent a vacuum was not limited. Since there was a limit to the height water pumps could raise water, however, the pressure of the air on the reservoir of water that was being pumped had to be the real explanation and not the *horror vacui*.[70]

Only in a very few cases, such as cupping jars, did Schott reject air pressure and adhere to the fear of the vacuum as the explanation. Therefore, somewhat ironically, by employing air pressure to explain virtually all the experiences traditionally used to assert the impossibility of a vacuum, Schott undermined the mass of everyday experiential evidence that he had earlier claimed overwhelmingly proved the impossibility of the vacuum. Even if the Jesuits' speculative philosophical arguments against the vacuum still stood, the air pump had robbed them of the experiential evidence they had long relied upon to deny the vacuum. If Schott was aware of this problem, he made no attempt to reconcile the conflicting claims of experience and experiment.

In its peregrinations the air pump was many different things to the people it encountered. For Guericke it was a "visual demonstration" of the vacuum. For the princes at Regensburg it was a curious novelty. For Jesuit intellectual authorities it was a threat that had to be tamed by being treated as an experience supporting the plenum. For Boyle it was a means to exploring the characteristics of air. For Schott it was at different times several of these things. For him it began as one thing but became something else. Initially it was a novelty, a marvel that could be used in the Jesuit program of winning influence with an important patron and raising the prestige of the Society through cultural endeavor. Schott also shared the view of his fellows that the pump was a threat to be tamed.

Schott illustrates a marked shift in the way instruments could be perceived and used: the pump changed from an experience to an experiment. This enabled him to preserve the *horror vacui* but also to adopt experimental natural philosophy. Torricelli's mercury tube did not end the abhorrence of the vacuum. Nor did Guericke's air pump or Boyle's. Nature still abhorred a vacuum. But it did not have an immediate, or efficient, causal role in the phenomena displayed by these devices in those cases where air pressure performed that role.

As had been the case in astronomy, Jesuit mathematicians now claimed that with their devices they could make reliable claims about the physical world. Schott had officiated at the wedding of mathematics and philosophy in the happy union of experimental philosophy. Yet it would take some time before Physics professors were completely won over to recognizing this union as legitimate. Certainly the case of astronomy showed they were beginning to accept mathematical observations as reliable foundations for physical claims, but even by the time Schott encountered the air pump the issue had not been completely solved. There remained several obstacles to the integration of instruments and mechanical devices into the Physics curriculum.

Jesuit Physics in the Later Seventeenth Century

Despite Kaspar Schott's engagement with experiment, the philosophical curriculum continued to follow the cycle of the triennium required by the *Ratio studiorum* over the latter part of the seventeenth century.[71] Within the year of Physics, *physica generalis* expounded on the subject matter of Aristotle's *Physics* and *physica particularis* on his other books of natural philosophy. Individual novel doctrines were adopted, particularly those dealing with the heavens, but the overall structure of the curriculum, the function of natural philosophy, and the method of its instruction did not change significantly.

Certainly, the career structure of Jesuit professors contributed to this conservatism, as they did not have the time to develop expertise in natural philosophy. Also, their conception of natural philosophy as preparation for theology focused the curriculum on certain kinds of questions. The lack of skilled mathematicians (there was hardly one of any note in Germany in the seventeenth century after Schott's death in 1666) shut down a conduit that had brought novelties into the Jesuit colleges. Of course, the hardening of attitudes toward the new natural philosophy exhibited in the condemnation of Galileo and the Ordinatio of 1651 must have contributed. So while it may have been possible to creatively navigate the provisions of censorship, it seems that few Jesuit natural philosophers had the desire to do so in the later seventeenth century. Not only were there no talented mathematicians, but philosophers such as Cornaeus and Arriaga, whose work had consistently tested the tension between liberty and obedience, novelty and orthodoxy, were also absent.

The German Jesuits' reception of the pneumatic instruments such as the mercury tube and the air pump illustrates this unwillingness to engage with novelties. Despite Schott's own enthusiasm for experiment, for philosophy professors matters were somewhat different, and it would take some time before experiment became a standard part of the Jesuit Physics curriculum in Germany. The air pump, or any other experimental device, confronted the Physics professor with several problems he had to solve if he wanted to include it in his lectures.

First, since natural philosophy was a speculative science, it had little room for devices or instruments. Whereas Jesuit mathematics classes traditionally covered hands-on training in practical tasks (such as the design of sundials), such activities were absent from Physics classes. Thus the idea of actually performing experiments in Physics lectures was not obvious. Furthermore, air pumps were very expensive and hard to find in the seventeenth century.[72] Second, the air pump grew out of the mechanical tradition, a discipline that historically was part of mathematics. In effect, Guericke had been trying to solve a physical problem—the existence or nonexistence of the vacuum—through mechanical means. It was an endeavor that challenged disciplinary boundaries as the Jesuits knew them. The final problem was where to situate a treatment of air pumps and other pneumatic and barometric instruments in the Physics curriculum. Since mechanics traditionally was not taught in Physics, a place would have to be created for mechanical devices like the pump and other instruments.

Melchior Cornaeus's treatment of the air pump in his *Curriculum philosophiae peripateticae* illustrates the problems experimental instruments presented natural philosophers.[73] Open to novel approaches to natural philosophy, including using elements of mathematics in Physics, Cornaeus had to solve the problem of where to insert the air pump in the traditional philosophy curriculum. He could not simply treat it as an "experientia" demonstrating the impossibility of a vacuum and put it in his treatment of the book 4 of the *Physics,* namely space, time and the vacuum, along with traditional experiences such as cupping jars, because its sheer novelty and remarkable effects meant that it was not an experience—that is, "phenomena known and evident to all." The air pump required detailed consideration and dissemination before it could be considered to be evident.

Nevertheless, Cornaeus believed that the pump contributed to the discourse surrounding the possibility of a vacuum, and he solved the problem by placing his account in a lengthy appendix after the section of his compendium that discussed the vacuum, entitled "Certain Mechanical Theorems Deduced from a Consideration of the Vacuum." His use of the term *mechanical theorems* in a philosophy text is quite noteworthy and unusual for the time. Nevertheless, the need for an appendix reveals that devices such as the air pump stood at this time outside the regular Physics curriculum.[74]

Cornaeus was, however, virtually unique in addressing the air pump in a Physics text. Of course, he and the other Würzburg Jesuits were the only Jesuits who had access to an actual pump. But in the prevailing climate of conservatism, and with few mathematicians to introduce the works of Boyle, Guericke, and Schott to the colleges, there was little desire in Germany even to consult the virtual air pumps these works presented.

Several Physics texts did, however, discuss the mercury tube, which had after all been around for a long time and had been subjected to more widespread discussion than the pump. Universally these treatments continued the initial Jesuit response of presenting the mercury tube as a further experience demonstrating impossibility of a vacuum, not as an experiment exploring characteristics of atmospheric air, as Schott had come to see it and the pump.

In 1659, only two years after Cornaeus's and Schott's work on the air pump, a Jesuit professor referred to the Torricellian mercury tube in his Physics lectures, although without naming its inventor. The professor, Johann Serrarius, displaying the prevalent conflation of the terms *experience* and *experiment,* described it as an *experimenta* demonstrating the impossibility of the vacuum—which would have shocked the tube's inventor, who thought it did the exact opposite. Significantly, the experiment occurs in his treatment of the material covering book 4 of the *Physics* in the section on the vacuum—not in the section on the element of air in *De generatione.* Serrarius's goal was still to refute the possibility of the vacuum, not to explore the properties of air. He simply gave the mercury tube a place alongside cupping jars and watering cans as another experience demonstrating the impossibility of a vacuum.[75] Serrarius made no mention of the air pump.

Another approach was simply to ignore such novelties and thereby avoid the difficult issues they brought with them. As late as 1693–94 an anonymous lecture manuscript, probably from the University of Mainz, made no mention of hydraulic/pneumatic experiments. It made no reference to Schott, Boyle, Guericke, or Torricelli. In response to the question "How does nature impede the vacuum?" the professor stated, "Nature impedes the vacuum either through the ascent of heavy things or the descent of light things, or through the rarefaction of bodies or by another means, otherwise bodies would break through the fear of the vacuum." In containers from which all the air had allegedly been removed, either some small quantity of air actually remained or enough air penetrated through pores to prevent a vacuum.[76]

This adherence to a strictly traditional Aristotelian-peripatetic explanation thirty years after Guericke and Boyle's work on air pressure stands in striking contrast to the rapid integration of telescopic discoveries into the Jesuit mathematics curriculum in Germany in the early 1610s. The willingness to engage with novelties that had marked Jesuit natural philosophy at the start of the century had faded by its end. The new century, however, was to reverse this trend.

PART 3

The Eighteenth Century

CHAPTER 8

Censorship and *Libertas Philosophandi* in the Eighteenth Century

THE STRUCTURE AND CONTENT OF THE JESUIT PHYSICS CURRICULUM changed fundamentally in many ways over the eighteenth century. From the essentially static period of the later seventeenth century, the process of change accelerated. The development was not the result of the structures of censorship simply being abandoned. Censorship continued over the eighteenth century, but, like the curriculum itself, it was transformed.

The Leibniz–des Bosses Exchange

In the middle of 1707, Gottfried Wilhelm Leibniz (1646–1716) got into an argument with his Jesuit correspondent, Bartholomaeus des Bosses. The theology professor at the Jesuit college at Hildesheim had written to the philosopher in nearby Hanover a year and a half earlier, opening a correspondence that lasted until Leibniz's death in 1716. Leibniz, the last of the German baroque polymaths, touched on many of his myriad interests in their exchanges: the Chinese rites controversy, opinions of recently published books, gossip from Rome, monads and transubstantiation, and more. Both participants sincerely attempted to engage with the other's thought, and each treated the other with respect, even affection, but the conjunction of three topics that had been raised in their letters provoked an outburst from Leibniz: the continuing polemics over Jansenism, the scope of the Church's infallibility, and the authorship of an obscure philosophical work that Leibniz liked.

His somewhat heated remarks appear to mark Leibniz, often regarded as the first philosopher of the *Aufklärung,* as an outspoken advocate of philosophical liberty. He was frustrated by the condemnation of certain Jansenist propositions the preceding year, particularly since it was not clear that all disputing parties agreed on what Jansen meant or said, or indeed what each other meant or said. He exclaimed, "If only distinguished men both in your order and elsewhere would either lay aside or restrain their passions with which good men are not seldom burdened. I could not praise either the acts of the Sorbonne professors toward your brothers or the acts of your brothers toward the memory of Jansen. I think such condemned propositions are like nose wax, since nobody knows how variously the nouns *possibility* and *necessity* are understood."[1]

Liebniz's sophisticated understanding of the shifting and slipping referents of philosophical terms convinced him that polemics over such topics were inherently futile. Des Bosses, in contrast, appears as the supporter of the Church's absolutist pretensions in matters of religion. He replied that while the Church can err in matters of fact, what was at stake in the Jansenist controversy were matters of fact relating to the doctrine of the faith, in which the Church could not err. He "kissed [Leibniz] as a friend of peace and the common good, for wanting both sides to avoid hateful things," but sometimes it was necessary to refute them.[2]

In the same series of letters, Leibniz and des Bosses also discussed a book that Leibniz had recently read with great interest: *Philosophia vera theologiae & medicinae ministra* (Philosophy, the true servant of theology and medicine), supposedly published in Cologne in 1706 under the name Aloysius Temmik (or Temmick). Des Bosses informed Leibniz that Temmik was actually a pseudonym for a Jesuit who had died two years earlier.[3] The Jesuit provincial and Roman revisers had rejected the work, but through the efforts of an extern it had been printed, not in Cologne, as the title page stated, but in Würzburg, where the bishop had forbidden further distribution of the work. Des Bosses later identified the author's real name as Aloysius Kuemmet.[4]

Des Bosses did not like the work at all for several reasons. For one, the author stated that all forms except humans' were merely modes of matter. But since des Bosses could not understand how a mode of a merely passive thing such as matter could exert an active force [vis], he concluded that this would attribute all action to God alone and thereby "overthrow all philosophy." To paraphrase, How could there be any *natural* philosophy if everything was the direct result of the supernatural?[5] Leibniz thanked des Bosses for the information but heatedly responded, "Even if I do not approve of the author's primary opinions, I wish that the liberty to philosophize be granted to learned men, even yours, which causes emulation and stimulates talents. In contrast, they are killed by the servitude of the mind. Nor can you expect anything from those to whom you grant nothing." This, he continued, was responsible for the decline of learning in Italy and Spain. Besides, what Temmik wrote was openly supported by many scholars in France.[6] Returning later in the same letter to the Jansenist controversies, Leibniz exclaimed,

> I believe that persecutions for opinions that do not teach crimes are the worst possible. Upright men should not only abstain from them [i.e., persecutions] but abhor them and should work so that those over whom we have some authority are deterred from them. Honors and privileges that are not owed may be denied those who support opinions that seem to us to be annoying. I do not think it is permitted to snatch them away, or even more to rage with proscriptions, chains, the galleys, and even more severe and evil measures. For what else is this than a kind of violence, from which you cannot be safe except through the crime of abjuring what you think is true?[7]

This may look like a clear case of a spokesman for the Enlightenment condemning the barbarity of an earlier age. But Leibniz's position was not quite as straightforward as it may at first appear. Des Bosses was probably rather taken aback by Leibniz's outburst. The previous year the Jesuit had sent Leibniz some examples from a list of opinions taken from Descartes and Malebranche's work recently forbidden by General Michelangelo Tamburini. Des Bosses, sure that Leibniz would agree with the list, wrote, "I congratulate our censors for finding in you a like-thinking guide."[8] Leibniz, no great admirer of Descartes, did approve of the list and recalled the earlier list of 1651. "I like to hear about these censures, whether yours or others', and I do not condemn them," he wrote. "For this matter pertains to formulas for speaking cautiously and avoiding offenses that are not necessarily deserved."[9]

Leibniz, whose lifelong ecumenical goal was to restore harmony among Christians and to reconcile the New Philosophy with the Aristotelianism of the schools, liked such lists because they helped prevent the quarrels and bitter recriminations so prevalent in early modern intellectual and religious culture. Of course, Leibniz was not particularly successful in preventing quarrels and was himself involved in one of the most vicious controversies of the time when he engaged Isaac Newton in the infamous dispute over priority for the invention of the calculus. But Leibniz, like the Jesuits, was faced with the dilemma of granting sufficient liberty while ensuring that it did not lead to disorder and fruitless quarrels. Also Leibniz, like the Jesuits, felt obliged to denounce philosophical views that led to unacceptable theological conclusions. In his correspondence with Samuel Clarke, Leibniz argued that the God of Newton's natural philosophy was an unacceptably imperfect one who had to constantly intervene to keep his inefficient universe running.[10]

While the Jesuits and Leibniz were grappling with similar problems, they appealed to different criteria to determine the acceptability of a particular doctrine, which necessarily meant they reached different conclusions. Leibniz did not accept that the Catholic Church had the authority to determine philosophical truths, writing to des Bosses, "[I] do not think the Church has any infallibility, except in preserving salutary dogmas taught long ago by Christ. The rest pertain to teaching, where reverence is necessary but not assent. If Rome defined that there were no antipodes, if it damned the motion of the earth today, should we think that it was to be held infallible?"[11]

Leibniz's rhetorical questions presumed a negative answer, but of course the Church had condemned the motion of the earth in its verdict in the trial of Galileo, a ruling that was still in effect at the time and to which Jesuits such as des Bosses were still obliged to adhere.

Libertas Philosophandi in the Eighteenth Century

There were remarkably enduring continuities in all aspects of Jesuit censorship of natural philosophy in the eighteenth century: in the justifications for the need for censorship such as the discourse of uniformity, in the hierarchy's instructions governing censorship, in the practice of censorship, and in the professors' attempts to negotiate and even circumvent censorship. In short, censorship was just as contested a process as it had been in the preceding century.

The Jesuits were by no means unique in the Age of the *Aufklärung* in limiting the freedom to philosophize. Just as Leibniz's attitude toward censorship was ambivalent, so was that of another outspoken proponent of philosophical liberty. Christian Wolff (1679–1754), easily the most influential philosopher of the first half of the eighteenth century among Catholics as well as Protestants in Germany, had a personal interest in the question of liberty; he had suffered for his opinions, being expelled in 1723 from his position at the University of Halle and sent into exile overnight by King Friedrich Wilhelm I of Prussia for his lectures on Confucius, which, his Pietist critics at the university claimed, led to atheism.

Wolff was a great defender of philosophical liberty because of his complete faith in the philosophical method. He argued in his 1728 *Preliminary Discourse on Philosophy in General* that nobody who philosophized according to his philosophical method could err. Nor could such a philosopher defend what was contrary to revealed truth or virtue.[12] Thus it was "absurd to prevent someone from philosophizing because he uses philosophical method. . . . There is no danger to religion, to virtue, or the state if full freedom to philosophize is given to those who philosophize according to the philosophical method."[13]

But again Wolff's position was not that distant from the Jesuits' (even if they did not recognize it). For example, Wolff argued that whoever used the philosophical method could distinguish probability from certitude and "does not pass off a hypothesis as a demonstrated truth." Consequently philosophers would not undermine a correct reading of Scripture. Quite remarkably, he continued,

> This same thing was recognized by the Roman Curia when it permitted the use of the Copernican system as a hypothesis to explain and compute celestial motions. However, before its truth was as evident as the roundness of the earth, it should not be used dogmatically so that Sacred Scripture would have to be explained according to it. Those who contradict revealed truth and who teach things whose legitimate consequences are contrary to revealed truth do not philosophize according to the philosophical method. For they use uncertain principles and weak demonstrations.[14]

The enlightened, Protestant philosopher stood shoulder to shoulder with the Holy Office. In decreeing that heliocentrism could be treated only as a hypothesis, it had followed the proper philosophical method, since its truth was not evident in 1616 or 1633. Furthermore, even though adherence to the philosophical method prevented error, it was quite clear that some, such as Spinoza, had deviated from the method to produce errors. Thus Wolff granted the state the power to limit the freedom to philosophize to "prevent opposition to religion, to virtue, or to public life."[15] He left open who was to judge whether a conclusion had been reached properly by following the philosophical method. Presumably this Lutheran would not have granted this authority to the Holy Office in his own day or to the Pietists who had succeeded in expelling him from Halle. His method was aimed at preempting censorship by showing that philosophy did not need it. His was a censorship in theory if not in practice.

While the Jesuits had more in common with Wolff than they realized, they were suspicious of his enthusiastic endorsement of *libertas philosophandi.* At a disputation at Würzburg in 1737 under Josef Braun, Christian Wolff's appeal for liberty was rejected: "Wolff errs most gravely and philosophizes most badly . . . by writing: 'If complete *libertas philosophandi* is granted to those who philosophize with the philosophical method, one should not fear any danger for religion, virtue, or the state.'" Such an insolent appeal contradicted experience, reason, and Scripture. Had there been no heretics among the philosophers, Braun wondered?[16]

In the Society of Jesus, the foundation that had made censorship necessary, namely the need for philosophy to conform to theological truths, remained. A few Jesuits still openly argued for the ultimate authority of Scripture in all areas of knowledge, including the natural sciences, even after the middle of the century. Michael Froehling's (1717–85) *Philosophical Meditation on What Is Permitted to the Philosopher Concerning the use of Scripture* (1753) dealt with the vexed and ancient problem of *libertas philosophandi.*[17] Froehling granted that liberty was a good thing but asserted that many people did not have a correct notion of philosophical liberty. In philosophy two extremes were to be avoided, namely too much credulity and impudence (*Frechheit mentis*). "Impudent" liberty did not follow the truth of Scripture. Since Scripture came from God, it was the font of all truth. Relying on the familiar Jesuit trope of freedom through obedience, Froehling stated that "we rightly damn that impudent liberty, which does not want to follow the Truth, while it convinces itself to be confined by the false, and fashions chains for Reason, not considering that it struggles to the apex of truth most freely through obedience, but to the thickest darkness of errors and ignorance by its own arrogance."[18] A stirring statement, but clearly the Jesuits were not, and never had been, turning to Scripture for all the answers to philosophical questions. Philosophy had to look elsewhere and consequently still had to be supervised.

Formal Censorship in the Eighteenth Century

The Society's hierarchy continued to operate the mechanism of censorship it had inherited to enforce the boundaries between those disciplines, even if some fine-tuning of the machinery was required from time to time. The Society continued to admonish its members to adhere to orthodox opinions. Complaints also continued to reach Rome from the provinces lamenting the spread of excessive liberty in teaching philosophy at the Society's schools.

At the end of the seventeenth century in 1696 the Fourteenth General Congregation responded to such complaints and to requests that General Piccolomini's 1651 Ordinatio be updated to deal with more recent novelties. It reaffirmed that "[t]he Society is greatly repelled by any opinion containing novelty and laxity in moral matters" and ruled that a further list of proscribed theological and philosophical opinions be drafted.[19] To carry out the congregation's wish, General Thyrsus Gonzales wrote a letter to the whole Society urging it to maintain solid doctrine. Relying on well-established tropes, Gonzales reaffirmed the Society's "singular aversion to and horror of all inventions of new opinions that have more splendor than truth."[20] He asked for the provinces' suggestions on the matter and formed a commission to draw up the list, which was eventually published under General Michelangelo Tamburini (1648–1730, general from 1706) in 1706.[21]

Cartesian philosophy was the chief target of the list. The first proposition was that most fundamental to Descartes's method: the human mind can and should doubt everything, other than that it thinks, therefore it exists. There were many more Cartesian propositions on the list. Among the most significant were:

28. There are no corporeal substantial forms distinct from matter.
29. There are no absolute accidents.
30. The Cartesian system can be defended as a hypothesis whose principles and postulates cohere correctly among themselves and with its conclusions.[22]

This was in effect an explicit, official prohibition of Cartesian matter theory at Jesuit schools, even as a hypothesis.

The Sixteenth General Congregation of 1731–32 once again warned against excessive liberty and love of novelty in philosophy. Although it praised experimental philosophy, it also reaffirmed the Society's commitment to Aristotle, particularly in natural philosophy, and again instructed the general, Franz Retz (1673–1750, general from 1730), to draw up a list of propositions that professors would be prohibited from teaching.[23] The list appeared in 1732 and, like the 1706 list, concerned itself, among other subjects, with matter theory. Its first three propositions once again banned Cartesian and atomist views on this subject.[24] Yet neither the 1706 nor the 1732 list stopped professors from devoting considerable attention to all aspects of Cartesian philosophy in their lectures and texts.

Despite the promulgation of these later lists, there is ample evidence to show that the prohibitions of the 1651 *Ordinatio pro studiis superioribus* did not quietly fade away. They were certainly still legally valid within the Society; the 1732 list explicitly quoted the *Constitutions,* which specified that opinions banned by previous generals remained off limits. In fact, the 1651 Ordinatio continued to be cited by both censors and authors in preference to the later lists well into the eighteenth century. The revisers general constantly referred to it well into the eighteenth century.[25] Its influence also extended into the classroom. In his Physics course held in 1743 at the University of Würzburg, Professor Edmund Voit (1707–80) cited the 1651 Ordinatio at the conclusion of his discussion of numerous propositions.[26] For example, after categorically denouncing the Cartesian and atomist view of natural body and demonstrating the superiority of the peripatetic system (with the argument, among others, that it was the only system compatible with transubstantiation), Voit cited proposition 19 of the 1651 Ordinatio, which prohibited the opinion that mixed bodies (other than humans) do not have substantial forms and that their appearances derive from various mixtures of atoms.[27] Voit also cited during the course of lectures the Ordinatio's propositions concerning causes (5 and 6), the relationship of quantity, matter, and substance (23 and 24), the composition of the continuum (25 and 26), and several others. In all of these cases, Voit provided an explicit refutation of the opinions banned by the Ordinatio. Naturally Voit had to discuss the forbidden opinions in order to refute them, often in considerable detail, but his treatment was by no means even-handed; he left no doubt as to the falsity of the listed opinions. Adherence to the Ordinatio continued ninety-two years after it was first promulgated.

The Jesuits did not simply imagine the threat to their eucharistic theory. Their view of the fundamental principles of natural body and the implications of these for the Eucharist were hotly disputed. At the heart of these polemics lay the validity of the doctrine of absolute accidents. In Paderborn, the site of a Jesuit college, a certain Carpophorus dei Giudice published in 1718 a work that rejected absolute accidents as philosophically absurd. They were not *de fide* and were completely incapable of accounting for the eucharistic species. The author's rather unusual name may have hidden the identity of a Jesuit.[28]

He would not have been the only Jesuit disgruntled with the orthodox view. Aloysius Temmik's *Philosophia vera theologiae & medicinae ministra,* the 1706 text that Leibniz told Bartholomaeus des Bosses he found so interesting, also made a stinging critique of absolute accidents. Des Bosses told Leibniz that the author was actually a Jesuit named Aloysius Kuemmet. There is no record of an Aloysius Kuemmet, but a

Jesuit named Caspar Kuemmet (1643–1706) had taught and published on Hebrew and sacred Scripture in Mainz and Würzburg—the book's true place of publication, according to des Bosses.

At Ingolstadt in 1727 the Jesuits' former student and now professor of medicine Johann Anton Morasch published the first part of a philosophy textbook that followed the traditional structure of the material but openly advocated atomism. In the *Philosophia atomistica* Morasch not only devoted considerable space to refuting the peripatetic doctrine of absolute accidents but had the temerity to directly confront the theological question of whether they could be reconciled with transubstantiation.[29]

This opened the so called "atomist war." The Jesuit Georg Hermann responded with a peripatetic refutation of atomism that focused on forms and accidents.[30] Morasch weighed back in anonymously with a vindication of atomism that once again did not shy from the eucharistic implication of atomism and absolute accidents.[31] Morasch's former student and fellow professor of medicine Franz Anton Ferdinand Stebler sought a middle ground in 1740 with a pair of graduation orations entitled *Aristoteles atomista.* Less hostile than Morasch, he skirted the theological issues and contented himself with arguing that nothing in atomism was incompatible with Aristotle. Nevertheless, he began the work with epigrams not only from Morasch but also from Temmik, a hint that not only librarians such as Leibniz had come across this work. Although Stebler avoided mention of the eucharistic species, the theological implications of his work would have been clear to any Jesuit.[32]

The Practice of Censorship

There were also continuities in the strategies that Jesuits employed to operate within the strictures of censorship. Whether des Bosses's attribution of the *Philosophia vera* to Caspar Kuemmet was correct, he certainly considered it conceivable that a fellow Jesuit would anonymously publish a book that had been expressly forbidden. But even without resorting to rank disobedience, Jesuit philosophers could still cover a vast range of opinions simply by using the tried and tested techniques of assigning degrees of probability to various opinions.

This fact has not always been appreciated. Contemporary and modern sources credit Anton Kleinbrodt (1668–1718) with introducing experimental physics to Ingolstadt, where he taught the triennium from 1701 to 1704. The Bavarian learned journal *Parnassus Boicus* published a eulogy for him stating,

> What he dictated into the quills of his students admittedly followed scholastic doctrine in part, but his natural philosophy and physics are in general simply furnished with a rational choice between the old and new teachings in philosophy, particularly according to the opinions of Descartes, Helmont, Entius, Sturm, etc. In fact, under this worthy man experimental philosophy was seen virtually for the first time in the philosophy faculty at Ingolstadt, and his physics tracts filled with much learned and Latin eloquence display that most sufficiently.[33]

For some reason, modern accounts follow Bernhard Duhr in claiming that Kleinbrodt was dismissed from his chair in 1704 for teaching Cartesian philosophy.[34] But this claim owes more to modern misconceptions about Jesuit censorship and what it meant to teach a proposition. Kleinbrodt did not actually adhere to Cartesian opinions: he taught Cartesian philosophy only in the sense that he discussed it, just as Jesuits had "taught" forbidden opinions since the sixteenth century. His 1703/04 Physics lectures survive and, despite the eulogy's claims, are in many respects not particularly novel.[35] Like other Jesuit lecturers, Kleinbrodt gave the opinions of many competing authorities, often praising them but ultimately rejecting all except the approved opinion. His treatment of the position and motion of the earth conforms in every way to the traditional Jesuit handling of the issue.[36] His *Mundus elementaris,* a series of three lengthy Physics disputations derived from his lecture course, received the imprimatur of the deans of both the theological and philosophical faculties, who were unstinting in their praise of the work's style and content; the Jesuit dean of philosophy particularly admired the way Kleinbrodt had joined the old with the new.[37] The dissertations clearly stated that there was prime matter that was not composed of atoms but was an incomplete substance. Furthermore, they argued that there were substantial forms not only in humans but, in explicit contrast to Descartes, in animals, plants, and inanimate objects also.[38] Although Kleinbrodt's work was replete with eclectic influences, including Cartesian, on the essential elements of Jesuit physics it was entirely orthodox.

Kleinbrodt's departure from Ingolstadt following his teaching of Cartesian and other nonperipatetic doctrines was not the result of a dismissal for such teaching. No eighteenth-century sources as Mederer and the *Parnassus Boicus* claimed this. Nor is there anything in the surviving documentation to indicate that this occurred.[39] Kleinbrodt did not leave the university under a cloud. After completing the triennium, he went to Dillingen to teach mathematics, quite the normal career progression within the Society.[40] Other than his earlier interest in experimental philosophy, what Kleinbrodt taught was little different from what other Jesuits were teaching. A survey of the main schools of thought on the constitution of matter, covering both the Cartesians and the atomists, was a standard element of Jesuit Physics instruction and did not in any way signal a deviation from the permissible.

Nevertheless, the censors were kept busy ensuring that professors assigned the correct degree of probability to crucial propositions. The revisers of the Upper German province were extremely dissatisfied with a manuscript by Joseph Falck and compelled him to rewrite significant passages. This case illustrates the continuing influence of the Ordinatio but also confirms that provided the Ordinatio was upheld and certain fundamental doctrines were asserted, Jesuits could address contemporary developments in natural philosophy.

The work in question was Falck's *Mundus aspectabilis* (The visible world), a philosophical compendium eventually published in 1737.[41] After teaching philosophy and mathematics at Freiburg and then mathematics at Ingolstadt, Falck (1680–1737) was summoned in 1715 by Elector Maximilian Emmanuel of Bavaria to be his sons' philosophy tutor. Falck later became confessor to Charles Albert, the son who succeeded as elector and later enjoyed a brief, ill-starred reign as emperor. Falck remained at the court in Munich, dying in 1737 in Copernican fashion just as the first printed pages of his book were delivered to him.[42] The manuscript, finished in 1724, was originally titled *Philosophy Taught to the Most Serene Dukes of Bavaria.* This close association with the ducal house and the authority that would thereby be attached to the views expressed in the work attracted the scrutiny of the provincial censors; the revisers' reports on the manuscript are the most numerous and detailed preserved in Munich. There were at least two rounds of censorship in Germany and another in Rome. The first German readings occurred in 1724 and 1725, the second, of the revised manuscript, considerably later, in 1736.[43]

Among the first group of censors, Horatius Burgundius and Leonard Tschiderer (1683–1752) complained that Falck did not adhere to the *Ratio studiorum's* decree on following Aristotle, at least on substantial questions.[44] They identified an essential core of Aristotelianism that Jesuits had to follow: first, a prime matter distinct from atoms; second, substantial forms in both animate and inanimate bodies; and finally, absolute accidents or qualities. Burgundius and Tschiderer wrote that Falck did not use his modal terms properly when dealing with these matters: he did not designate adequately which propositions were true, false, or probable. They complained that Falck said that "the Peripatetics admit *not improbably,* that there are substantial forms in mixed inanimate things. This is too little. . . . The atomists infer *not improbably* that natural causes produce nothing except motion. The words *not improbably* are to be omitted [underlining in original]." Falck's wording made a peripatetic doctrine less certain than it should have been and an atomist doctrine more certain than was permitted.

Similarly in the case of substantial forms, which were vitally important because of their connection with transubstantiation, the censors noted: "He defends the peripatetics' opinion on substantial forms as only probable. Such words, which give a nod to the contrary opinion as being equally probable, are better omitted." On the Copernican system, they wrote: "[Falck] says some doubt remains. It seems this should be omitted, for there is no doubt about the falsity of the Copernican system." Heliocentrism still had to be treated as a hypothesis: a mathematically convenient but physically false construct. The censors' verdict was unambiguous: "If this philosophy is to be printed it must first be emended so that the system of the peripatetics, especially its substantial opinions related above, is defended and the opposing systems and arguments are seriously refuted, not ambiguously but clearly and positively, which would not be difficult if the author wanted to." Perhaps the corrections took so long because Falck did not really want to make them.

The censors' reports also reveal that the earlier tensions within the Society between mathematicians and natural philosophers continued. Among the 1736 group of censors, the Ingolstadt mathematician Nicasius Grammatici (1684–1736) wrote that Falck had completely integrated the (now lost) recommendations of the Roman censors and that he "defends well the substantial opinions of the Aristotelian system as it is customarily taught in this province."[45] But Anton Mayr (1673–1749), a theologian, philosopher, prefect of studies at Ingolstadt, and himself the author of a philosophical compendium, repeated the complaints that had been made by Burgundius and Tschiderer eleven years earlier. Mayr thought Falck's work would be useful "because by reading it our members will be able to understand the doctrine of the moderns and afterwards will know how to attack them." But he was concerned that Falck still favored the opinions of the atomists too much by not refuting them adequately.[46] While it was important to know the opinions of the moderns, it was more important to know why they were wrong.

Although virtually all censors pointed out that Falck's many concessions to atomist doctrines created problems for the vital issue of transubstantiation,[47] they also took issue with Falck on purely philosophical questions, such as that of absolute lightness. Joseph Mayr (1671–1743) was not the only one to note that there were problems in Falck's treatment of the issue of heaviness and lightness because he appeared to follow the opinion forbidden by the 1651 Ordinatio.[48] He suggested that if Falck added that absolute lightness was at least possible outside the sublunar world, then he could avoid appearing to follow the forbidden opinion. The censor was forced to confess that he did not know what the reason for the prohibition was, or even in what exact sense the proposition was prohibited, revealing the distance Jesuit Physics had moved since

1651. Nevertheless, recommending conservatism, he advised, "[I]t is safer to avoid these things, even as a way of speaking."[49]

In the published version of the work Falck continued to devote much attention to the views of moderns such as Descartes, the atomists, and Copernicus. He described the Cartesian view of body and the heliocentric system in great detail. However, Falck also included what the censors wanted to hear; he left no doubt as to which opinions were correct and which were suspect. He vigorously upheld those fundamental peripatetic principles necessary for the physics of transubstantiation. He declared that the Cartesian view of body could not be defended by Catholics because of its incompatibility with the Eucharist. "The truths of faith must be preferred to all else," he proclaimed.[50] Falck still treated numerous physical experiments, including many from hydrostatics and pneumatics. On the issue of absolute lightness, he continued to say that air and even fire had inherent heaviness but, in deference to Mayr, was also careful to note that this did not rule out the possibility of absolutely light things existing beyond the sublunar realm, for example in the effluvia of celestial bodies—hardly a significant concession to the authorities. When he explained hydrostatic experiments, he still employed the concept of relative lightness and density.[51]

The Waning of Censorship

As late as 1763 Franz Neuff (1727–77), professor of logic at Mainz, warned that orthodox faith still faced a wide range of enemies such as Jansenists, atheists, deists, "indifferentists," and "pseudopoliticists." Consequently in some philosophical matters there could be no liberty, since "it is foreign to the sane mind to want to impugn those things we know to have been revealed by God with arguments sought from the laws of nature."[52] Any disagreement between Scripture and philosophy was more likely to lie in human ignorance than in Scripture. Therefore, attempts to explain away significant discrepancies between Scripture and philosophical claims, such as those made by the Copernicans, had not yet attained the status of perfected demonstrations. In particular, nobody had yet shown the causes by which the celestial bodies were moved. Here Neuff mentioned Kepler and Tycho but quite astonishingly made no mention of Newton, who had provided a plausible account of the causes of planetary motion over seventy years earlier.[53] If Newton's account was not adequate, then Neuff did not attempt to explain why. Neuff's treatment must have seemed dated to many of his colleagues.

For there were powerful advocates for change in the Society, among them the Dalmatian Roger Joseph Boscovich (1711–87), who worked primarily in Italy and was the preeminent Jesuit scientist of the eighteenth century. Certainly Boscovich held that there was a connection between good philosophy and true belief. In the dedication of his *Theoria philosophia naturalis* to Cardinal de Migazzi, the archbishop of Vienna, in 1758, Boscovich remarked that "a contemplation of all the works of Nature is in complete accord with the sanctity of the priesthood," for the mind easily passed from contemplating nature to honoring its divine founder and "His infinite Power & Wisdom & Providence." A religious superior must take care that in "the study of the Wonders of Nature, improper ideas do not insinuate themselves into tender minds; or such pernicious principles as may gradually corrupt the belief in things Divine, nay, even destroy it altogether, & uproot it from its very foundations. This is what we have seen for a long time taking place, by some unhappy decree of adverse fate, all over Europe."[54] Yet Boscovich's admonition did not specify any particular false philosophy or any system that should be taught. In many ways the passage is indistinguishable from the kind made by all natural philosophers and natural theologians around Europe in the eighteenth century.[55]

Boscovich urged the Church not to tie itself to any particular philosophical thesis, particularly when it seemed to be incorrect, as in the case of geostatic cosmology.[56] Boscovich also warned against excessive attachment to any particular metaphysics of matter. In 1760 Boscovich wrote from Paris to his brother Baro in Rome that if one suggested there that "a material thing has an effect, a force, an ability to change place," this encouraged unbelief and materialism. To such charges, Boscovich responded that "the greatest harm that can be done to religion is to connect it with the things in physics which are considered wrong even by a great number of Catholics. The youth then, convinced of the opposite, do not say religion is the truth and consequently such and such a thing in physics is true, they say that such and such thing in physics is wrong and consequently religion is wrong, and I think that in this way a lot of harm has been done to religion here and is still being done."[57]

The Society's hierarchy was aware of such pressures for change and to some degree acknowledged them. The general congregation officially proclaimed the fundamental compatibility between Aristotelian and experimental philosophy in 1731 and again in 1751. But there were members of the Society who wanted it to go further. The committee on studies made up of delegates to the 1755 General Congregation considered a (now) anonymous inquiry from one province asking whether the 1651 list of propositions should be preserved "in this age." The request singled out the twenty-third proposition, namely whether quantity was distinct from matter or substantial form, as well as those propositions concerning the motion of the planets and heavens, indicating that the author was interested in a reconsideration of the ban on atomism and heliocentrism.[58]

In response, the committee decided that nothing on the 104-year-old list should be changed. Their justification for their inaction echoes their predecessors' discussion over one-and-a-half centuries earlier, when lists of forbidden proposition were first considered: maintaining the list showed respect for General Piccolomini and the Ninth General Congregation; drawing up a complete new list would be an enormous undertaking; the Society would seem to be dissimulating by now allowing things it had prohibited for years; and there was the danger that people would assume that anything not specifically banned would automatically be permitted. The committee did, however, recognize that many opinions excluded from Jesuit schools in some provinces were openly taught in others. In fact, it acknowledged that things were taught in many provinces that were specifically banned by the 1651 Ordinatio. Rather than ruling against such practices, the committee merely recommended to the general congregation that provinces that had their own lists of opinions should continue to follow them.

In practice this devolution of authority upon the provinces meant that the enforcement of the rules governing the choice of opinions was left up to the discretion of local superiors. Thus enforcement reflected the attitudes of the individual provincial, rector, or prefect of studies, just as it had in the sixteenth century. But even conservative superiors did not enforce a narrow Aristotelianism. Georg Hermann (1693–1766), provincial of Upper Germany in 1754–59 and again in 1765–67, had a reputation for conservatism. He had fought in the "atomist war" against Morasch. Mederer wrote that he was "a man of ancient manners and mind who approved of nothing that was new." Mederer noted that Hermann was a throwback, always being "a sworn follower of the more rigid peripatetics rather than his own predecessors," among whom Mederer named Kleinbrodt and Falck.[59]

In 1755 Hermann decreed that the lists of forbidden propositions promulgated by the generals were to be followed in both method and substance under threat of strict punishment.[60] But even this reputedly strict provincial did not require anything beyond an adherence to a fairly small core of peripatetic physics. In terms very similar to those used by the reviewers of Joseph Falck's textbook, Hermann listed these doctrines as "matter and substantial form in the peripatetic sense, the true production of things de novo, the existence of many absolute accidents."[61] This left vast areas of physics completely open. This was a space that Jesuits authors freely took advantage of. The texts to which Hermann gave his imprimatur as provincial included Joseph Mangold's *Philosophia rationalis et experimentalis* (1755), by no means a "rigidly peripatetic" work.[62] And in the succeeding decade, even this small core of Aristotelian doctrines would be discarded from the curriculum.

By the time of the Suppression, the Jesuits had reached a position akin to Wolff's and Boscovich's. Unlike generations of his predecessors, Benedikt Stattler, the most prominent Jesuit philosopher in Germany on the eve of the Suppression, had no concerns about *libertas philosophandi.* He admitted that the practice of philosophy was linked to religion and the public good and that the ultimate goal of his philosophical compendium was the defense of religion. But he claimed that his desire for truth, his reverence for religion, and reason granted him *libertas philosophandi.* Like Wolff almost half a century earlier, he insisted that the proper method in philosophy would keep him from straying. No longer did theology or the censors have to play that role.[63]

CHAPTER 9

The Spread of Experiment

THE RECORD OF THE CYCLE OF PUBLIC ACADEMIC LIFE AT THE JESUIT colleges—the magnificently engraved sheets of theses defended, the disputations, the academic orations—can give the impression that little changed in the Jesuit philosophical curriculum over the course of the century between Kaspar Schott and Melchior Cornaeus's encounter with the air pump and the middle of the eighteenth century. Certainly many historians have thought so. The theses that the Jesuits' students publicly defended at the University of Prague, for example, were repeated decade after decade virtually without change. They made no reference to any area of experimental natural philosophy until the 1740s. Nor was there a chair for experimental philosophy at Prague until 1750.[1]

Elsewhere the public record reveals that the natural philosophical curriculum at Jesuit colleges underwent significant changes as the Jesuits slowly but with increasing enthusiasm integrated experiment into their natural philosophy. Printed handbills advertising the graduation ceremonies at the University of Dillingen provide the topics of the orations delivered at the celebration (although unfortunately the texts of the orations have not survived) and reveal the Jesuits' steadily increasing concern with experimental natural philosophy and their conviction that it was possible to reconcile this new form of philosophy with that of Aristotle. As early as 1717 the orator asked if the barometer could be trusted, a question hinting at the fundamental epistemological issue raised by instruments.[2] In 1718 the topic of the graduation oration was "Does the experimental physics of the moderns surpass the speculations of the ancients?"[3] In 1734 the orator asked whether the French theologian and philosopher Emmanuel Maignan's charge that the peripatetics were completely devoted to speculation and not at all to experiment was just.[4] In 1743 Professor Joseph Zwinger (1705–72) offered a tribute to Torricelli on the centenary of his death.[5] Two years later the topic was the general question of whether familiarity with experiments helped philosophy.[6] The year after that the orator considered what benefits to philosophy could be expected from the air pump in particular. He also considered the question that was fundamental in determining the role of experiment in philosophy: "Can experiments decide physical questions?" That experiment had already achieved that crucial status in physics was further confirmed by the third topic for that year, when the orator asked if philosophy could even survive without experiments.[7] In 1749 the speaker offered an evaluation of the Jesuit fathers' ability in experimental philosophy and considered what was still necessary to perfect it.[8] The next year, with Berthold Hauser (1713–62), the foremost German Jesuit authority of his age on experimental philosophy, presiding, the speaker turned to the favorite Jesuit topic of the compatibility of experiment and Aristotelian philosophy. "Was Aristotle a *Physicus Experimentalis?*" he pondered, and further, "Must the conjunction of rational and experimental philosophy be sorrowful?"[9]

Moreover, if one looks beyond the public face of the university at texts that more accurately represent actual classroom instruction, it is clear that across the German assistancy fundamental developments were occurring beneath the surface of what may appear to be static. While the institutional underpinnings of experimental philosophy were slowly established, the Jesuits solved the question of where a practical, mechanical activity such as experiment fit into a traditionally speculative discipline such as philosophy. By the middle of the eighteenth century, *experiment* had become synonymous with *natural philosophy* for the Jesuits.

Experiment Institutionalized

If the German Jesuits were reluctant or unable to grapple with the experimental philosophy in the later seventeenth century, the situation changed fundamentally in the eighteenth. A crucial signpost is the work of Anton Kleinbrodt at Ingolstadt early in the new century. Significantly, in his 1704 Physics lectures Kleinbrodt dealt with the Torricellian tube and the air pump not as a part of his consideration of the vacuum but later in the course, in the section of *physica particularis* that discussed the attributes of the elements, including air. Thus his prime purpose was no longer to use the instruments as *experientia* against the vacuum but to show how they could explore the properties of air—that is, as experiments. Repeating the conclusion that Schott had reached after reading Guericke and Boyle, Kleinbrodt maintained that all the phenomena that had been traditionally explained through the *horror vacui,* such as the inability of pumps to raise water over thirty feet and the tendency of smooth marble blocks to stick together, could be better explained through an "elastic virtue" in the air. Kleinbrodt nevertheless maintained the impossibility of a vacuum and, relying on an argument that Schott and his contemporaries had used half a century before, stated that at the top of the Torricellian tube there was not an empty space but rather a "spirit of mercury."[10] Kleinbrodt's teaching had considerable influence. His students included not only later Jesuit professors but also Johann Adam Morasch, future professor of medicine, founder of the anatomical college at the university and atomist opponent of Jesuit natural philosophy.

By the 1710s the arguments explaining pneumatic phenomena through the spring of the air had become firmly integrated into the Jesuits' peripatetic curriculum. At Würzburg in 1712 Professor Johannes Seyfrid's (1678–1742) students argued the thesis that a vacuum was not naturally possible and that whatever "is said to happen through the fear of the vacuum can be explained peripatetically, through the gravity and elastic power of the air and other bodies."[11] The extent to which experimental phenomena were becoming a standard part of the Physics curriculum is shown by Professor Franz Ellspacher's (1680–1748) 1714 dissertation, devoted entirely to the barometer and its associated phenomena. He followed the by now standard Jesuit line that there was no vacuum in the barometer and that the phenomena were caused not by the *horror vacui* but by the pressure of the air, which was absolutely and not merely relatively heavy.[12] Dissertations discussing the barometer were now common among German Jesuits, probably because very little equipment was needed to replicate the standard repertoire of mercury tube experiments in comparison with the paraphernalia of the air pump. Ellspacher's work was followed several years later in 1718 by a similar dissertation also dedicated to the barometer by Paul Zetl (1679–1740) at Ingolstadt.[13] By this stage no Jesuit Physics texts still asserted that the phenomena were directly caused by the fear of the vacuum.

Although professors did not explicitly articulate it, this change embodied a fundamental shift in perceptions of the relationship between the natural and the artificial. These dissertations, although devoted to mechanical devices—mechanical in the sense that they created a space that was in many regards unnatural—were not mathematical dissertations. Rather, they grew out of the Physics curriculum; Ellspacher's title specified that the barometer was being subjected to *philosophical* inquiry, while Zetl's claimed that the barometer would be explicated through *physical* reasoning. The traditional distinction between the natural and the artificial was thus being effaced. Another manifestation of this process, almost unnoticeable to modern eyes, was that the treatment of simple machines (levels, wedges, pulleys, etc.), which—according to the *Mechanical Questions* attributed to Aristotle—moved bodies against their natural inclinations, shifted around this time from mathematics to the Physics curriculum and its texts. An innocuous change, perhaps, but one that further signaled the end of the Aristotelian distinction between natural and artificial motions and indicated that Jesuit natural philosophy was increasingly treating all motions the same way.[14] Certainly residual elements of older distinctions between natural and artificial were to linger in the Physics curriculum into the middle of the century in a somewhat confused mixture. For example, one often encounters references to natural motion and natural places in treatments of motion that are strongly influenced by Newtonian mechanics. In short, elements of peripatetic physics endured.

Also necessary for the institutionalization of experiment were the material underpinnings of experimentation. These were only slowly acquired. In the early eighteenth century, Jesuit colleges had "museums" or cabinets, crammed with all the exotic items and artifacts so typical of these collections in a period of transition from cabinets of curiosities to the modern museum.[15] A 1707 inventory of "physical experiments, mathematical instruments, and other rarities held in the philosophical room" of the University of Würzburg reveals the chaotic variety of such collections.[16] Astrolabes, magnets, fossils, a magic lantern, four burning lenses, crabs and seashells, assorted horological devices, scales, an armillary sphere, Kircher's Pantometrum, plants and seeds, pictures of Jesuits, saints, and obelisks, and much more cluttered its cabinets and walls. Many of the items had been there since the time of Schott, who had died forty years earlier, including a bookcase with his manuscripts, now unfortunately vanished without a trace. Most of the instruments were mathematical—that is, designed for measuring—rather than physical—that is, designed for creating phenomena. Attached to the ceiling were various experimental instruments, including a barometer and a thermometer, but Guericke's air pump, brought to Würzburg by Johann Phillip von Schönborn fifty years earlier, had disappeared. The mathematical and horological instruments may well have been used in mathematical lectures on astronomy and sundials. However, the collection could not be described as a well-stocked experimental museum that could provide demonstrations or instruction.

However great the Jesuits' approval in principle of experimental natural philosophy and of teaching the results of experiments performed by others, this did not translate directly into a willingness to actually perform experiments with specialized apparatus. Since Schott had attempted to replicate Guericke's experiments in 1656 and Kleinbrodt taught experimental philosophy at Ingolstadt in the first years of the eighteenth century, one reads with some surprise in Mederer's *Annales* of the University of Ingolstadt for the year 1729 that "[i]n support for experimental physics at the beginning of this year, the philosophical faculty had a pneumatic pump built at its own cost for 150 florins. And this seems memorable at least because until this time that instrument had not been seen in this academy."[17] Over seventy years after the Jesuits first encountered the air pump, after numerous statements of support for experiment, the Jesuits at Ingolstadt had not yet acquired that instrument, so fundamental to experimental natural philosophy. Clearly there was a substantial gap between teaching and defending theses relating to experiments and actually performing the experiments themselves.

There are several reasons for this. Apart from any philosophical or pedagogical reasons for not performing experiments, there were fundamental material reasons. Even if one wished to perform experiments, obstacles stood in the way of acquiring instruments. For one, relatively few craftsmen in Germany were capable of manufacturing instruments such as the air pump until well into the eighteenth century. Another major reason was cost: it can be said that air pumps were the "big science" of the seventeenth and eighteenth centuries. Although many improvements were made in their design throughout this period that made them easier to use and improved the quality of the vacuum they produced, they remained expensive instruments, indeed among the most expensive of natural philosophical instruments.[18] A 1753 catalogue of the stock of instrument maker G. F. Brander's (1713–83) workshop in Augsburg listed three models of air pump. The cheapest, a perpendicular, single-barreled cylinder with a barometer conveniently attached to measure the pressure inside the receiver, cost 150 to 175 florins, including the necessary glass receivers. A second, with an inclined cylinder, cost 250 florins, and a third, which followed Willem Jacob 'sGravesande's (1688–1742) design, cost 350 florins. In comparison, microscopes cost 30 to 40 florins, although some astronomical instruments such as the more complex quadrants could cost up to 800 florins.[19] Such costs made air pumps major capital items. However, despite the cost, for any institution that was serious about conducting experiments an air pump was perhaps the most necessary instrument; a survey of late-eighteenth-century instrument collections in France reveals that along with an electrical machine, an air pump was found in every collection.[20] Thus it is significant that the Jesuit colleges were without air pumps for so long.

Although the Jesuits had individual experimental instruments, it is clear that their natural philosophical curriculum was not built around instruction through demonstration. They simply did not have the resources to support such pedagogy. This situation changed dramatically toward the end of the first half of the century. The case of Ingolstadt is best documented. In 1746 Elector Maximilian III Joseph of Bavaria ordered that Jesuits teach the biennium at Ingolstadt just as they did at their other Bavarian colleges. To make the philosophical curriculum relevant to all students, not merely those going on to theological studies, he insisted that experimental philosophy be cultivated along with mathematics and, in particular, Church and secular history.[21] Nevertheless, these instructions, like others issued by the Bavarian government to the University of Ingolstadt in the first half of the eighteenth century, made very few specific demands on the philosophy faculty, particularly in comparison to the other faculties. For example, the state established and funded a chair in history to be occupied by a Jesuit, but the state made very little direct input into the Physics curriculum beyond general admonitions to teach experimental physics and devote more attention to mathematics. The most heated battles being fought at the Catholic universities in the eighteenth century were not about how natural philosophy should be taught and for the most part did not touch upon the arts or philosophy faculties.[22]

The Jesuit faculty quickly responded to the elector's command with an outline of the new curriculum. They stated that if they were to teach experimental philosophy as they had been commanded, they would need the instruments that academies in other lands had. Funds for such resources were at present completely lacking. Furthermore, if a "Collegia Experimentalia" were to be established, it would need its own auditorium where experiments could conveniently be done.[23] This confirms that even though individual instruments such as the air pump that Mederer reports was purchased in 1729 were on hand, there was little institutionalized support for experimental philosophy at the university.[24]

This soon changed. The philosophical faculty's "diary" notes that the philosophy faculty was actively teaching "Physica Experimentalis" in 1748.[25] In 1753 the philosophy faculty at Ingolstadt again asked the elector for funds for experimental apparatus. Their request made clear the status that they accorded experimental philosophy:

> Your Electoral Serenity has Yourself realized through Your most wise reason and own experience that a thorough Philosophy cannot be made truly comprehensible without either the instruction of experiences [experienzen] or the actual conviction of the senses. Our philosophical faculty has most eagerly decided to present to its academic auditors a natural science [Naturs-Wissenschaft] established ac-

> cording to the true taste of current learning and amply provided with experiences. Since verbal instruction only engages the understanding of the listeners and has not left the impression which the genuine approval of the eye could achieve, for the promotion of a thorough philosophy and thereby of the flourishing of your most humble university we send to Your Electoral Serenity our most humble request to allocate an annual amount from the academic treasury aiming exclusively at the public benefit of philosophy.[26]

It is clear from the professors' esteem for the pedagogical power of experimental devices that they regarded them primarily, if not exclusively, as tools of instruction, not of "research." Whereas traditional philosophy engaged only the understanding, experiments also engaged the senses and could win assent through the more forceful visual impressions they made on the audience.

The elector decided to contribute one hundred florins for the purchase of instruments, an amount that would not have been sufficient for the purchase of a major instrument such as a pump from Brander's catalogue but would have covered the cost of several smaller pieces.[27] Just over a year later the faculty reported to the elector the benefits brought by the instruments purchased with the initial installment of funds and requested further support. They praised the generosity of his first contribution, which confirmed him as "the highest patron of such sciences." The fathers had bought many "experimental machines" with which experiments could be placed before the eyes of the auditors in public lectures and be made more comprehensible. It was certainly a good start, but, they reminded the elector, it could by no means be regarded as "a complete philosophical cabinet [*Armarium Philosophicum*]." A better collection, the fathers argued, would win the elector greater applause and improve the fame and usefulness of his university.[28] Thus the professors based their appeal on the pedagogical efficacy of experimental philosophy but also on the reciprocal benefits of the patron-client relationship. By fostering experimental philosophy the elector would win prestige and his university fame. This appeal was successful; it earned another two hundred gulden for the purchase of experimental instruments.[29]

Such developments were not limited to Ingolstadt; the chronology is similar at other Jesuit colleges. Throughout Germany they rapidly developed collections of instruments from the 1740s. At Innsbruck in the Hapsburg lands, Ignaz von Weinhart established an experimental "Armarium" in 1742 with his own instruments. Significantly, Weinhart was a professor of mathematics, confirming that the impulse toward experimental natural philosophy was strongly supported by mathematicians. For 1751–61 the state assigned 200 florins annually to the purchase of instruments for the experimental instruments and 150 florins annually thereafter. Weinhart worked with local artisans to improve instruments such as the air pump, which he supplied to other Jesuit universities such as Ingolstadt, Würzburg, and Dillingen. Weinhart also performed experiments before visiting dignitaries such as Emperor Franz I, Archduke Peter Leopold, and Cardinal Giuseppe Garampi.[30]

In Freiburg the philosophy faculty acquired a barometer and a thermometer in 1745. The diary of the philosophy faculty noted from 1755 onwards the frequent purchase of physical instruments, mainly electrical and hydraulic. Experiments were publicly performed at Freiburg also; two students were rewarded with thirty kreutzer in 1746 for helping to display the air pump to a visiting general. In 1756 the faculty applied to the university for funds to build a physico-mathematical museum, and when these were not granted, they built it anyway with Jesuit funds. The faculty did receive seventy-five florins per year for three years to buy instruments, including an air pump and Magdeburger hemispheres. Most of the instruments purchased came from Innsbruck, where they presumably were manufactured under Weinhart's supervision.[31]

The college at Bamberg bought an air pump and then an electrical machine in 1747.[32] At Mainz a list of instruments "greatly necessary for experimental physics" was drawn up. It included such favorites as the pneumatic pump with its receivers as well as the Magdeburger hemispheres.[33] The syllabus of the philosophical biennium at the University of Mainz for the 1747/48 academic year advertised that in his treatment of "special philosophy" Blasius Henner would cover the system of the world, the motion of the stars, the mixed elements, and psychology and would "disclose each in turn insofar as possible through artificial and natural experiments."[34]

Reforms in other states followed a course similar to those at the Bavarian university in Ingolstadt: that is, as part of a broader program of educational reform, the government made a vaguely articulated general statement of support for experimental philosophy—which in any case came well after the Jesuits had begun to integrate experimental philosophy into their Physics curriculum—followed some time later by concrete measures to support the teaching of experimental philosophy. The Schönborn dynasty, for example, was responsible for a series of reforms at the University of Würzburg. In 1731 and 1734 Bishop Friedrich Karl von Schönborn (1674–1746) introduced new *Studienordnungen.* As at Ingolstadt, these required the teaching of experimental physics but did not specify the content of the curriculum to any degree. The 1731 *Studienordnung,* for example, stated that in Physics the *curiositates eruditae* were to be granted the appropriate attention and that "particularly in that which belongs to the books on heaven and earth, the elements, motion, duration and time, and psychology [animastica], appropriate regard should be paid to experimental philosophy."[35] Yet the university had already produced numerous disputations and dissertations treating various aspects of experimental natural philosophy, in particular the mercury tube barometer.

At Würzburg as at Ingolstadt, it was not until some years later that the state took concrete steps to provide institutional support for the teaching of experimental philosophy. On 2 September 1749, Bishop Carl Philipp von Greiffenklau decreed the establishment of a chair of *Physica Experimentalis* that was to be filled by a professor from the Society of Jesus who was to receive a salary of two hundred gulden.[36] The newly appointed occupant, Blasius Henner (1713–59), took his responsibilities seriously, undertaking a tour of other universities with Anton Nebel (1711–54) and Heinrich Pfeffer (1714–post 1770), the mathematics professors from the Universities of Würzburg and Mainz respectively. Although they mainly visited Protestant universities, Henner found his visit with "Father Frantz," the professor of experimental physics at Vienna, particularly useful. This was Josef Franz (1704–76), one of the most prominent Jesuit scholars of the day, who taught mathematics, astronomy, and physics, constructed an astronomical observatory, tutored Emperor Joseph II in philosophy, and was rewarded for his efforts with a magnificent funeral. So while their taking this trip indicates that the Würzburger Jesuits felt the need to "catch up," it also shows that they had no qualms about learning from and visiting Protestant institutions and that they felt that Jesuit institutions and faculty compared favorably with the Protestant ones they visited. During his tenure as professor of experimental physics Henner also wrote a textbook on the field.[37]

The growing institutional support for experimental physics required a fundamental change in the career structure of Jesuit philosophy professors. Traditionally professors rotated through the triennium or biennium. But once chairs for experimental physics had been established, their occupants spent considerable time in them. Blasius Henner occupied the chair at Würzburg for ten years from 1749 until his death in 1759. He was succeeded by his brother Georg Henner (1732–?) from 1759 until 1770. The third occupant of the chair was Ambros Egell (1732–1801), whose tenure began in 1771 and continued past the abolition of the Society until infirmity forced him to resign in 1797. Virtually all the major Jesuit colleges, in particular those at universities, acquired a chair in experimental physics before the suppression of the Society.

Physics was acquiring greater autonomy within the philosophical curriculum. Ever since the time of Christoph Clavius, there had been Jesuit mathematics professors who enjoyed lengthy tenures in their chairs, in quite striking contrast to Physics, which was taught by junior professors cycling once or twice through the triennium or, later, biennium. Physics had not been considered to be a discipline requiring specialists. The establishment of chairs of experimental philosophy that were occupied virtually permanently by a single professor was an acknowledgment on the part of the Society that experimental philosophy required its own dedicated practitioners. Another sign of the growing autonomy of physics, particularly experimental philosophy, within the larger curriculum was the appearance of textbooks covering only the year of the curriculum that covered physics, as opposed to the compendia that treated the entire philosophy course. By the mid–eighteenth century, textbooks devoted solely to natural philosophy were published. In addition to Blasius Henner's textbook, examples of this genre include Karl Scherffer's *Institutionum physicae* (1752–53) and Josef Redlhamer's *Philosophiae naturalis* (1755), both published in the Austrian province.[38]

While we have focused on the mercury tube and the air pump, the Jesuits increasingly addressed those areas that attracted the attention of the experimental community in the eighteenth century progressed: optics, light, colors, electricity. Numerous works either cover all of these fields or focus on a particular one.

Aristotle and Experiment

By the middle of the eighteenth century it would have been difficult to describe Jesuit natural philosophy as scholastic. The Jesuits themselves did not use the term to describe their work, except in the most generic sense of "taught in a school or university." They often referred to their philosophy as peripatetic, indicating an adherence to the core of propositions discussed last chapter, but after the middle of the century some began to consciously reject that term or used it to define the kind of philosophy they did not teach. Where, then, was Aristotle? From the beginning of the Jesuits' engagement with experimental natural philosophy, they did not recognize any inherent dichotomy between it and Aristotelian natural philosophy. They frequently justified their teaching of experiment with what was by the eighteenth century a well-established trope: were Aristotle alive today, he would himself be practicing experimental philosophy.

As in other questions, the attitude of the German Jesuits toward experimental natural philosophy could be buttressed and encouraged by authoritative textbooks emanating from the Collegio Romano. Giovanni Baptista Tolomei, whose philosophy textbook was reprinted in the Upper German province, taught that Aristotle had been eminent in the empirical disciplines and "in all kinds of experiments in nature."[39] Unfortunately, he lamented, the peripatetics had become slaves to Aristotle's subtlety—that is, his speculative disciplines—and had neglected the empirical aspects of his work. This had deservedly earned them scorn. This was a very common sentiment among the self-professed adherents of the Enlightenment, and one increasingly shared by the Jesuits: Aristotle's true, empirical philosophy had been corrupted by the Middle Ages. Only by returning to experiment would the peripatetics be able to escape this path. Moliere could not have put it better himself when Tolomei wrote:

> Henceforth the barbs and insults, which the antiperipatetics are accustomed to hurl will cease: [for example,] that to learn the physics of Aristotle it is doubtless sufficient to commit a few general and unexplained words to memory: for example, from such a cause such matter is arranged, and induces such dispositions and such a form is educed; that from such a form such accidents emanate, that such qualities are diffused by such a medium, that there are such modes; that such an effect is such because it proceeds from such a cause, and whenever the phrase should be changed [that it is sufficient] to add occult qualities through which such a nature works such effects; and so through suchnesses of this kind they ridicule their listeners and further obscure nature itself.[40]

Experiment restored true natural philosophy after centuries of medieval obscurantism.

Over the first half of the eighteenth century the fundamental compatibility of experimental philosophy with Aristotelian philosophy became the Society's official policy and was expressed in the decrees of its hierarchy. In 1731, responding to fears in the provinces that the Society was restricting itself "exclusively to speculation and metaphysical subtleties," the Sixteenth General Congregation officially acknowledged the position experimental natural philosophy had attained within the Society. While it once again denounced excessive freedom in the formulation of philosophical opinions, the congregation granted that "the more attractive style of experimental physics is allowed for purposes of explaining natural phenomena." The congregation further explained its ruling by stating that there was no opposition between Aristotelian philosophy and "the more attractive style of learning in physics . . . with which the more notable natural phenomena are explained and illustrated by mathematical principles." In fact, the congregation specifically stated, there was complete harmony between them. Nevertheless, it reminded the Society's members that the founders of the Society had adopted Aristotelian philosophy as the one most supportive of theology and that consequently it must continue to be taught. The "peripatetic system" of the natural principles and constitution of bodies was not to be omitted from the philosophical classroom.[41]

The essence of this decree was confirmed over twenty years later by the Seventeenth General Congregation in 1751. It referred to the precedent set by the Sixteenth General Congregation and reaffirmed the fundamental compatibility between Aristotelian and experimental physics. Once again, the purpose of philosophy as a preparatory study for theology was emphasized: so that abuses might not arise in experimental philosophy, "those questions and concepts should be transmitted with accuracy which, though they are not and should not be theological in themselves, nevertheless smooth and strengthen the approach to scholastic theology." To maintain philosophy's propaedeutic role, the syllogistic method was to be used in the classroom through the use of questions and disputations. In explicating the decree, General Ignatio Visconti wrote that "there is no material that cannot be treated by the syllogistic method."[42] Experiment and syllogism were complementary approaches to philosophy. In contrast, a dominant thread in the English experimental tradition that began with Bacon and continued on through Boyle and the Royal Society consciously rejected syllogism and disputation as fruitless squabbling that could lead only to dissension, not knowledge. The conviction within the Society that experimental philosophy was not only compatible with Aristotelian philosophy but actually truer to the real Aristotle than mere speculative science was mirrored in the German provinces.

What the integration of Aristotelian philosophy and experimental physics meant in practice can be seen in Edmund Voit's (1707–71) 1743/44 Physics lectures at the University of Würzburg, where he used Aristotle's four causes to save the concept of the *horror vacui.* Since it was no longer possible to state in the face of all the evidence that the fear of the void was the efficient cause of all the phenomena formerly ascribed to it, Voit declared that it was their final cause: all the phenomena did occur in order to prevent a vacuum and all the disruptions to the fabric of the universe that it would entail, but the means through which this was achieved, and hence the efficient cause, was atmospheric air pressure.[43]

While they adopted novel elements into their Physics curriculum, the German Jesuits insisted on maintaining traditional elements of their pedagogy, as the General Congregation of 1730 had demanded. In response to the elector's insistence on the implementation of the biennium and experimental philosophy at Ingolstadt in 1746, the philosophy professors warned against going so far as to introduce the method used in non-Catholic universities and advocated keeping the syllogistic method, particularly for public disputations. They maintained that the primary role of philosophy was to prepare students for theology (a sentiment that the professors of the other higher faculties naturally did not share), which used syllogism and required that incoming students be well versed in its use, for without it they could not dispute properly. Furthermore, non-Catholics, the professors maintained, hated syllogism, as it unfailingly revealed the errors of their theology. However, they recommended some departures from the traditional method. Dictation should cease, and instead students should use one standard textbook. The professors recommended the one written by the Jesuit Anton Mayr, which they believed struck the right balance between ancient and modern philosophy as well as between faith, reason, and experiment.[44] Mayr's 1739 (1673–1749) *Peripatetic Philosophy Conformed to the Principles of the Ancients and the Experiments of the Moderns* was the first of a number of compendia covering the whole of the philosophy curriculum published by the German Jesuits around the middle of the eighteenth century.[45] As its title indicates, it was an explicit attempt to blend peripatetic and experimental philosophy.

In fact, adherence to the syllogism was virtually the only sense in which the mid-eighteenth-century Jesuits could be termed scholastic. If *scholastic* means being familiar with the scholastic doctors, then the term is not appropriate. Even in the late seventeenth century—that is, before they had wholeheartedly adopted experimental philosophy—the German Jesuits had moved away from the sources of high scholasticism. In 1682 Kaspar Knittel's (1644–1702) *Pleasant and Curious Aristotle* had sought to describe the conflicting schools of thought on the main topics of scholastic philosophy. Yet Knittel cited Thomas alone of the medieval doctors (with one or two references to the Dominican Durandus) and even then almost entirely through other Jesuit textbooks, virtually never from Thomas's works themselves.[46]

This development accelerated in the eighteenth century. Often dissertations or disputations whose titles seemed to fit squarely within the scholastic tradition actually contained treatises on various aspects of experimental physics. In fact, by the mid–eighteenth century, Jesuit natural philosophers had become completely ignorant of the scholastic doctors. The sources they employed included none of the high scholastics such as Aquinas or Scotus, and even Suárez and other major figures of the sixteenth-century scholastic revival had faded from use. In their place one finds seventeenth- and eighteenth-century mathematicians and experimental philosophers—many of them Jesuits but by no means all. A manuscript dating from around the middle of the eighteenth century outlining the philosophical curriculum of the colleges of the Upper German province contains a lists of recommended authors:

> Regnault, Zanchi, Falck, De Lanis, Deschales, Grimaldi, Tolomei, Fabri, Schott, Kircher, Casati, Gautruche [i.e., Jesuits] . . . Nollet, Muschenbröek, Newton, Keill, Voltaire, Philosophia Burgundica, Rohault, 'sGravesande, Euler, Boyle, Bayle, Müller's Collegium Experimentale, Vatter, Chauvin, Fortunatus a Brixia, Corsin, Pourchot, Sturm, Wolff. In many matters the commentaries of the Academy of St. Peterburg; the Historia of the Royal Academy, especially in the index recently published, the Trivoltienses; for electricity Tallabert, Doppelmayr.[47]

There is hardly a figure on the list who can be considered a late scholastic author, much less a medieval scholastic. The closest was probably the French Jesuit Pierre Gautruche, whose compendium of philosophy, *Institutio totius philosophiae,* was published in 1653.[48] Although this was a list solely of authorities for the whole of philosophy, the preponderance of mathematicians and natural philosophers is overwhelming. The list contained numerous Jesuits, both seventeenth century (De Lanis, Casati, Fabri, Schott, Kircher, Grimaldi) and more recent (Tolomei and Falck). There are numerous experimental philosophers and popularizers (Musschenbroek, 'sGravesende, Boyle, Sturm), as well as mathematicians (Euler) and those who transcend simple classification (Newton, Wolff, Bayle). There are even several whom one would perhaps not expect to be there, such as Voltaire—one suspects the Jesuits' students were not reading *Candide.* Conspicuous through their absence are Descartes and Leibniz. The Society had proscribed their philosophies several times, although this did not mean, as we have seen, that their opinions were banished to the realm of silence. And even Aristotle himself was not mentioned among the *auctores.*

Cardinal Tolomei wrote at the end of the seventeenth century that philosophers had to abandon the path of medieval philosophy, which had ruined Aristotelian philosophy. By midcentury, the German Jesuits were effacing any tracks showing they had ever trodden that path. Josef Redlhamer (1713–61), professor at the University of Vienna, began his 1755 textbook of natural philosophy with a brief overview of the history of natural science.[49] After first dealing with the ancients he proceeded to the Middle Ages, when the Arabs, in particular Averroes, obscured true philosophy because "in place of truly physical questions they substituted abstract, contentious metaphysics, thereby neglecting experiments."[50] Redlhamer omitted any mention of the medieval Latins, in effect erasing the Jesuits' own scholastic heritage. Instead, he jumped from his denunciation of the Arab philosophers to Bacon, "the founder of experimental physics," rather a flattering description for someone who never performed an experiment. There then followed a genealogy of modern physics that included the standard figures, such as Descartes, Gassendi, Boyle, Galileo (who appeared in a short note appended to Boyle), Torricelli, Guericke, then the Jesuits Kircher, Schott, and Fabri, and finally Newton and several of the Dutch natural philosophers. After listing their achievements, Redlhamer employed a familiar trope: if Aristotle returned from the dead and saw all these things, then "overcome in mute amazement he would correct many of the things he had written, throw out others and add others and through love of truth freely emend others."[51] Aside from the erroneous detour provided by the Arabs, the course of the history of science followed the straight path of the development of experimental science. The Jesuits, in Redlhamer's telling, were integral to this process. His readers could be forgiven for supposing that the Jesuits had never done any other kind of natural philosophy.

Air Pumps as Emblems

Steven Shapin has said that air pumps had an emblematic character in the early modern period, as they were models of the right way of proceeding in experimental philosophy.[52] One well-known example is Robert Boyle's air pump in the background of Wencelaus Hollar's frontispiece to Thomas Sprat's *History of the Royal Society* (1667). The Jesuits depicted this view of air pumps magnificently at the University of Dillingen, revealing the extent to which experimental philosophy and pneumatics in particular were emblematic of philosophy as a whole. The ceiling frescos of the magnificently decorated assembly hall of the Marian Congregation at the University of Dillingen, painted in 1762 by Johannes Anwander, allegorically depict in each corner one of the four faculties of the university, philosophy, law, medicine, and theology. Philosophy, a female figure with winged temples and the caption "experientissima," sits surrounded by the paraphernalia of the various branches of mathematics and natural philosophy: maps of lands and fortifications, a globe, rulers, a compass and set-square, a telescope, a pair of parabolic reflectors. In pride of place on a bench before the figure of Philosophy is an air pump (figure 7). It is a single-cylinder pump whose cylinder lies almost horizontally. Designed by Wolferd Senguerdius in 1681, the first example of this model was manufactured in the workshop of Samuel van Musschenbroek. Due to its ease of use it was widely copied and manufactured in Europe well into the eighteenth century. G. F. Brander's workshop in the nearby city of Augsburg produced numerous examples of this model; thus the frescos may depict an actual instrument used at the university.[53] In the glass bell–shaped receiver is a bird, anticipating Joseph Wright of Derby's famous depiction of the air pump of 1768 and confirming that the urge to put a small feathered or furry creature in the pump's receiver was universal. Through its central position in the fresco, the air pump clearly was the most powerful embodiment of natural philosophy.

However, the air pump had more than just an "emblematic character"; it and other experimental instruments were used as actual emblems in Jesuit architecture. The seventeenth century has been called the "golden age" of Jesuit emblematics, during which the German provinces produced numerous works on emblems.[54] Emblems were integrated into art and architecture as well as literary works on theology and philosophy. Emblems generally consisted of an *inscriptio*—that is, a motto—a

FIGURE 7. Ceiling fresco from the Goldener Saal in the former Jesuit university of Dillingen. Each of the faculties is allegorically represented in a corner of the ceiling. Here Philosophia sits surrounded by mathematical and experimental instruments, including most prominently, an air pump. In the glass bell receiver is a bird. Author's photograph.

pictura or illustration, and a *subscriptio* or explanatory passage. Often the *pictura* could be omitted, as was the case with many emblem books that contained no illustrations at all. Similarly, when emblems were employed in architecture or painting, the *subscriptio* (or even the *inscriptio* if the emblem was sufficiently well known) could be omitted. The generation and development of emblems required training to master the symbolic vocabulary of the emblem. Students at Jesuit colleges were trained in the use of emblems in rhetoric class, the fifth year of the humanistic curriculum. The notebook of Hubert Houben, a student in Professor Derckum's rhetoric class at the Jesuit college in Cologne in 1722, shows that students were still trained in the eighteenth century to link a pictorial image with a line of Scripture and a lyrical "Allusio ad Emblema."[55]

Jesuit and other natural philosophical and mathematical texts, particularly in frontispieces, often celebrated patrons in allegorical ways. Furthermore, natural historical and even natural philosophical phenomena were often used for emblems: for example, the heliotrope flower that turned to follow the path of the sun was an allegory for the way humans should turn to God.[56] Galileo designed a emblematic medal for Prince Cosimo de Medici around the device of the magnet that punned on the name Cosimo and the cosmos. However, throughout the seventeenth century, little use seems to have been made of instruments as emblems. One of the best-known seventeenth-century encyclopedias of emblems, Philippo Picinelli's *Mundus symbolicus,* contained no emblems drawn from instruments; the only emblem using a machine was one based on the pulley.[57] Late-seventeenth- and early-eighteenth-century German Jesuit emblem books also contain few emblems employing machines or instruments.

But the spread of experimental devices meant that they were increasingly available to become part of the vocabulary of emblematics. A striking example of the early use of emblems drawn from natural philosophical instruments is a medal cast to commemorate the amicable division of the Duchy of Brunswick-Wolfenbüttel in 1704, when Duke Anton Ulrich bestowed his title and the bulk of the territory on his eldest son August Wilhelm and the smaller County of Blankenburg on the younger son Ludwig Rudolf. The front of the medal depicts Otto von Guericke's Magdeburger hemispheres, held together by atmospheric air pressure after the air inside the hemispheres had been pumped out. The motto *Non vi* (not by force) refers to the strength of the attraction between the hemispheres, which teams of horses could not overcome. The reverse shows the two hemispheres lying separated with the inscription *sed arte* (but by skill). Just as the hemispheres could not be separated by brute force but only by turning a small valve that allowed air to enter, so the territory could not be divided by violence but only by a wise and peaceful solution.[58]

While the air pump fresco on the ceiling of the Goldener Saal at the University of Dillingen was not strictly speaking an emblem, as it lacked an *inscriptio* and a *subscriptio* and did not depict a particular virtue or moral lesson, elsewhere in the same building there is striking emblematic usage of experimental instruments. The ceiling of the university's Experimental Museum is decorated with a novel series of emblems, most of which are not found in any emblem book. Around the edge of the ceiling, surrounding the central image of the Holy Spirit, which shoots rays outwards, are twelve emblems consisting of stylized gold depictions of experimental devices accompanied by mottoes. They are:[59]

1. Parabolic incendiary mirrors	*Exemplo accendit* (it enflames by example)
2. Bell in air pump receiver	*Docet absque tumultu* (it teaches without commotion)
3. Telescope	*Coelestia spectat* (it regards the heavens)
4. Double-cylinder water pump	*Date et dabitur* (give and it will be given)
5. Triple siphon/fountain	*Communicat omnibus* (it imparts to all)
6. Microscope	*Abdita pandit* (it reveals the hidden)
7. Chain hanging from magnet	*Attrahit et imbuit* (it attracts and imbues)
8. Diffracting prism	*Varietas delecta* (diversity delights)
9. Barometer/thermometer	*Servit tempori* (it serves throughout time)
10. Series of pulleys	*Totum aecuat singula* (the whole equals each one)
11. Balance/scales	*Ponderat omnia* (it weighs all)
12. Suspended-boy electrical trick	*Animat dum recreat* (it animates while it renews)

Because of its central place in the design, the subject of the emblems, the unnamed "it," is the Holy Spirit. Thus in the workings of these devices one can see the chief attributes of the Holy Spirit, although the lack of a *subscriptio* or explanatory passage leaves the exact intention of the designer somewhat open. The identification of most of the devices is clear. Some of the *picturae* had long histories, in particular the balance, the magnet, and even the telescope. Others were completely novel. Number 12 shows a boy insulated in a static electricity experiment that was frequently depicted in works on electricity and performed as a salon trick (figure 8). This remarkable emblem shows how a contemporary fad could be integrated into the emblem genre. Number 2 is somewhat obscure; it could represent a furnace, but the motto hints at the inability of sound to pass through the air pump's evacuated receiver. The emblem would therefore be a depiction of a bell inside an evacuated space, the test intended to demonstrate the presence of some corporeal substance in an allegedly empty space first devised by Athanasius Kircher.

FIGURE 8. An emblem from the emblem cycle on the ceiling of the Experimental Museum at the University of Dillingen. This depicts the kind of static electricity experiment so popular in the eighteenth century. Its motto is "Animat dum Recreat" (it animates while it renews). Author's photograph.

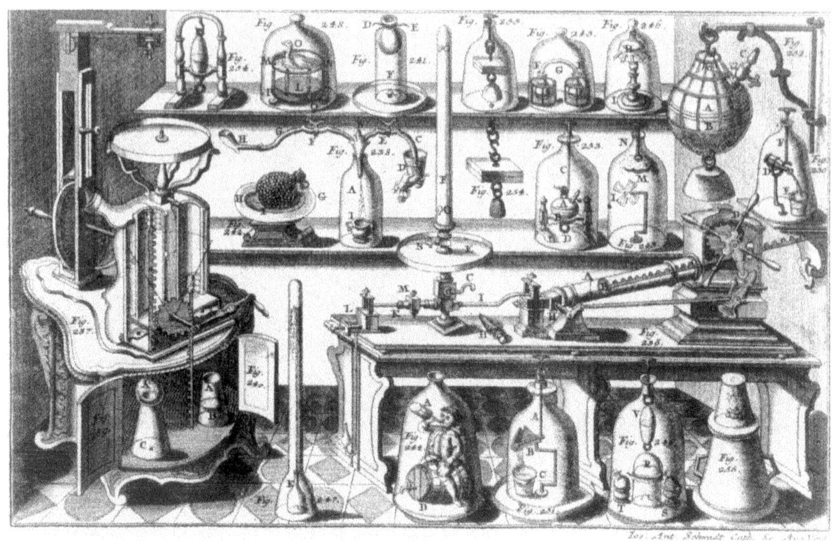

FIGURE 9. Instruments depicted in Berthold Hauser's *Elementa philosophiae.* These show two anatomized air pumps along with a vast range of receivers. Hauser's work contained similar engravings of electrical, magnetic, and mechanical instruments. It is unlikely that the Jesuit colleges at the universities of Dillingen and Ingolstadt where Hauser taught actually had all of these elaborate and expensive instruments. Courtesy Stadtbibliothek Mainz. Signatur: III l 4° 355h.

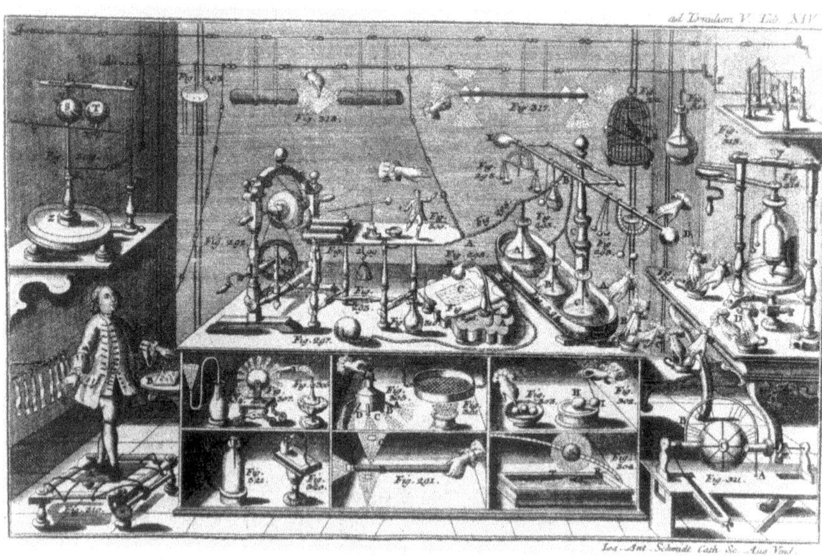

FIGURE 10. Electrical instruments depicted in Hauser's *Elementa philosophiae* (1755–64). Courtesy Stadtbibliothek Mainz. Signatur: III l 4° 355h.

The series was probably designed and created under the supervision of Berthold Hauser, professor of experimental philosophy at Dillingen for ten years from 1752 until his death in 1762. Hauser published a massive, multivolume philosophy textbook, *Elementa philosophiae ad rationis et experientiae ductum conscripta, atque usibus scholasticis accomodata* (1755–64), that provided a detailed treatment of experimental philosophy.[60] Several volumes contained many engravings of experimental instruments, including many of those used in the emblems (figures 9 and 10). Unlike emblems in churches or emblem books, the emblems on the ceiling of the Experimental Museum were not designed for or presented to a general audience and would have been seen only by professors, students, and visiting dignitaries. Nevertheless, they show that experiment had become as much a part of the symbolic vocabulary of Jesuit professors and students as the characters of the Bible and the gods and heroes of ancient mythology.

Nevertheless, one should end with a cautionary note. The magnificent engravings in Hauser's huge philosophy compendium depicted a vast range of instruments from the utilitarian to the flamboyantly baroque. Yet they can be regarded only as wishful fantasy rather than as an accurate depiction of the contents of Jesuit experimental cabinets. The collections at all the Jesuit colleges were limited by funds. Although the states might give money to purchase instruments, the colleges had neither the money nor the personnel to maintain large collections adequately. In 1755 the philosophy faculty at Ingolstadt started taxing graduating masters and bachelors to raise money for the experimental cabinet, but this raised only a trivial amount. When Elector Max III Joseph visited in 1763, everything in the collection was in poor condition, and when Professor Steiglehner, the first non-Jesuit to teach experimental philosophy at Ingolstadt, took the first inventory of the collection in 1781, he described almost all the instruments as broken, unusable, rusted, or damaged and wrote them off as junk. Again, the poor state cannot be attributed solely to the Jesuits; Steiglehner's predecessor, the (ex-)Jesuit Ignaz Helfenzrieder, had constantly sought funds for the cabinet without success.[61] So it must be said that even though the Jesuits vociferously supported experimental philosophy, the material resources devoted to it remained limited. The modern image of students working at the laboratory bench replicating experiments is just as anachronistic for the later eighteenth century as it is for the seventeenth.

CHAPTER 10

The Jesuits and Their Contemporaries in the *Aufklärung*

BY THE EARLY EIGHTEENTH CENTURY THE RELIGIOUS AND POLITICAL order in Germany was very different from that in the later sixteenth century. True, the Holy Roman Empire was still a patchwork of states of widely varying sizes, and a politically unified Germany was more remote than ever. But the confessional issues that had torn Germany apart in the century before 1648 had been settled. Wars continued to sweep across the empire throughout the eighteenth century, but they were not fought for religious reasons; Catholic France's alliance with Protestant Sweden against the Catholic Hapsburgs in the Thirty Years' War forever ended whatever meaning confessions had in wars in western Europe. While eighteenth-century Austrian or Bavarian princes may well have been as devoutly Catholic as their sixteenth- and seventeenth-century ancestors, the confessional age was over. Catholic princes had originally endowed Jesuit colleges in order to train clergy to help stem the Reformation by confirming the faith of their Catholic subjects or reconverting their Protestant ones. But Catholic bishoprics and dynasties such as the Hapsburgs and Wittelsbachs, which were struggling for their very existence against the spread of Protestantism in the second half of the sixteenth century, no longer faced that danger. German states now faced being swallowed or dismembered by their larger neighbors or by France, which steadily nibbled at the western border of the empire.

In this environment the function of higher education changed. The states now sought bureaucrats who were trained in *Cameralwissenschaft* or *Polizeiwissenschaft,* the sciences of public administration, and could run the territorial governments efficiently or technicians who could promote economic and industrial development. Their training was to be based on the disciplines of law, history, and applied mathematics. Those who wanted to reform the German states increasingly questioned the value of the philosophical education that the Jesuits provided and challenged their hold over the faculties of arts with increasing vigor.

Some have claimed that the Jesuits, wedded to the *Ratio studiorum* and a conception of philosophy purely as preparation for theology, resisted these trends and modernized their philosophical curriculum only when pressured by Enlightened reformers, for example. But this reduces the complexity of the situation to caricature. When the states provided the resources, the Jesuits happily added history to the philosophical curriculum, even though there was no provision for it in the *Ratio.* As this chapter will show, with regard to the natural philosophical curriculum at least, simple dichotomies of Enlightened versus obscurantist do not adequately describe the Jesuits' interactions with their contemporaries in the Age of *Aufklärung.*

The Enlightenment Critique

Censorship, restrictions on philosophical freedom, the subordination of philosophy to theology, and an insistence on maintaining elements of peripatetic philosophy made the Jesuits a prime target for those who saw the Society's philosophy as incompatible with the values of the Enlightenment, or its German version, the *Aufklärung.* Not surprisingly the Protestant, north German centers of the *Aufklärung* launched many attacks on the Jesuits and their philosophy.

The Leipzig scholar Jakob Brucker (1699–1770) excoriated scholastic philosophy as a degenerate form of true Aristotelian philosophy in his massive *Critical History of Philosophy* (1742–43). He saw the recent history of philosophy as a heroic effort to reveal the harm caused by scholasticism's "vanity, futility, and inane subtlety and [by] those things that have risen out of that adulterous offspring." The only reason scholastic philosophy had survived so long was its inseparable connection with scholastic theology, without which "the Church could hardly wage war for the received dogmas."[1]

Ironically, his critique of scholastic philosophy mirrored the Jesuits' own rejection of it. But while the Jesuits believed that they had abandoned it (and even overlooked that they had ever taught it), Brucker regarded them as scholastics still. The colleges of the Society of Jesus, enslaved in obedience to the Roman curia, had always been a particularly fertile field for "the weeds of the scholastic pseudophilosophy."[2]

Despite Brucker's heated language in condemning the Jesuits in general, when he considered individual Jesuit philosophers his view was considerably more nuanced. Arriaga, he said, "not only consulted the observations of the moderns but also used and approved many of them after repudiating the opinion of the scholastics," although Brucker maintained that Arriaga had always adhered to the method of the scholastics. The Conimbricenses should be praised for their erudition, but it was deformed by scholastic shrewdness and subtleties.[3] In short, while Brucker claimed that scholasticism in general was a diseased philosophy and that the Jesuit variety in particular was no different, when he considered individual Jesuit authors, his judgment was considerably tempered.

But even more significant than his assessments of individual Jesuits was Brucker's silence on the vast majority of Jesuit authors. For example, the only German Jesuit he discussed was the rather obscure Dillingen professor Franz Rassler (1649–1734), who, Brucker correctly noted, wrote in 1685 a work on Empedoclean corpuscles and another work on the vanity of scholastic philosophy (although Brucker did acknowledge that Schott had been the first to publish Guericke's experiments).[4] He did not mention a single German Jesuit natural philosopher of his own age; Brucker knew little about Jesuits in the seventeenth century and was completely ignorant about what they were doing in the eighteenth. What makes this omission all the more striking is that the *Acta Eruditorum*, the learned journal published in Brucker's home of Leipzig, showed considerable awareness of the Jesuits' contributions to scholarship.[5] Volumes published in the 1680s and 1690s included both original contributions by Jesuits, such as Adam Kochanski's mathematical articles in the 1682, 1685, and 1686 volumes, and reviews of numerous texts written by Jesuits.

Catholic *Aufklärung*

Such negative evaluations of Jesuit scholarship are to be expected from Protestant Germany. More surprising is the extent to which much modern scholarship shares these prejudices. Traditionally the historiography of the *Aufklärung* has addressed its Protestant manifestations. In this regard intellectual history has fared no better than political history in escaping the distorting gravitational pull of Frederick the Great and his Prussia.[6] More recent scholarship has sought to define the major characteristics of the Catholic *Aufklärung* in order to determine how it differed from its Prussian and Protestant forms.[7]

In general the Catholic *Aufklärung* was not anticlerical. However, one of its most visible aspects was its hostility toward the Jesuits. According to Richard van Dülmen, this was a defining characteristic of the Catholic Enlightenment.[8] Like the Protestants, Catholic adherents of the *Aufklärung* condemned the Jesuits for instilling obscurantism in their students. Count von Stadion, the chief minister to the elector of Mainz, claimed that "[a]s long as the Jesuits keep control of the lists of sins of great men and by means of the schools can beat or coax infantile prejudices and dogma into the heads of young people, they will always enjoy the support of the majority."[9] In the Catholic territories also, many regarded the Jesuits as incorrigible scholastics.

Motivated by feelings of inferiority toward their Protestant contemporaries, the Enlightenment's Catholic partisans sought to reform education in their lands. This resulted in many bitter struggles for control over the faculties of philosophy and theology throughout Germany. Attempts at reform sought to break the Jesuits' virtual monopoly over appointments in theological and philosophical faculties. The reformers generally met with little success in this regard, and the Society's privileges, particularly in the philosophical faculties, remained essentially intact until the abolition of the Society in 1773. While elements of the educated elites might have despised the Society, the Jesuits continued to enjoy great support from princes and the populace as well as from influential elements in Catholic society such as cathedral chapters.[10]

When the reformers focused on curricular changes, they made more gains. The most prominent reformer of Catholic education in the second quarter of the century was Johann Adam Freiherr von Ickstatt, who, although Catholic, attended Protestant universities and became a pupil of Christian Wolff. Ickstatt made his name by implementing reforms at the University of Würzburg, although they had little effect on the philosophy faculty. Then in 1746 Elector Maximilian III Joseph of Bavaria made Ickstatt, his former teacher, director of the University of Ingolstadt.[11] Ickstatt had many, often bitter, disagreements with the Jesuits there. But as Hammerstein has shown, these struggles were on the whole concerned more with disciplines such as history, ethics, and theology than with physics and mathematics.[12] In the latter fields Ickstatt's suggested reforms seldom went beyond generic requests that experimental philosophy and mathematics be taught.

Andreas Gordon, O.S.B.

Studies of the Catholic Enlightenment have traditionally portrayed the Jesuits as stubborn opponents of reform. For many fields that lie outside the scope of this study, this may well be correct. However, one should be wary of accepting the polemics waged between the Jesuits and their critics at face value. Terms such as *Enlightenment* are not all-or-nothing categories. In fact, the Jesuits often had much in common with the self-styled proponents of the *Aufklärung* with whom they sparred.

One Catholic natural philosopher who engaged in a bitter dispute with members of the Society of Jesus in the mid-1740s was the Scottish Benedictine Andreas Gordon (1712–50). Gordon studied at the Benedictine university at Salzburg and from 1737 taught natural philosophy at the University of Erfurt. Gordon was known for developing various electrical machines and for publishing a three-volume philosophy compendium, *Philosophia utilis et jucunda* (A useful and pleasant philosophy) in 1745. In 1748 he was elected corresponding member of the academy in Paris. Gordon's biographer Ludwig Hammermayer rightly points out the important contribution of Gordon and his colleagues in the Benedictine order to the diffusion of experimental physics in southern Germany. But his portrayal of Gordon as an Enlightenment paladin fighting Jesuit obscurantism adheres too closely to Gordon's own perceptions and rationalization of the affair.[13]

At the small university in the confessionally divided town of Erfurt, the Benedictines shared teaching duties with Protestants. The Jesuits were excluded from the philosophy faculty. The local Jesuit college taught only the humanities curriculum and was not allowed to teach philosophy. This situation created tension between the two orders. Gordon clearly liked to needle the Jesuits. For example, he wrote that he had performed an experiment at the University of Erfurt in 1742 in which he made water rise five feet up a tube by evacuating the air in the top of the tube with an air pump. Gordon claimed that an unnamed Jesuit present "began to declare publicly that the manifest cause of this phenomenon was the fear of the vacuum, which I contradicted in vain." When the pump could raise the mercury only twenty-seven inches, "[h]aving been asked by me what the cause of this phenomenon was, the said Jesuit responded, 'Mercury fears the vacuum less than water,'" which provoked laughter from all present.[14] Gordon's anecdote may be true, but no Jesuits were still publishing such claims about the *horror vacui* in the 1740s.

Certainly competition for students figured in the tensions. In 1745, after several of his students left him to study at Jesuit universities, Gordon attacked the Jesuits' philosophy, which he termed scholastic, in an academic oration.[15] But fundamental differences in their philosophies lay at the root of the dispute. While Gordon's philosophical compendium, *Philosophia utilis et jucunda,* drew on a wide range of sources, Hammermeyer admits that it showed "a remarkable lack of ability or interest in theological-metaphysical speculation."[16] Gordon deliberately limited philosophical investigation to the perceptible and measurable. The Jesuits responded after Gordon published a second critical oration.[17] However, it seems perplexing to say, as Hammermayer does, that they sought to defend "pure Aristotelian-scholastic doctrine" while putting up a "fighting retreat" in the face of the Enlightenment's advances. The Jesuits certainly did not reject all aspects of modern physics. One of the Jesuits who responded to Gordon's oration advocating the elimination of scholastic philosophy and the adoption of the new physics was Joseph Pfriemb, a Jesuit professor of philosophy at Mainz and an enthusiastic advocate of experimental philosophy who had written a series of graduation odes on the standard Jesuit theme of "The Happy Marriage of Theoretical and Experimental Physics."[18] The Jesuits, then, did not object to Gordon's teaching experimental philosophy.

Pfriemb and his fellow Jesuit Lukas Opffermann (1690–1750) were motivated by real animosity toward Gordon, however, and sought to have him tried before the archbishop-elector of Mainz, in whose territory Erfurt lay, claiming that his ethical teachings had heretical implications. The authorities did not conduct a trial, and after the dispute became public, the local Jesuits and Benedictines were summoned before the city's governor and told to drop their feud.[19] The exchanges between Gordon and the Jesuits Pfriemb and Opffermann were marked by the usual features of early modern polemics: accusations of willful misquotation, of dishonest use of sources, of unjustified claims to precedence. On several questions they argued past each other. Neither side passed up any chance to hurl insults at the other. Pfriemb criticized Gordon's departure from good manners but was himself no more civil in his characterization of Gordon's oration as a monster and an abortion.[20]

But the polemic was essentially based on fundamental disagreement over the correct relationship between philosophy and theology and proper criteria for determining the truth of philosophical claims. Gordon was outspoken in his rejection of scholastic philosophy, claiming that it agreed with neither reason nor experience. Furthermore, he wondered, how could a philosophy composed by a pagan three hundred years before Christ be of use in explicating the Gospels? In fact, scholastic theology itself was tainted by many questions that did not support the faith.[21] Among the examples that Gordon presented to support his claim that scholastic philosophy was stupid was the hylemorphic composition of bodies from matter and form, one of the few remaining doctrines in the peripatetic core of Jesuit philosophy.[22] In place of "barbarous" scholastic philosophy, Gordon advocated embracing mathematics and experiment. The only criterion for establishing the truth of philosophical claims was the extent to which they agreed with reason and nature.

The Jesuits considered Gordon's attempt to completely free philosophy from theology to be unacceptable. For Pfriemb the history of dangerous philosophical doctrines showed that philosophy could not be constrained by reason alone; the path to incorrect doctrine was prepared by false philosophy. Pfriemb upheld the Society's use of Aristotelian and peripatetic philosophy, in part because it stimulated young minds but primarily because it was the philosophical system most compatible with the faith—Aquinas had made Aristotle Christ's friend, he wrote.[23] In short, Pfriemb stood squarely in the Jesuit tradition that reason and experience were in themselves not enough to ensure the truth of a philosophical opinion and that the opinion also had to conform to orthodox theology. Thus Gordon neglected to teach, and indeed openly ridiculed, elements of scholastic philosophy that the Jesuits considered essential for the defense of orthodox theology, in particular hylemorphism.

It is easy to understand Gordon's frustration with his Jesuit rivals. The archepiscopal authorities' unwillingness to hold a trial shows that they did not share the Jesuits' fears, but by instructing Gordon to refrain from teaching metaphysics and theology they showed some degree of sympathy toward the Jesuit position. Gordon also exaggerated the backwardness of Jesuit physics, claiming that his rivals stated that all novelties were dangerous and that they had not accepted any new opinions in physics other than the weight of the air. But he identified real problems in Jesuit instruction, such as the frequent changes in professors. He also stressed the need to perform experiments in public lectures, rather than merely in private. By the time of the polemic with Gordon, however, the Jesuits had themselves begun to remedy this situation.

The Bavarian Academy of Sciences

In the face of the Jesuits' dominant position in the universities, the Enlightenment's self-proclaimed advocates created their own forums. The first major academy of sciences in Catholic Germany was the Bavarian Academy of Sciences (Bayerische Academie der Wissenschaften), founded in 1759 as the culmination of several attempts to foster intellectual life in the Catholic lands. In keeping with the character of the Catholic *Aufklärung,* the Bavarian academy was not anticlerical. Indeed, one early plan to found an academy originated among the Bavarian Benedictines. Another early attempt was led by the Augustinian Eusebius Amort, who actually invited Jesuit involvement by trying to recruit to the project Josef Falck and Nicasius Grammatici. After this effort failed, Amort founded the learned journal *Parnassus Boicus,* which included many Jesuit contributions.[24]

But once such visceral Jesuit haters as Georg Lori, a student and protégé of Ickstatt, took over and successfully completed the efforts to found a Bavarian academy, opposition to the Jesuits became "almost the academy's raison d'être."[25] Lori sought to establish the academy outside the traditional educational institutions, in particular the University of Ingolstadt. Lori insisted on keeping all preparation for the founding of the academy secret from the Jesuits. Particularly crucial in Lori's view was that the academy be exempt from the university's censorship and have the privilege of censoring its publications itself. As he recounted with great satisfaction numerous times in his correspondence, when the elector wanted the university to retain censorship authority over the new academy, Lori threatened to simply abandon the project if this privilege was not granted.[26] Lori won his freedom from he termed the Jesuit "yoke." The first director of the philosophical class (i.e., section) of the academy was the Benedictine Ildefons Kennedy, who knew Andreas Gordon from the Scottish Benedictine monastery in Regensburg, so there is no doubt that the circle around Lori were aware of Gordon's confrontations with the Jesuits.[27]

Clearly the Jesuit monopoly over appointments in the philosophy faculties at Catholic universities rankled Lori. He frequently referred to them as the "Monopolists" and "Solipsists." In return Lori refused to admit them to the academy, even though they were not explicitly excluded from membership. When the Benedictine Ulrich Weiß inquired if Jesuits would be admitted to the academy, Lori tersely replied, "The Jesuits are not members because they are scholastics and Jesuits."[28] But Lori's rejection of the Jesuits was a case of cutting off one's nose to spite one's face. In response to Lori's solicitations for recommendations of potential members, scholars around Germany nominated Jesuits. Anton Roschmann in Innsbruck suggested Ignaz Weinhart, the professor of mathematics and experimental philosophy at the University of Innsbruck.[29] Johann Georg von Stengel wrote from Mannheim that the Jesuit astronomer Christian Mayer would be a good addition to the philosophical class.[30] Even Lori's own mentor Ickstatt, the director of the University of Ingolstadt, who had himself been reconciled to the Jesuits after considerable initial difficulties with them, acknowledged the Jesuits' leadership in mathematics and experimental philosophy when he recommended the Jesuit mathematician Georg Cratz for membership of the academy in 1759, writing to Lori, "Accept our professor of mathematics the reverend Father

Cratz. He must share astronomical and physical observations for which the fathers have excellent instruments. Once we have survived our childhood things will go well, I believe. Initially we have to go in ribbons and allow ourselves to be led."[31]

Despite Ickstatt's acknowledgment that the Jesuits' talents and resources were essential for the development of the academy, Lori refused all recommendations of Jesuits, justifying his arbitrary decision by an appeal to liberty: "No one is incapable of entering the academy, but the regulations, which do not suffer despots, must be followed exactly. Our constitution is very democratic. In democracies, all tyrants are hated. Does Your Excellency not know those people who have ruled over scholars and science like a sultan over the Muslims? I fear Greeks bearing gifts."[32] But when he boasted that the Jesuits would not be able to make fun of "us" anymore, he seemed to be motivated by resentment rather than a noble attachment to liberty.[33]

The result was that the fledgling academy suffered from a major shortage of members qualified in mathematics and the natural sciences.[34] Nevertheless, Jesuits did win the academy's prizes: Cratz for a work on the distance and weight of the moon in 1761/62, Benedikt Stattler for a work on hydrostatics in 1770/71, and Johann Evangelist Helfenzrieder for his work on dykes and water flow in 1772/73.[35]

From its creation the academy focused on scholarship and publication, following the eighteenth-century distinction between university and academy. Although the academy conducted some public lectures, such as Ildefons Kennedy's series on experimental physics, it did not enter into competition with the University of Ingolstadt in teaching. In 1763 it began publication of its transactions (*Abhandlungen*). In 1784 Lorenz Westenrider published an official history of the academy's first eighteen years. Much like Sprat's *History of the Royal Society,* this text functioned more as a programmatic statement than as an account of the academy's actual deeds.[36] Naturally it repeated Lori's narrative of liberation from the yoke of Jesuit scholasticism, even though the Society of Jesus no longer existed. Kennedy quickly set about developing a collection of instruments, purchased mainly from G. F. Brander in nearby Augsburg. Not surprisingly, among the very first instruments was an air pump. For the academy Brander's workshop made a particularly nice one with fashionably curved legs and glass panels instead of wooden doors.[37]

The Bavarian Academy of Sciences inspired imitation. In 1763 the Academia Theodora-Palatina was founded by Johann Daniel Schöpflin in Mannheim, the capital of the palatinate. Here the academy's statutes specified that only laymen or secular clergy could be admitted, which necessarily excluded Jesuits. This barred the palatinate's most accomplished scientist, the Jesuit astronomer and mathematician Christian Mayer (1719–83), professor of mathematics and astronomy at the University of

Heidelberg and court astronomer and director of the electoral observatory. Mayer was a well-known figure in eighteenth-century science: he traveled to Paris and met with prominent members of the Parisian Academy of Sciences, such as Laland and Cassini, and was invited in 1769 to St. Petersburg to direct observations of the transit of Venus. Mayer, who had already been rejected by Lori from the Bavarian academy, was understandably annoyed at the pettiness that excluded him from the Mannheim academy.[38]

The Theresian Reforms

The war that followed the accession of Maria Theresa to the Hapsburg throne in 1740 brought the monarchy to the brink of collapse as other European states, in particular Prussia under the ambitious and unscrupulous Frederick II, sought to exploit Austria's weaknesses. While the province of Silesia was lost, never to be regained, the Hapsburg monarchy's desperate fight for survival ushered in an age of fundamental reforms of the state and society. Although historians increasingly question terms such as *absolutism*, these reforms assaulted local privileges in an effort to harness the human and material resources of the monarchy's culturally, ethnically, and geographically diverse territories to the interests of the state. The universities did not escape the state's attempts to subordinate their interests to its own. In many respects, however, these reforms displayed the limitations that characterized eighteenth-century absolutism, particularly in the Hapsburg lands. For example, universities remote from Vienna successfully delayed implementing reforms that they felt unduly undermined their autonomy. And, importantly, much-touted educational reforms often simply mirrored the curricular changes that the Jesuits were themselves already introducing to their colleges.

In charge of education reform in the Hapsburg monarchy was Gerard van Swieten. Although he personally admired individual Jesuits, Van Swieten despised the Society of Jesus. In 1752 Van Swieten produced a new *Studienordnung* for the faculties of theology and philosophy that was approved by Empress Maria Theresa. However, Van Swieten had been assisted in the formulation of the new *Studienordnung* not only by Cardinal-Archbishop Johannes Josef Graf Trautson, the *Studienprotektor* of the University of Vienna, but also by a Jesuit, Ludwig Diebel (1697–1771), who had been a professor of mathematics, philosophy, and theology.[39] Initially the *Studienordnung* applied only to the University of Vienna, but within several years it was applied to all the universities in the Hapsburg monarchy (Prague, Graz, Innsbruck, Olmuc, and Freiburg).[40]

As part of the larger reforms aimed at revitalizing the state through rationalization, the *Studienordnung* attempted to eliminate excessive holidays, wasteful ceremonies, and costly baroque pomp. It appointed a director to supervise every faculty and sought to replace the traditional bachelor's and master's exams with a *Staatsexamen.* Furthermore, it was concerned with developing a curriculum that promoted commerce and economic productivity and as such emphasized "utility." Yet like reforms in Bavaria and Würzburg, when it came to discussing the curriculum, this one was very general and did not descend into specifying detailed content. Dictation was forbidden. As part of the program of utility, useless *Subtilitaeten* were forbidden, particularly in metaphysics.[41] The only time the decree referred to specific doctrines was when it stated that "the unfounded doctrine (which cannot be confirmed by any experience) of peripatetic matter and form is henceforth completely forbidden"[42] and demanded that classes "be applied to the true experimental physics." The *Studienordnung* also decreed that philosophical studies be limited to a biennium.

In many ways the effects of the *Studienordnung* on the Jesuit colleges were minimal. The biennium had already been adopted at many colleges—over thirty-five years earlier, in fact, at Freiburg.[43] As for the rejection of the scholastic doctrine of matter and form and the introduction of experimental physics, the former doctrine was slowly fading away of its own accord, and the latter did not require an imperial decree, as it was already practiced at Jesuit institutions.[44] The Jesuits could, however, use the decree to support their own educational program. The philosophy faculty at Freiburg wrote a response to the new *Studienordnung* in 1753, welcoming its requirement that experimental physics be adopted into the curriculum. It recommended that a special fund be set up to acquire the necessary instruments. The faculty's diary reported from 1755 the purchase of new instruments and the performance of electrical and hydraulic experiments. In 1756 the philosophy faculty built a *physico-mathematisches Museum* with the Jesuits' own money after the university refused to provide the necessary funds. From 1765 a public *Kolleg* was held on Wednesday afternoons performing physical experiments for a curious crowd. Naturally, among the first instruments purchased were an air pump and Magdeburger hemispheres.[45]

The Jesuits did resist implementing some parts of the 1752 *Studienordnung.* For one, they found its division of the material absurd, with a three-month and an eight-month semester in which some professors were overburdened at particular times while others had nothing to do.[46] The non-Jesuit faculties of law and medicine resisted the reforms as strongly as the Jesuits.[47] However, the failure of one of Van Swieten's reforms is of great significance. He argued in a 5 November 1757 letter to the empress that the Jesuits' presence had been a catastrophe for the University of Vienna and for every university where they had taught. Since by neglecting their duties the Jesuits, he claimed, had not upheld their side of the Pragmatic Sanction of 1623, which had granted them the faculties of arts and theology, the agreement was now void.[48] But Van Swieten was unable to persuade the empress to abandon the core element of the Pragmatic Sanction and allow appointments in the faculty of philosophy to be open to competition. From 1765, the "Normale Viennense" was in effect throughout the monarchy, which required the Jesuit provincial to submit three names for every vacant chair to the imperial commission supervising studies for its decision. Moreover, the provincial could not remove a professor without the state's permission.[49] Nonetheless, the Society of Jesus continued to occupy all the chairs in philosophy until the Suppression.

While adherents of the Enlightenment might have railed against Jesuit education, they were also bitterly divided among themselves over how universities should be reformed. The absolutist reforms in the Hapsburg lands under Emperor Joseph essentially envisioned the universities as professional schools producing public servants (in the most literal sense). After the Suppression they abolished several institutions, put the rest under strict state supervision, and imposed a standardized curriculum. Ironically, the Josephine reforms were even further from the northern Protestant reformers' conception of the university than the Jesuit colleges had been.[50]

Protestant and Catholic Universities

The Jesuit colleges were subjected to bitter criticism from both Catholics and Protestants during the eighteenth century. While it is not our intent to rule in these polemics and judge whether the Jesuits or their critics were more justified, we do need to contextualize such criticism. We cannot assess the Jesuit universities against the measuring stick of the modern research university—valid comparisons can be made only with the Jesuits' contemporaries.

The criticisms that adherents of the Enlightenment leveled at Jesuit colleges were directed against all German universities in the eighteenth century and indeed often at the very institution of the university.[51] The first fundamental problem with German universities was their excessive number, which resulted in small enrollments and consequently small numbers of professors. This was a function of Germany's political fragmentation. Additionally, virtually all German universities were chronically underfunded, and as salaries were linked to enrollment, talented men were discouraged from academic careers. Small institutions also meant that professors had to be generalists rather than focusing on particular fields. Jesuit universities' enrollments were similar to those of their Protestant contemporaries, with a few exceptions, such as Göttingen.[52] Second, universities were often in provincial towns, away from the major cities with their courts and other cultural centers, further contributing to the intellectual stagnation of both university and city, which could not creatively stimulate each other.

Third, universities were hotbeds of misbehavior of all kinds, including excessive drinking, brawling, dueling, sexual violence, and assaults on professors, that make the behavior of modern undergraduates pale in comparison. Such activities disrupted academic activities and discouraged parents from sending their sons to universities. This was perhaps the chief criticism leveled at universities in the eighteenth century. Again, Jesuit universities were no different from their Protestant contemporaries and suffered a similar decline in enrollments over the eighteenth century.[53]

Two responses can be made to modern charges that the Jesuit universities neglected research in favor of teaching. First, this criticism is based on a modern conception of the function of the university, one that developed after the Prussian university reforms of the early nineteenth century, and is thus anachronistic. Second, all German universities shared the view of the university as a teaching institution. Even the University of Göttingen, widely regarded in historical literature as the precursor of the modern research university, saw itself primarily as a teaching institution. Göttingen did have an academy of sciences of which the university professors were members, but this dualism confirmed the pedagogical function of the university.

Furthermore, the kinds of texts published by Göttingen professors were little different from those published by Jesuit professors across Germany: textbooks, short treatises in various academic genres, and works for general readers. Specialized original contributions were relatively rare, even at Göttingen.[54] Also, the number of publications did not differ greatly between professors at Göttingen and those at the more important Jesuit colleges, such as Ingolstadt and Würzburg.

Certainly the career structure of Jesuit professors did not allow them to develop expertise in natural philosophy. When from the mid–eighteenth century individual Jesuits did occupy chairs of natural and experimental philosophy for extended periods, the benefits were immediate. Protestant professors did not move among institutions to the same extent as Jesuits did. But as the chairs in the various fields of arts or philosophy were the lowest paid, they were often briefly held and used as stepping stones on the way to more lucrative positions in the higher faculties.

One should again be wary of making gross distinctions between Catholic and Protestant institutions regarding the content of instruction. The Catholic institutions generally lagged—both objectively and in their own perceptions—behind the Protestant contemporaries in the teaching of modern approaches to law, politics, and government along with their supporting disciplines such as history, but in the natural sciences the distinction is not quite as clear-cut.[55]

Historians have repeatedly pointed to Germany's first course in experimental natural philosophy, taught at Altdorf in 1672 by Johann Christoph Sturm and published in several volumes.[56] Other Protestant universities gradually introduced courses in experimental physics. Yet universities did not provide substantial support for experimental natural philosophy. Professors had to buy their own instruments (not an option open to Jesuits, who had sworn a vow of poverty), and if institutions did acquire an experimental cabinet, it was usually as the bequest of a deceased professor. Even the University of Göttingen did not have a cabinet of instruments until it purchased Professor Georg Christoph Lichtenberg's collection in return for an annual pension of 200 thaler.

Instruction in natural and experimental philosophy varied greatly among Protestant universities. At the University of Erlangen, where Johann Christian Arnold was professor of biblical Physics for many years, experimental physics was not taught.[57] A visitation of the University of Giessen in 1720 indicated that experimental physics was being taught.[58] At Göttingen, founded in 1737, natural philosophical lectures included experimental demonstrations from the outset. The rather eccentric Lichtenberg (1742–99) taught experimental physics from 1769 as an extraordinary professor and from 1775 as an ordinary professor, but he taught in his own home, using instruments he had purchased himself.[59]

Furthermore, one should be a little skeptical of the rigorousness of instruction in experimental philosophy in this period, when it was regarded as being just as much entertainment as anything else—something Lichtenberg understood as well as anyone. One of his students recalled:

> People used to say that with him instead of studying physics, you had a course on experimentation, that he did not have students as much as spectators. It is true that he did not concern himself overmuch with formal proofs. He brought them in almost by chance. But he knew a great deal and realized that only a few were capable of understanding, or at least were patient enough to listen to mathematical explanations, for he could see how bored people became when he brought in a little mathematics during the first hours of a course.[60]

And if the Jesuits did not give their students "hands-on" training in the laboratory, Schimanck notes that this did not become practice at German universities until the later nineteenth century.

There was a range of approaches to natural philosophy at universities in eighteenth-century Germany. While followers of Christian Wolff dominated the Protestant universities by the middle of the century, they were united solely by their praise of his method.

Wolff did much to introduce Newton to Germany, but he was himself quite eclectic. Among his followers were eclectics, strict Newtonians, and eclectics who favored Newton but did not exclude other approaches. In contrast, Pietists inspired by Christian August Crusius were skeptical of Wolff's rationalism, particularly his claims about the power of human reason. The Jesuits found much of use in Wolff's vast corpus. The Principle of Sufficient Reason (undoubtedly borrowed from Wolff rather than directly from Leibniz) occurs again and again in Jesuit metaphysics.[61] They also repeatedly cited throughout the century his textbooks of experimental philosophy. On the other hand, Berthold Hauser's preface is indebted to Crusius's sentiments on the dependence of the world on the will of the Creator. In short, the situation of universities in eighteenth-century Germany cannot be reduced to one in which a monolithic Catholic bloc faced a similarly unified Protestant one.[62]

CHAPTER 11

The Transubstantiation of Physics

IN THE FIRST HALF OF THE EIGHTEENTH CENTURY, BENEATH THEIR declarations of the correct relationship between theology and philosophy and their often polemical exchanges with philosophers outside the Society, the Jesuits' natural philosophical curriculum was slowly changing. Yet while they adopted experimental philosophy, they upheld a core of peripatetic doctrines, several of which were essential for their defense of the physics of the Eucharist. However, in the last quarter-century of the Society's existence before the Suppression of 1773, the pace of change accelerated greatly. Jesuit philosophers came to openly espouse atomism, for example. Consequently their treatment of fundamental questions such as the physics of the Eucharist changed radically. While some Jesuits propounded a natural philosophy that drew on an eclectic mix of sources, others became strict followers of Newton. All, however, were strongly influenced by the foremost Jesuit scientist of the age, Roger Joseph Boscovich.

Restructuring the Curriculum

In the middle of the eighteenth century the Jesuit philosophy curriculum underwent a thorough restructuring. Certainly this occurred in considerable part in response to the demands of the German states that wanted a shorter and more practical philosophical curriculum. In implementing the states' demand for a philosophical biennium in place of the traditional triennium, the Jesuits had to reassess the curriculum and remove material that would be reserved for a third year that only students going on to theological studies would take. But these external pressures really required only a shortening of the curriculum, not a restructuring—that came from the Jesuits themselves. There were two particularly strong influences on the new curriculum. The first was the influence of Christian Wolff, which was felt throughout all of Germany, whether Catholic or Protestant. The second was the Jesuits' increasing distance from scholasticism and its attendant metaphysical questions. Contributing to this restructuring was the fact that the textbooks used at the Jesuit colleges in Germany were now being written by German Jesuits, whereas in the sixteenth and seventeenth centuries these had been written by Jesuits in other lands. This meant that the authors were much more familiar with Wolff's work than with, say, Spanish or Italian authors. They were also able to respond to the demands of the German states.

The format of these textbooks reveals two important features. The first is the dominance of natural philosophy, particularly experimental philosophy, within the overall philosophy curriculum. Rather than being one-third of the curriculum, it now formed well over half of it. Anton Mayr's *Peripatetic Philosophy Fashioned According to the Principles of the Ancients and the Experiments of the Moderns,* published in Ingolstadt in 1739, already displayed a clear preponderance of natural philosophy: while the entire first volume (448 pages) was devoted to logic, metaphysics received only a 90-page appendix to the fourth and final volume. The remaining 1,616 pages covered physics and natural philosophy.[1] This imbalance became a feature of all Jesuit textbooks. The Jesuits were more than ever an order of natural philosophers.

The second development in these later textbooks was an extremely significant shift in the order of the material. The traditional progression of logic, then physics and associated natural philosophical matters, and finally metaphysics was replaced in the middle of the century by an order derived from Christian Wolff. In Wolff's schema, metaphysics followed logic but preceded physics.[2] The restructuring did not occur overnight but took hold throughout the middle of the century. Mayr's compendium still followed the traditional scholastic order of logic, physics (both general and particular), and metaphysics, as did the textbook by the Austrian Jesuit Anton Erber (1695–1741), *Philosophical Textbook Composed by the Scholastic Method,* published in 1750.[3] However, Joseph Mangold's (1716–87) *Rational and Experimental Philosophy Adapted to Present-Day Students* (1755), published in three volumes, each about 500 pages, no longer adhered to the scholastic division of the material but adopted the Wolffian structure.

The first volume treated logic and metaphysics, while the second and third were devoted to *physica generalis* and *physica particularis* respectively.

The largest of the philosophical compendia was Berthold Hauser's massive, eight-volume *Elements of Philosophy Composed According to the Guidance of Reason and Experience and Suitable for Academic Usage,* published over the ten years following 1755. Totaling 5,885 pages, it both adopted the new structure and displayed the dominance of natural philosophy.[4] The first three volumes covered logic and metaphysics. The fourth volume treated *physica generalis* in 924 pages, while the final four volumes devoted over 3,200 pages to *physica particularis.* This division of the course, which concentrated on natural philosophy while the other parts of the philosophical curriculum, particularly metaphysics, diminished, was retained in later textbooks and was replicated in other academic textual forms such as disputations, lists of theses, and lecture notes.

Just as important as the gross divisions of the curriculum was the constitution of those divisions. The parts of metaphysics were now ontology, cosmology, psychology, and natural theology. Much of the first half of the traditional Physics curriculum, *physica generalis,* which had concerned itself with the topics covered by the first four books of Aristotle's *Physics,* such as matter, quantity, and form, was removed from physics and into metaphysics. Considering that questions formerly treated in the Physics course, such as the distinction between substance and accidents, were increasingly moved to the ontological section of metaphysics, the size imbalance between physics and metaphysics could be regarded as being even greater, as some of the content of metaphysics actually dealt with physical questions in their broadest sense. For example, Benedikt Stattler's (1728–97) *Philosophy Explained by the Method Appropriate to the Sciences* (1769–72) consisted of eight volumes: one of logic, four of metaphysics (ontology, cosmology, psychology, and natural theology), and three of physics.[5] However, since ontology contained chapters on quantity, extension, space, motion, and time (i.e., the subject matter of Aristotle's fourth book of the *Physics*) and cosmology dealt with motion, heaviness, cohesion, and impenetrability, much of metaphysics actually dealt with traditional natural philosophical questions. In sum, the Jesuit philosophical curriculum was first and foremost a course in natural philosophy.

Atoms and Corpuscles

By the third quarter of the eighteenth century, the German Jesuits were becoming increasingly open to matter theories other than hylemorphism. The Jesuits' strenuous objection to Descartes's theory of body, which reduced all physical phenomena to the action of particles of matter in motion, had been one of the defining characteristics of their natural philosophy. But this did not mean that seventeenth-century Jesuit natural philosophers were incapable of appreciating the blossoming mechanical philosophy. Nor did it prevent them from using conceptual devices such as corpuscles to describe the action of bodies. Jesuit professors and authors such as Rodrigo de Arriaga had used terms such as *corpuscles* and *particles.* Even after the Ordinatio of 1651, Jesuits continued to use these terms. In the 1650s and 1660s Kaspar Schott often referred to the motion of corpuscles when explaining how pneumatic devices worked. But for him and his colleagues, they were quite distinct from atoms. Corpuscles and particles were in no way solid, indivisible pieces of matter.

The revisers general granted this instrumental understanding of corpuscles their approval in 1674 when they were asked whether it was permissible to teach that rarefaction and condensation occurred through the intromission and emission of corpuscles. Even though earlier revisers had rejected the proposition, they responded that regardless of whether the proposition was true or false or in conformity to Aristotle, it should be permitted, since it was now taught sufficiently commonly. However, they continued, one should not take corpuscles to be atoms, which were prohibited in part by propositions 18 and 19 and in part by propositions 37 and 38 of the 1651 Ordinatio, nor could one admit that the intromission and emission of corpuscles filled or left empty spaces.[6] The revisers' world did not consist of atoms in the void.

This ability to talk about particles and corpuscles provided they were carefully distinguished from atoms granted Jesuit natural philosophers the space to exploit the explanatory power of the mechanical philosophy. The eclecticism that marked the Jesuits' approach towards experimental philosophy was apparent here also. A model in this enterprise was the Italian Jesuit Giovanni Baptista Tolomei, whose efforts to integrate scholastic and modern philosophy elicited Leibniz's admiration.[7] His influential textbook, widely read by the German Jesuits, stated that whatever was good in other systems could be fitted into the peripatetic one, which could account for Gassendi's Epicurean atoms and Descartes's subtle substance.[8] As an example, Tolomei gave an explanation of how wood burned that, he claimed, was "most peripatetic." His account featured a mixture of flying particles of wood, viscose and watery humors, and an accidental form called heat that resulted from the fire. Despite his use of particles, Tolomei's goal in physics was still peripatetic: to understand the constant process of the generation and actualization of forms.

The ambivalent nature of the German Jesuits' attitude toward mechanical philosophy in the eighteenth century is clearly illustrated in a 1745 oration by the Jesuit Philipp

Friderich (1709–48), professor of Physics and mathematics at the university of Mainz.[9] Friderich's high regard for experimental philosophy was revealed in a trope, familiar since the Renaissance and well used by the Jesuits: if Euclid, Archimedes, and Ptolemy could rise from the dead and see the microscope, the telescope, and Torricelli's, Boyle's, and Guericke's pneumatic devices, they would approve and award their inventors the "palm of erudition." But they would only smile to themselves if they could see the "exotic character of Moderns who peddle their method and toss around philosophical liberty."[10] Particularly bad were those who attempted to explain nature through mechanical analogies; here Friderich singled out Descartes.

Despite his denigration of mechanical philosophy, in the same year Friderich also wrote a reconciliation of Descartes and the peripatetic natural philosophy and in fact defended Descartes against some of his detractors, particularly the Calvinist ones. Friderich stated that "[w]e do not reject Descartes's mechanism completely but subordinate it to the peripatetic system."[11] That is, while Descartes's view of body and accidents could not be reconciled with the Eucharist, his corpuscles and molecules could be integrated into a peripatetic system that had been shorn of "metaphysical subtleties," which, Friderich wished, could be sent back to Spain whence they came. In fact, Descartes had done a great service in cutting down the forest of useless questions and "beating the scholastic dust out the cloaks of the ancient philosophers."[12] Here Friderich joined in the Enlightenment denunciation of medieval, scholastic philosophers, who had allegedly corrupted ancient philosophy—another common trope of the time that the Jesuits enthusiastically adopted.

After the middle of the century, however, this instrumental view of corpuscles was replaced by an increasingly realist one. This paralleled a development in which many questions with implications for the Eucharist, such as the constitution of natural bodies, were moved out of the Physics curriculum and into Metaphysics. With the Eucharist no longer hanging over discussions in Physics, more room opened up to hold opinions that were previously forbidden. Maximus Mangold (1722–97) abandoned the peripatetic view of body in his 1763 textbook entitled *Modern Philosophy Adapted to Public Lectures.* Although the title indicated it was designed for the academic philosophy curriculum, the terms *peripatetic, speculative,* and *scholastic* were conspicuously absent from the title. In fact, Mangold overtly adhered to atomism in his physics, stating that "for me it is correct to agree with Gassendi. And so I think that the prime matter of all natural bodies is homogeneous atoms, or minimal particles, ultimately indivisible, which as soon as they are created by God are compacted by this same Creator himself into certain very small molecules indivisible through natural forces."[13] Mangold's atoms were not mere explanatory devices that complemented peripatetic prime matter and substantial forms; for him bodies actually consisted of indivisible particles. He was without doubt an atomist.

The Physics of the Eucharist Transformed

Until midcentury, the treatment of the physics of the Eucharist retained essentially the same character that it had acquired by the early years of the century. For example, Anton Nebel's 1747 philosophy disputation at the University of Würzburg, which demonstrated the existence of absolute accidents by appeal to theological truths, differed little from the work of Georg Saur, his predecessor at Würzburg forty years earlier.[14] In his 1755 textbook, Joseph Mangold adopted the Wolffian order of the curriculum, namely logic followed by metaphysics and then physics. His metaphysics was further divided into ontology, psychology, and natural theology. In addition to treating questions that had always been part of metaphysics in the older curriculum, such as being, essence, and existence, his ontology now included the bulk of *physica generalis,* namely substance and accidents, causes, the distinctions between natural, supernatural, and artificial, and the concepts of time and space. In effect, these once physical questions were reclassified as metaphysical. As a consequence of this shift, any treatment of the Eucharist also moved to metaphysics.

Despite the restructuring of the material, Mangold's treatment of the question of the Eucharist still differed little from earlier Jesuit work on the subject. Mangold claimed that the moderns agreed with the peripatetics that substance was "being existing in itself" (*ens per se existens*), with the exception of Wolff, whose definition of substance contradicted all received opinions. Descartes's assertion that the essence of body was three-dimensional extension was, for Mangold, the source of many serious errors. From the received definition of substance, the essence of body could not consist in that which could be separated from body, but three-dimensional, or "actual," extension could be separated from body, as was clear from the mystery of the Eucharist. The Council of Trent had ruled that the whole Christ—that is, his essence—was present in each part of the sacrament, no matter how small. Thus the Corpus Christi could have no extension. The unacceptable consequence of Descartes's view was that Christ was not wholly in the Eucharist.[15]

But Mangold was also conscious that in a philosophical work such an appeal now had to be justified. He noted that some philosophers, such as François Bayle, had argued that since the mysteries of faith were beyond common laws of nature—that is, the realm of philosophers—they should be left to theologians. Yet Mangold regarded this argument as feeble, for whoever subscribed to the orthodox faith also had to accept the principles it was based upon—and the faith stated that Christ was wholly within each particle, something that Descartes's physics could not explain.[16] Mangold admitted that even Suárez had been forced to confess that it was difficult to defend the view that quantity was something distinct from body without making an appeal to the Eucharist. Nevertheless, that appeal did demonstrate the necessity of a distinction between quantity and body.

After his discussion of substance, Mangold turned to accidents but made an important distinction: in physics he would treat the arguments for absolute accidents that were taken from nature; in ontology he would only treat those arguments derived from the Eucharist. Thus he separated arguments derived from nature, which were in the realm of physics, and those derived from supernatural phenomena, which, though just as real and compelling, belonged in metaphysics. As with many earlier treatments of absolute accidents and the Eucharist, Mangold conducted a historical excursus into what the Council of Trent had actually meant by the term *species*. Like his predecessors, he concluded that the council had understood *species* to be nothing other than "real or absolute accidents existing without a subject."[17] Joseph Mangold still fit wholly within a tradition that stretched back over a century to Thomas Compton Carleton.

Yet in the following decade this tradition would come to an end in works such as Maximus Mangold's openly atomist compendium. In keeping with the dominant style of German philosophy at the time, both Catholic and Protestant, Maximus Mangold's work was thoroughly eclectic, as his treatment of substance and accidents shows. As in his brother's work, they were no longer covered in Physics but in the section on ontology in Metaphysics. But while Joseph had claimed that all philosophers basically agreed on the definition of substance, Maximus thought the discrepancies among philosophers on this question were nothing short of amazing. Rather than attempting to determine the correct definition of substance, he invited each reader "to choose the one that he judges to be clearest and more consistent with those principles that he follows in physics."[18] Eclecticism ruled; the matter was simply one of personal taste. The brevity of his treatment—only two pages—shows how unimportant he considered the whole matter to be. Furthermore, he devoted only a single paragraph to accidents, in which he made no mention at all of the Eucharist or Trent.

When Mangold considered the essence of body, he noted that it exhibited two predicates above all others: impenetrability and extension. In response to Descartes's claim that the whole substance of body consisted of three-dimensional extension, Mangold argued that the essence of body regarded physically could not be *extensio actualis* or three-dimensionality. Extension was preceded by impenetrability and it was from this that actual extension came because impenetrable bodies could not occupy the same place. Therefore, impenetrability had to be the primary predicate of body. Mangold then turned to the Eucharist to argue once again against Descartes's view of body and extension. In a single sentence Mangold stated that the Eucharist showed that extension was separable from body; in it the same body remained but its extension did not. Mangold concluded by stating that "body considered physically is correctly defined as substance naturally impenetrable to other substance similar to itself." Here the Eucharist was once again deployed to argue for a particular philosophical viewpoint. But Mangold referred to it in a single sentence, a stark contrast to earlier voluminous treatments. Furthermore, even though it was once again used to refute Descartes's equation of body and extension, Mangold did not use it to argue for any peripatetic concept such as absolute accidents but rather to support impenetrability as the prime principle of matter, a project quite foreign to earlier generations of Jesuits, for whom the principles of body were prime matter and substantial form. Now the discussion was not between scholastics and mechanical philosophers but between varieties of mechanical philosophy, on whether extension or impenetrability was the fundamental attribute of body.

The Jesuits' move away from scholastic philosophy had to affect their theology. As Edward Schillebeeckx wrote in his account of Trent and the Eucharist, one can never grasp the faith in a pure state, set aside from any other way of thinking. The extent to which one generation's mode of thought is historically conditioned can become apparent only to a later generation.[19] The Jesuits' thinking, in particular their philosophy, had changed significantly in the two centuries that separated Trent from the decades preceding the Suppression. As they moved away from scholastic philosophy, they became aware of the contingency of Trent's language and no longer felt compelled to interpret Trent's term *species* as they had two centuries earlier. Thus, while Maximus Mangold's philosophy differed greatly from earlier Jesuits', it was not in conflict with contemporary Jesuit eucharistic theology. Whatever else changed in the Jesuits' natural philosophy, they still claimed to teach nothing that conflicted with the truth of the faith. Rather, what had changed was Jesuit eucharistic theology itself.

As scholastic categories such as matter and form, substance and accidents disappeared from Jesuit philosophy, it was no longer meaningful or even possible for them to discuss the old questions of whether quantity was an accident or whether the substance of body consisted of extension. As their conceptual categories changed, eighteenth-century Jesuits became increasingly aware of the historical contingency of the Council of Trent's statements; the fathers at Trent had spoken in the philosophical idiom available to them. That idiom was increasingly meaningless, or at least foreign, even to Jesuit philosophers. Sixteenth- and seventeenth-century Jesuits, raised on a diet of late scholasticism, could hold only that by *species* Trent meant accidents. After the middle of the eighteenth century, when Jesuits no longer read even Suárez, let alone medieval scholastics, they could see Trent's *species* only as a generic term for the external appearances, which they could account for with the philosophical tools of their own day. And if Maximus Mangold could encourage his readers to adopt whichever philosophical definition of substance they liked best, then they could also adopt whichever account of the change of substance—that is, transubstantiation—they liked best.

This is precisely what the Jesuit theologians at the University of Würzburg did when they compiled the theological textbook commonly known as the *Theologia Wirceburgensis.*[20] The section on the sacraments was written by Thomas Holtzclau. When considering the question of how Christ existed in the sacrament, he accepted that the eucharistic accidents existed without subject in the sacrament but simply rejected the view that there were absolute accidents really distinct from matter. Now the Jesuits were employing an argument that opponents of absolute accidents had made: *accidents* was a relatively recent usage developed first in the eleventh century—that is, a historically situated term.

Instead, Holtzclau favored the theory of intentional species, in which all the various appearances of the bread and wine were placed directly in the senses by God, as proposed by the French Minim Emmanuel Maignan in the mid–seventeenth century.[21] Generations of earlier Jesuits, including those at Holtzclau's own university, had criticized this theory. This occasionalist theory argued that the term *species* or *accident* used by the councils referred to the appearances of the bread and wine, which were impressed on our senses by the action of tiny bodies. What caused these species and their consequent effects on our senses was the direct action of God. Christ himself with divine power reflected the light to make us see white, affected the nerves in our tongue to make us taste wine, and directly stimulated our nasal membranes to make us smell the bread. Of course, Holtzclau insisted that this view was quite in accordance with Thomas and the Council of Trent. Regardless of whether he was correct, this certainly was not the view his predecessors had held to be *de fide.*

Carlo Sardagna, professor of theology at the Jesuit college in Regensburg, did not address the question of absolute accidents in his treatment of the Eucharist. His main concern was to uphold what we can call a minimalist interpretation of Trent: the real presence occurred through a transubstantiation that replaced the substance of the bread and wine while leaving only their species. Although he used the terms *species* and *accidents* interchangeably, he ignored the question of what the species were philosophically. While the real presence still required transubstantiation, transubstantiation no longer required scholastic absolute accidents.[22]

On the eve of the Suppression the Jesuits were unable to understand, let alone defend, what their predecessors regarded as untouchable orthodoxy. Georg Vaeth's (1731–73) philosophical dissertation entitled *Concord of the Ontological Truth about Accidents with the Theological Truth about Eucharistic Species,* written at Würzburg in 1772, went beyond the *Theologia Wirceburgensis* and Sardagna to explicitly reject the absolute accidents that the Society of Jesus had steadfastly defended against all threats for generations. Rehashing an old argument with new terminology, Vaeth claimed that substance was "in itself the stable subject of contingent determinations."[23] Accidents could have no existence independent of the subject of which they were determinations; nothing would be a greater contradiction if they could. There would be "thought without the thinker, pain without the sufferer . . . , motion without the mobile, extension without the extended, quantity without large or small."[24] Accidents could have neither their own essence nor their own existence. The idea of an absolute accident was therefore absurd. Vaeth's distance from the earlier tradition of Jesuit philosophy is clear. In fact, his source for the "peripatetic" opinion he rejected was Berthold Hauser, whose work had only appeared a decade earlier and who was himself far from being a true scholastic.

Like his predecessors, Vaeth explored the history of the dogma of transubstantiation, but he denied that history showed Church rulings such as the condemnation of Wyclif at the Council of Constance affirmed the existence of absolute accidents, as "a theologian raised within the family of peripatetics" would state.[25] He insisted, much like Descartes, that the councils generally did not intervene in philosophical matters except in those rare cases that directly contradicted the truths of the Catholic faith. This was not the case, Vaeth claimed, with absolute accidents—whether absolute accidents existed posed no danger to the Catholic faith, so it was neither necessary nor fitting for the Council of Constance to decide on this question.[26] Trent had not been trying to enshrine any particular school of philosophy but had merely wanted to affirm that Christ was truly present in the Eucharist and in each and every part of the Eucharist and that the whole of the bread and wine was completely converted.[27] This was a complete break with Vaeth's Jesuit predecessors, who had stated that the existence of absolute accidents was *de fide.*

But Vaeth did not break completely with the traditions of Jesuit philosophy. He still held that philosophy should be able to account for theological truths. Vaeth believed that neither Descartes's theory nor the exposition of the atomists provided an adequate account of the Eucharist. It is not surprising that as a student at Würzburg he preferred Maignan's account, namely that "the eucharistic species, insofar as they are conferred on the consecrated host, are nothing other than the divine actions of Our Lord Christ, substituted for the absent actions of the bread and wine."[28]

Although it did not endorse Descartes's account of the eucharistic species, the Frenchman would have been pleased with Vaeth's work. Generations of Jesuit theologians, philosophers, and censors had bitterly opposed Descartes. But when Vaeth denounced absolute accidents as philosophically absurd and concluded that they were incapable of accounting for transubstantiation, the Jesuits finally granted Descartes's wish that the doctrine of real accidents be condemned as foreign to rational thought.

Newton and Boscovich

The history of eighteenth-century science is often narrated as the story of the dissemination of Newtonian mechanics and, more broadly, of the effort to use it as the model for all branches of natural inquiry. Although such a view oversimplifies the history of Jesuit science in Germany, broadly speaking Newton's trajectory among the German Jesuits in Germany, from a state of essential anonymity to eclectic acceptance and finally to an overwhelming victory, is similar to his path on the Continent in general.

Before the start of the eighteenth century there were virtually no references to Newton in Jesuit Physics texts. In the early eighteenth century very few Jesuit natural philosophers had the mathematical skill to read the *Principia* even if they had had the inclination to do so, so his influence in mechanics remained minor. The Jesuits first drew on Newton the experimentalist as their interest in experimental philosophy grew. As in the previous century, the Society's mathematicians usually acted as the conduit introducing new methods and materials to their colleagues who taught Physics. Also, popularizers such as 'sGravesande and above all Christian Wolff made much of Newton's work available to the Jesuits in a digestible form.[29] But until the mid–eighteenth century, Newton remained one source among many in the Jesuits' eclectic arsenal—a situation very different from Britain's Newtonian orthodoxy.

The Jesuits still regarded themselves as practicing *philosophy*—that is, a science of causes. Thus they held reservations about Newton's natural philosophy. Newton's unwillingness to hypothesize and commit explicitly to a particular account of matter meant that the Jesuits did not find his version of the mechanical philosophy to be as troubling as Descartes's or other atomist accounts. Yet ironically, Newton's reluctance to provide an ontological account of gravity presented the Jesuits with the greatest stumbling block. The Newtonian method presented the Jesuits with the same problems that it did French Cartesians: it did not offer causal accounts that they could regard as truly philosophical.

Jesuit textbook treatment of the tides illustrates this clearly. The Jesuits did not rely on Aristotle for the cause of the tides. In this matter he provided little help anyway; according to legend he had become so frustrated at his ignorance of the cause of the tides that he had thrown himself into the ocean in despair. Whatever Newton's followers might have felt, the Jesuits certainly did not think that he had convincingly solved the riddle of the tides. Quoting the Jesuit Causin, Berthold Hauser called this question the "sepulcher of human curiosity." A century earlier, Giovanni Battista Riccioli could already list seventeen different opinions on the cause of the tides.[30]

For the Jesuits, an adequate account had to provide an ontologically acceptable cause of the tides. Some still held theories that identified a cause unrelated to the moon. As late as 1761, the Jesuit Paul Offermann taught at Mainz that Galileo's, Descartes's, and Newton's theories of the tides were all wrong and that the most probable cause was that the sun excited marine effluvia, which created fermentation and swelling in the ocean. His evidence, however, seemed to rest on analogies with periodic fermentation in the human body.[31]

The majority of scholars, whether Jesuit or not, felt that there was an undeniable correlation between the moon and tides. But the mere identification of a correlation was not sufficiently philosophical. Hauser wrote that the core of the controversy was to explain the "virtue, action, or mode" by which the moon acted on the ocean to cause the tides. As with all topics, Jesuits such as Hauser outlined the main theories on the subject. They were quite familiar with Newton's theory that rested upon the gravitational attraction between the moon and the waters of the ocean. Certainly Newton's theory presupposed a moving earth, which even at this late date the German Jesuits did not accept.[32] Yet Hauser and the German Jesuits did not reject Newton's gravitational explanation of the tides merely because it implied a moving earth. Having become mechanical philosophers, they wanted a good mechanical account rather than one that rested upon Newton's occult force of gravity. Citing the Swiss mathematician Leonhard Euler, Hauser argued that Newton had simply despaired of finding a physical cause and assigned an occult one of the kind true philosophers rejected.[33]

Hauser did not adopt Descartes's vortices or Euler's attempt to combine them with Newtonian mechanics. Rather, he favored another mechanical account that was supported by the Viennese Jesuits Franz, Khell, and Zanchi and that he traced back to the French Jesuit Honoré Fabri. Tides were due to gravity, but gravity was itself a result of ethereal matter pushing down on the earth. As the moon revolved around the earth, it pushed down on the ether between itself and the earth, creating unequal pressure on the oceans and thereby upsetting their equilibrium (see figure 11). The tides were the result of the oceans' restoring their equilibrium. In addition to being quite compatible with the idea of a stationary earth, the theory had numerous advantages, which Hauser enumerated: it was simple, relied on natural operations, saved the phenomena, and was based on the laws of hydrostatics, the action of the ether, and the analogous behavior revealed by the barometer, the air pump, and other hydraulic machines. Therefore, Hauser concluded, it could be held as true.[34] Other Jesuits such as Maximus Mangold derived their treatment in large part from Hauser, although Mangold concluded in his usual fashion by throwing his hands up and suggesting that the reader choose whichever theory he liked best.[35]

Nevertheless, the German Jesuits did adopt Newtonian mechanics. Once again, this was a protracted process in which Newtonian mechanics blended with scholastic concepts. In 1740 Edmund Voit still made a distinction between projectile motion, caused by a force impressed extrinsically, and natural motion, caused by an intrinsic impetus.[36] Such appeals to natural motion (and, less commonly, natural place) remained common until the middle of the century.

Nonetheless, Newton's laws and basic equations of motion became a standard element of Physics texts. But many Jesuits remained critical of Newton's theory of universal gravity. Maximus Mangold, for example, noted that Newton's laws of motion were adequate for describing phenomena but that they did not adequately account for the underlying principles. Moreover, in the mid–eighteenth century, ironically, the Jesuits subscribed to Newtonian mechanics but denied the heliocentrism that it so elegantly explained.

Such incongruity could not last. Gradually the distinction between Newton's mechanics and his theory of gravity (and the heliocentrism it demanded) was erased. A major impetus behind the German Jesuits' eventual acceptance of both Newtonian mechanics and universal gravitation was provided by Roger Joseph Boscovich, whose natural philosophy of matter was extremely influential among the German Jesuits. Boscovich acknowledged that his theory drew on Leibniz's monads and Newton's forces. Certainly Boscovich did not imbue his points with the consciousness that Leibniz seemed to, and at an ontological level, the attraction between his points was different from Newtonian force.[37] Nevertheless, he held that Newtonian universal gravity and mechanics accurately explained the motion of bodies except at extremely small distances. His theory represented all bodies as networks of nonextended points that attracted or repelled each other according to their distance from each other. At extremely short distances, points repelled each other with a force that increased asymptotically as distance decreased. This accounted for the most fundamental characteristic

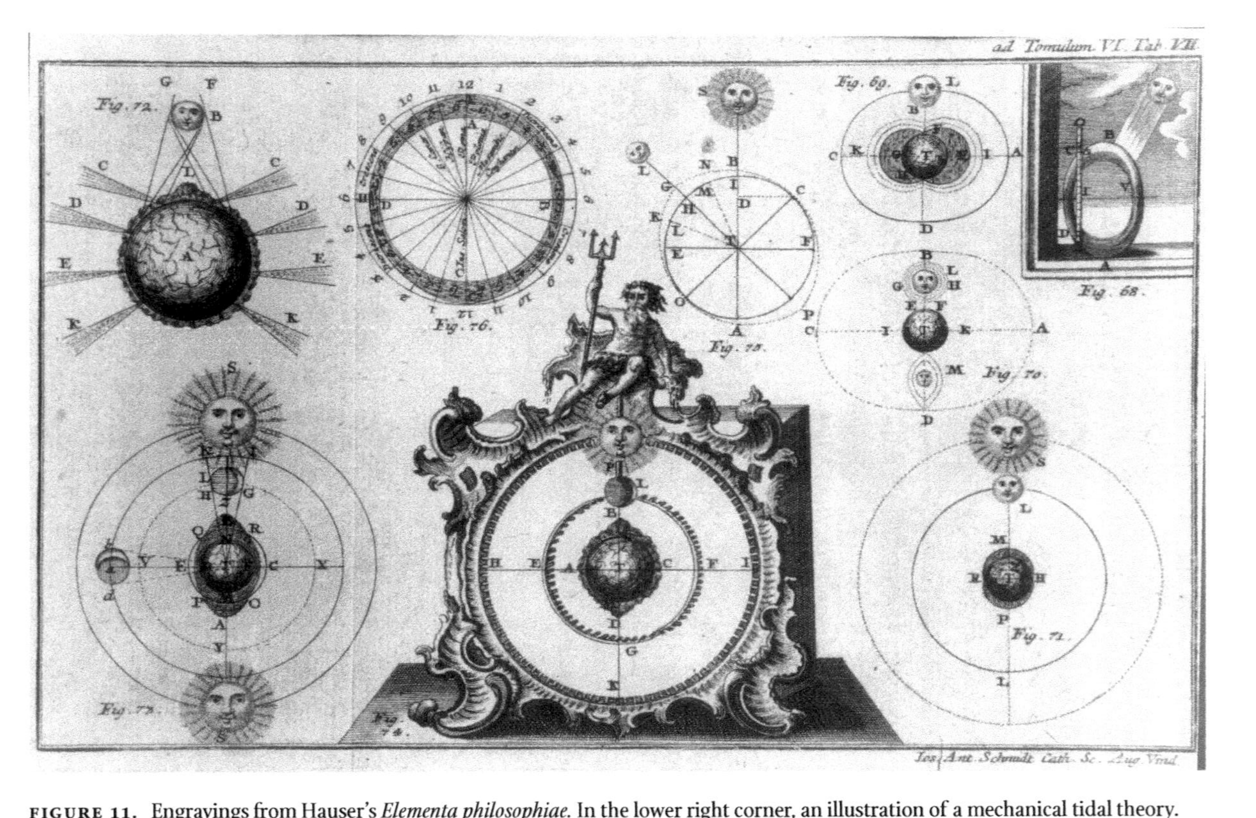

FIGURE 11. Engravings from Hauser's *Elementa philosophiae.* In the lower right corner, an illustration of a mechanical tidal theory. Note, however, that it still employs a geocentric model. Courtesy of Stadtbibliothek Mainz. Signatur: III l 4° 355h.

of matter—the impenetrability of bodies—and also the collisions that resulted from it. At greater distances, bodies attracted each other along a curve closely approaching Newton's inverse square law.[38]

Boscovich was still careful to avoid using the term *atoms,* consistently preferring *points* (*puncta*). This may have been to avoid the theological issues attached to atoms, but it was mainly to distinguish his theory from the conventional schools of atomism, which considered atoms to be solid, impenetrable, physical particles. Boscovich's points lacked extension and mass, the two attributes normally attributed to atoms. While his theory had little in common with conventional atomism, it shared even less with peripatetic physics. It made no mention of forms or accidents. In such a theory older concepts of matter, whether peripatetic or atomist, and the debates that coalesced around them, became less and less meaningful and relevant. Boscovich also rejected other elements of peripatetic physics; like Newton, he made the void an essential part of his theory.

The first edition of Boscovich's magnum opus, *Theoria philosophiae naturalis,* was published in Vienna in 1758 under the supervision of the Jesuit university professors there. A second, improved edition was produced under Boscovich's direct supervision in Venice in 1763. While it may have encountered opposition from some Jesuits, such as those in Paris who still adhered to peripatetic physics, the theory won widespread acceptance at Jesuit institutions in Germany in the final decade of the Society's existence.[39] Benedikt Stattler praised Boscovich as "the last of the heroes of philosophy."[40] Many of the Jesuit texts published in the German assistancy that were devoted to explicating his system were produced in Hapsburg universities, but Nicolas Burkhaeuser (1733–?) published a discussion of Boscovich's system at Würzburg in 1770.[41] Burkhaeuser wrote that his predecessor Christian Appel (professor of Physics in 1769) had been the first to teach the Boscovichian system at Würzburg.[42] Burkhaeuser's stated assumption that his readers possessed a knowledge of modern—that is, Cartesian and Newtonian—physics confirms that Newtonian mechanics was now a basic part of the curriculum.[43] In addition to those works dedicated primarily to explicating Boscovich's systems, there were many more dissertations that included a number of theses derived from his system. By the last days of the Society, theses on Boscovich's law of forces were a standard part of all Jesuit academic texts in Germany.

But Boscovich's natural philosophy also brought with it Newtonian mechanics and the mathematics it rested upon. The impact of Newtonian mechanics when rigorously pursued is revealed in the work of Jacob von Zallinger, an outspoken Newtonian in the last days of the Society. One of three brothers who entered the Society and distinguished themselves as scholars, Zallinger taught at several colleges in the

Upper German province and published several works on Newtonian physics.[44] His treatises *Law of Universal and Mutual Gravity* (1769) and *On the Physical Exposition of Mathematical Demonstrations* (1772) were thoroughly Newtonian, although unsurprisingly strongly influenced by Boscovich. The first volume of his textbook of natural philosophy, entitled *The Interpretation of Nature, or Philosophy Explained by the Newtonian Method,* appeared on the eve of the Suppression and the subsequent two volumes in the following years.[45] Zallinger also published an anonymous defense of Boscovich's system in German.

The term *Newtonian* in the title of Zallinger's textbook may appear odd. As was quite standard by now, the first volume covered logic and metaphysics, which included psychology and natural theology. It is hard to describe this volume as Newtonian in any meaningful sense. The second volume, however, subtitled "The Principles of Terrestrial and Celestial Mechanics," was Newtonian throughout. It began with Newton's rules for philosophizing, taken from the beginning of the third book of the *Principia.* Then, after a treatment of rectilinear and curved motion, it culminated in discussions of universal gravity and the "true," heliocentric system of the world. Thus universal gravity and heliocentrism were presented as the necessary corollaries of Newtonian mechanics. The third volume still contained much material that lay within the traditional scope of natural philosophy but well outside the much more limited scope of modern physics, such as a survey of the mineral, vegetable, and animal kingdoms. As with the first volume, it is hard to identify anything particularly Newtonian about it. So even though the textbook's title promised a Newtonian philosophy, only the middle volume could truly be called Newtonian. Yet for Zallinger the entire textbook was truly Newtonian, since he regarded mechanics as the most fundamental part of natural philosophy, and his mechanics was thoroughly Newtonian. The Newtonian study of motion now occupied the place of *physica generalis,* the starting point of all natural philosophy.

Zallinger explicitly dismissed rationalist schools of philosophy, such as Wolff's, that sought to construct all sciences along a geometrical model but merely described things as the mind imagined them, not as God had created them, and consequently had produced nothing certain or demonstrated. And in response to the hoary old charge that Newtonian natural philosophy did not identify real causes, he answered that uncertain disputes about the fundamental nature of forces or the first elements had never brought anything useful to philosophy.[46]

Zallinger's work can be seen as the culmination of the transformation of Jesuit natural philosophy in several further ways. First, his approach was unapologetically mathematical, and he dismissed those who said that this subjected physics to mathematics. Mechanics and universal gravity could be studied only with the tools of mathematics, but anyone who applied himself to their study could master them, as had all of his students over many years. Second, the two meanings of the term *mechanics* had now diverged completely. Zallinger used *mechanics* in the same way the modern physicist does—to describe the study of motion. The meaning is very different from that in, say, Kaspar Schott's *Mechanica hydraulico-pneumatica* (1657), which dealt with machines. The older usage rested on a distinction between natural and artificial motion, and between natural bodies and machines, a distinction that was beginning to blur in Schott's time and was further obscured with the entry of simple machines into the Physics course in the early eighteenth century. With Zallinger's wholehearted Newtonianism, this distinction was truly effaced, and one set of laws governed all motion.

Copernicus Vindicatus

Jacob von Zallinger was not alone in openly advocating heliocentrism. The brief period between the middle of the century and the suppression of the Society saw one more development that furthered the transformation of Jesuit physics when its professors were finally allowed to teach heliocentrism as a physically true account of celestial motions. In the eighteenth century, observational astronomy flourished at the Jesuit colleges in Germany. German Jesuit astronomers established excellent observatories in Ingolstadt, Würzburg, Mannheim, and elsewhere, participated in international astronomical expeditions, and were elected to foreign scientific societies.

In the classroom, however, Jesuit professors continue to reject the motion of the earth. Just as in the seventeenth century, they could teach heliocentrism in their schools by assigning it the status of a hypothesis—that is, a useful mathematical model that accounted for the phenomena in an entirely satisfactory way but could not be physically true, in this case because it conflicted with the literal reading of Scripture. Only the Tychonic world system could be treated as a thesis. This approach continued throughout the first half of the eighteenth century: the Copernican model remained a hypothesis, although Jesuit professors treated it in increasing detail. In fact, Jesuit Physics texts paid considerably more attention to the Copernican model than to the Tychonic and certainly did not hesitate to identify the latter's weaknesses. In the seventeenth century Jesuit professors provided physical objections to a moving earth, but in the eighteenth, as the Jesuits increasingly adopted Newtonian mechanics, their only remaining objections were based on the question of stellar parallax and Scripture.

This did not necessarily mean that Jesuit mathematicians and philosophers were crypto-Copernicans who merely paid lip service to the rules. Eusebius Amort's 1737 obituary for the Jesuit astronomer Nicasius Grammatici of the University of Ingolstadt stated that the lack of any observable stellar parallax was the decisive reason for why Grammatici regarded the Copernican system as only a hypothesis. Grammatici had designed a planetolabium (a mechanical model of the planets) according to the Copernican system, due to its greater ease of operation,

> but Father Nicasius did not intend thereby to realize this hypothesis, since there was still no irrefutable demonstration available that would compel us to depart from the obvious meaning of Scripture. For he confirmed on the contrary the immobility of the earth with his observation conducted for several years at my suggestion of the nebula in Ursa Major, in Orion, and in Collo Cygni, where the little stars of the third, fourth, and fifth magnitude that are so close to each other do not change their distance from each other and consequently present no parallax.[47]

Joseph Mangold's treatment of the competing world systems in his philosophy textbook written in the Upper German province in 1756 illustrates the situation just before Copernicus was taken off the Index. Mangold used a number of arguments that had been in currency ever since philosophers sought to refute the Copernican hypothesis on scriptural grounds in the sixteenth century. While the Copernican system could be treated as a hypothesis, Mangold advised that Scripture should be read literally when it did not oppose manifest reason, other pieces of Scripture, or the Church's tradition of interpretation. As there was not yet enough "sufficient reason" in favor of the earth's motion, passages of Scripture stating the earth's rest and the sun's motion should not be read in any way other than literally.[48] The argument was that of Cardinal Bellarmine in his letter to Paolo Antonio Foscarini, which allowed the possibility of a reinterpretation of Scripture only if the physical arguments in favor of the earth's motion became demonstrative.[49] The Leibnizian-Wolffian terminology of sufficient reason was a change from the preceding century, but the essence of the argument was the same.

Mangold was not simply ignoring all the developments in astronomy and physics that had occurred over the previous two centuries. Like Grammatici before him, he recognized that an observed stellar parallax would be a powerful argument for Copernicanism, and while he listed numerous astronomers who claimed to have observed stellar parallax, he questioned the accuracy of their observations. He also regarded Bradley's observation of the aberration of light as doubtful, since refraction varied in different weather conditions and instruments changed in heat and cold.[50]

Roger Joseph Boscovich's earliest works also advocated the Tychonic system. But as he came to recognize that Newtonian mechanics demanded heliocentrism, he sincerely tried to harmonize Copernican cosmology with the dictates of the Church. In a move reminiscent of Descartes, he suggested that by making a distinction between absolute and relative space one could assign motion to the earth and apply Newtonian physics to astronomy. He suggested that the earth moved around the sun in relative space but stood still in absolute space, while the sun moved with a motion opposite to that attributed to the earth in relative space.[51]

However, from the early 1740s onwards Boscovich pleaded that regardless of the truth, the interdiction on heliocentrism should be lifted, as it did more harm than good to the Church. In 1757, at Boscovich's urging, his friend Pope Benedict XIV removed Copernicus's *De revolutionibus* from the *Index of Forbidden Books.* It was not until 1822, however, that the Holy Congregation of the Inquisition permitted books teaching the motion of the earth to be printed. The German Jesuits, however, far from the power of the Roman Inquisition, did not wait until 1822 to change their teaching.

Dadic has written that the Jesuit colleges in the Hapsburg lands, which had been subjected to the Theresian reforms, immediately began teaching the motion of the earth as physically true after 1757. Interestingly, this was not the case throughout the German assistancy; 1757 did not bring about an immediate transformation in the Upper German province, for example. As an open advocate of Gassendist atomism, Maximus Mangold cannot be considered excessively conservative, yet his 1764 compendium differed little from texts published before 1757 in its treatment of the rival systems of the world. Because the Copernican system recommended itself through its simplicity and best fit the "astronomical phenomena," it could be defended as a hypothesis. But if Scripture was to be taken in its literal sense, then it could not be defended as a thesis—a status reserved for the Tychonic system alone.[52] Mangold did state, however, that the Copernicans responded that Scripture was not meant to teach physics. He left this argument unrefuted—perhaps a tacit acceptance of its validity.[53]

Whatever his beliefs on the purpose of Scripture, Mangold used the "Newtonian hypothesis" consisting of mutually balancing centrifugal and centripetal forces to explain the motion of the planets. Like many of his contemporaries in the Society, Mangold adopted Newtonian celestial mechanics without explicitly stating that the earth revolved around the sun. Other Jesuits continued to deny that the evidence commonly interpreted as supporting a moving earth necessarily implied this. In 1763 at the Jesuit college in Munich, Benedikt Stattler wrote that the sun did not differ in nature from the fixed stars, but he did not state that the sun was at rest in the center or that the planets revolved around the sun. As for physical evidence of the earth's motion, he insisted that the fact that the earth was an oblate spheroid depressed at the poles (which Newton had predicted and which had been confirmed by this time by measurements taken by scientific expeditions) did "not arise from its rotation around its axis."[54]

Nevertheless, Jesuits in the Upper German province did come to openly espouse Copernicanism. Five years after Stattler, Franz Sales Stadler openly argued at Munich that the earth rotated diurnally on its axis and revolved annually around the sun.[55] After the mid-1760s, dissertations from the University of Ingolstadt and the Munich college were unanimous in describing the earth moving on an elliptical orbit within a framework of Newtonian mechanics.[56]

Elsewhere Jesuits continued the long tradition of stating that Copernicus could be treated only as a hypothesis if one read Scripture literally. Nevertheless, such provisos did not keep them from adopting Newton celestial mechanics to describe the motion of the planets in a heliocentric system. At Würzburg in 1769 a disputation under Professor Christian Appel referred to the motion of the earth around the sun as a hypothesis but presented it as the only system of the world. It made no mention of any other possibilities.[57]

In sum, on the eve of the abolition of the Society, the transformation of the natural philosophy taught at its colleges in Germany had been completed. Jesuit Physics texts no longer contained any discussion of peripatetic natural philosophy. Even that small core of peripatetic theses that the hierarchy had insisted be maintained throughout the first half of the eighteenth century was gone. German Jesuits had turned to other theories of the fundamental composition and attributes of body, some adopting atoms but most favoring Boscovich's incorporeal particles. Considerations of topics such as the physics of the Eucharist no longer formed any part of the Physics curriculum. Once such questions had been transferred to Metaphysics, the theology of the Eucharist was divorced from any particular school of natural philosophy. After years of favoring the Tychonic over the Copernican system of the world, the Jesuit curriculum now taught a Copernican system based on Newtonian celestial mechanics. It is hard to say at what point Jesuit physics ceased to be peripatetic, but during the two decades preceding the end of the Society it lost its last trappings of Aristotelianism.

EPILOGUE

After the Suppression

THE SUPPRESSION OF THE SOCIETY OF JESUS WAS THE CULMINATION OF a long campaign by the Bourbon monarchies. It achieved its first major successes with the expulsion of the Jesuits from Brazil in 1754 and then Portugal itself in 1759. This was followed by the expulsion of the Jesuits from the French (1764) and Spanish (1767) empires. Finally, under grave pressure, Pope Clement XIV signed the bull *Dominus ac Redemptor* on 21 July 1773, suppressing the Society of Jesus throughout the world, an absolutely unprecedented act in the history of the Church.[1]

Many of the same anti-Jesuit elements that figured in the Bourbon lands were present in Germany, in particular the growth of national churches, which saw the Jesuits as ultramontane and foreign elements. Such concerns did not have to lead to expulsion, however. Bavaria, for example, insisted that the Jesuit province be coterminous with Bavaria, resulting in the establishment in 1770 of a Jesuit Bavarian province, separate from the rump of the Upper German province.

Certainly some significant personalities were very hostile toward the Jesuits. Maria Theresa's chancellor Prince Wenzel Anton von Kaunitz had considered pressing the pope to abolish the Society entirely as early as 1768. Aware of Maria Theresa's affection for the Jesuits, however, he did not want them to be brutally treated and exiled as they had been in the Bourbon monarchies.[2] But in general there was little active campaigning for a suppression of the Society of Jesus, and the bull took many, including the Jesuits themselves, by surprise. In its execution the Jesuits were generally not treated in any of the German lands as badly as in the Bourbon lands.[3]

Regardless of princes' and ministers' feelings toward the Jesuits, all were forced to acknowledge their importance in education. Even Kaunitz acknowledged that it was unavoidably necessary to keep on many Jesuits as professors, at least until Austrian higher education could be fundamentally refashioned.[4] Perhaps one of the greatest ironies of the Suppression is that Germany's most powerful Protestant prince, Frederick the Great of Prussia, delayed promulgating the bull the longest. With his conquest and annexation of Silesia from the Hapsburgs at the start of his reign, he had acquired a large Catholic population. Although initially no great friend of the Jesuits, he came to acknowledge their services to education.[5]

Frederick and Kaunitz were not alone in acknowledging the importance of the Jesuits in their territories' higher educational systems. At virtually all of the colleges and universities in Germany, the Jesuits were kept on in their former roles, but as secular priests. At the University of Cologne, all the former Jesuit professors were allowed to continue at their positions after 1773.[6] At Paderborn the bishop asked the Jesuit professors to stay on, maintaining their communal life and Roman dress: twenty of twenty-two agreed to.[7]

At Würzburg the chair for Aristotelian physics was transformed into one for theoretical physics and was occupied by the ex-Jesuit Nicolaus Burkhäuser, who taught until 1803. Ambros Egell had begun teaching experimental physics in 1771 and continued to 1797. Franz Huberti, professor of mathematics and astronomy from 1754, remained in his chair. Another ex-Jesuit, Franz Trentel, also taught these subjects as an adjunct professor from 1773 and then as a regular professor from 1775.[8]

In the universities in the Hapsburg territories, the Jesuits were largely evicted from their chairs—with the exception of mathematics and physics, where they could not be easily replaced. At Innsbruck ex-Jesuits continued to occupy the chairs of mathematics and physics for over a decade. Ignaz von Weinhart was followed in 1780 by Franz Seraph von Zallinger. A second chair of physics, established in 1770, was occupied by Franz Sales Stadler until 1776. He was succeeded by Jakob Anton von Zallinger, who stayed for one year and was replaced by his brother Franz. When he took Weinhart's chair in 1780, his was occupied by yet another ex-Jesuit, Joseph Stadler.[9]

Immediately after the Suppression, the Bavarian Academy of Sciences, which had steadfastly excluded Jesuits, whatever their qualifications, promptly elected two ex-Jesuits to membership. In the decade after the Suppression, nine ex-Jesuits who worked in either mathematics or the natural sciences were elected to the academy.[10]

Certainly the persistence of the Jesuits in chairs of mathematics and physics at the German universities can be seen as a confirmation of their expertise and the quality of their instruction. But ironically it also confirmed one of most persistent criticisms leveled against them since the sixteenth century. In an era when there were virtually no paid positions for mathematicians and natural philosophers outside the universities, the Jesuit monopoly over university chairs discouraged all those who did not wish to become priests, let alone members of a religious order, from embarking on a career in these fields. Consequently, when the Society was suppressed there was no pool of qualified candidates from which to choose their replacements. While the total suppression of the Society of Jesus was perhaps not necessary to end this monopoly and open up career opportunities in science to all (males at least), it did accelerate what was by the later eighteenth century a necessary and inevitable step.

CONCLUSION

THIS BOOK CLEARLY HAS NOT EXHAUSTED THE DISCUSSION OF THE German Jesuits' vast educational enterprise. Studies of patronage need to be extended to encompass the Jesuit colleges. The difficult task of reconstructing the economics of the Jesuit pedagogical mission and of academic publishing has hardly been undertaken. Many other fields of Jesuit classroom instruction require further study, particularly in the eighteenth century, including the experimental fields of perception and cognition, optics and electricity, and observational astronomy.

Nevertheless, this study has identified fundamental characteristics of Jesuit natural philosophy, in particular significant, long-lasting traditions. Among them is the survival of important elements of scholastic, or, better said, peripatetic, philosophy until after the middle of the eighteenth century. Slowly the scholastic tradition was overwhelmed through the workings of another Jesuit tradition, the integration of novelties. While censorship did cause problems for scholars within the Society in a number of ways, it would be a serious distortion to write the history of the Society's intellectual activities merely as a story of repression. Censorship was the concrete manifestation of the relationship between theology and philosophy, another tradition that was a constant feature of Jesuit physics. Although this relationship remained stable at an idealized, normative level throughout the period studied here, underneath there was room for considerable change in philosophical matters.

Any attempt to understand the Jesuits Physics curriculum must place it in a variety of contexts: the development of the natural sciences in early modern Europe, the culture of scholarship within the Society, and the institution of the early modern German university. None of these is in itself enough to account for the path Jesuit Physics took between 1630 and 1773. Although the relationship between theology and natural philosophy must be considered, this dynamic alone cannot explain all aspects of Jesuit natural philosophy. Many of the limitations placed on the curriculum were not necessarily due to the Society or its theology but were shared by universities in general at that time. Theology does not explain why the Jesuits did not perform air pump experiments in their classrooms until eighty years after Schott first published an account of them in 1657. Rather, the cause of the lag was institutional, lying just as much with the function and self-image of the early modern university as with the structure and method of the Jesuit curriculum. Also, the problem of integrating a practice such as experiment with speculative natural philosophy, a problem quite unrelated to theology, had to be solved first. Moreover, the delays at Jesuits universities were little different from those at Protestant universities in Germany.

In certain important ways the development of natural philosophy at the Jesuit colleges recounted here broadly resembles that described by the traditional narrative of the Scientific Revolution. The content of the curriculum was vastly different in 1773 from what it had been in 1600. It was no longer purely speculative but relied upon experiments. It was no longer purely qualitative but used mathematics. It no longer dealt explicitly with theological questions. Not only were its practitioners permitted to hold opinions that had previously been forbidden, but in many regards its concerns had moved well beyond those that had dominated throughout the seventeenth and early eighteenth centuries. Its practitioners had become increasingly professionalized. No longer was Physics taught by junior faculty as an obligatory step on the path to another career; rather, it was taught by scholars who dedicated their entire careers to the discipline.

However, the chronology of the standard narrative of the Scientific Revolution differs from this one. The changes took a long time—close to two centuries. This results in a view of the Scientific Revolution very different from that of the attenuated standard narrative that generally ends with an event such as the publication of the *Principia* in 1687. And whatever it was, the Scientific Revolution was not an all-or-nothing affair. Jesuit professors were able to pick those elements that fit both their theology and philosophy while strenuously rejecting others.

Is it possible, then, to say that these changes were revolutionary? Determining what exactly composed the Scientific Revolution is a vexed question.[1] Most historians would agree that the use of controlled observation such as experiments to conduct science, the mathematization of science or at least of natural philosophy, and the assumption that natural processes can best be accounted for through the action of some kind of particles rather than through an organic model such as the growth and decay of forms are among the most important factors marking the fundamental changes that the investigation of nature underwent in Europe from the mid–sixteenth to the mid–eighteenth centuries. The Jesuit Physics curriculum lacked all of these factors at the start of the period considered, although it was already beginning to come to terms with them. By 1773 Jesuit physics was characterized by all of them. Granted, it is difficult to classify a process that lasted nearly two hundred years as revolutionary, but one cannot deny that by the end of the process Jesuit physics had been transformed.

If by the 1750s and 1760s the physics taught in the Jesuits' German colleges had largely assumed the characteristics of physics taught at other European universities, it was not completely indistinguishable from science outside the Society. Jesuit scientists were still subject in some degree to constraints imposed on them by their Society. Also, they adhered to disputation as the core of their pedagogy, while many innovators consciously dismissed it as scholastic wrangling incapable of uncovering the truth. Nor had Jesuit natural philosophy achieved the contours of twentieth-century physics. It still differed considerably in many ways. For example, it included much material that was beginning to be separated from physics as disciplines such as chemistry, geology, biology, and physiology established their independence.

Thus with its starting point at a time when the Jesuit colleges in Germany taught a largely Aristotelian natural philosophy and its conclusion with them teaching Newtonian physics, this story may resemble the standard narrative of the Scientific Revolution, so often told as a story of improvement, progress, even conquest. But the purpose of this work has not been to retell the Scientific Revolution with this or any other trajectory. It is much more interested in everyday processes than the end points per se. While the story of the Scientific Revolution is primarily a narrative of discovery and invention, this is a story of dissemination, adaptation, and pedagogy. The activities that fell between the two end points are just as important as the destination reached.

For several decades historians have sought to tell a cultural history of science in which science is treated as a cultural product like any other, as cultural activity rather than as a privileged quest for transcendent truths. Hopefully this work has contributed to the enterprise of revealing the extent to which science is embedded in culture. Yet this is equally a social history of science, of the science of the everyday, of the centuries-long annual cycle of lectures, disputations, academic publications, and ceremonies, of the classroom in which young Jesuit professors tried to ascertain what was useful and valuable in the New Science and attempted to integrate it into their teaching. For these men—natural philosophers, theologians, mentors, priests, and often mathematicians all in one—the new science of Galileo, Newton, Huygens, and the other heroes of early modern science was just one of many factors that they grappled with in their intellectual universe, and not necessarily a major one at that. Good social history deals with institutions but in a way that brings them back to life. The Jesuit colleges were lively institutions that cannot be re-created just from statutes like the *Studienordnungen,* reformed again and again, or from princely decrees that have formed the fodder for so many institutional histories of universities. The structure and role of these institutions within the university, order, and state influenced how and what the Jesuits taught just as much normative documents.

Certainly one could level many criticisms against the Jesuit colleges and universities. While some critiques may have an appeal, it is difficult to identify what the deleterious results of the supposedly backward Jesuit science actually were. That it somehow stunted the intellectual, cultural, or even moral development of Catholic Germany? How could one demonstrate this? One could argue that the Jesuit educational institutions neglected their duties to the state and society by not instilling an adequate standard of mechanical and technical skills in the youth they taught. Yet most historians accept that natural philosophy and then science played little role in promoting useful technologies.[2] So the Jesuits colleges' alleged failure to adopt modern science is hardly responsible for Germany's failure to spark the Industrial Revolution. Even those who claim that through the creation of a culture of Newtonianism the Scientific Revolution did contribute directly to the British Industrial Revolution admit that universities played only a minor role in the development of this culture.[3] Another anachronistic assessment is to measure the Jesuit institutions against the yardstick of the modern university, or even against the University of Göttingen (although if one were to do this, several Jesuit colleges, such as Ingolstadt, Würzburg, and Innsbruck, would not measure up too badly).

Judgments about the Jesuit colleges will remain anachronistic unless we first determine what they sought to do, what the states expected of them, and how well they fulfilled this dual charge. While this study has focused primarily on the first of these questions, it has not neglected the second. Moreover, the Jesuits successfully fulfilled the task given to them by the German states. The Jesuits entered the teaching enterprise in Germany at a time when Catholic higher education was in precipitous decline due to the Reformation. Counteracting this situation was what the princes primarily wanted of the Jesuits, not an education in practical mathematics. Certainly the situation changed in the eighteenth century, as the confessional age was replaced by an era of competing states with absolutist pretensions. Again, the Jesuits sought to provide the education these states demanded, if not to all their critics' satisfaction—if this was ever a possibility.

From the end of the sixteenth century, when the Jesuits refounded natural philosophical education on the basis of an eclectic Aristotelianism, to the second "Golden Age" of Jesuit science after the middle of the eighteenth century, the story of Catholic natural philosophy in Germany is essentially the history of the Physics curriculum at the Jesuit colleges. Throughout this period they served at the whim of the princes who could have replaced them at any time had they been unhappy with their performance. Yet the Jesuits maintained their monopoly over the chairs of natural philosophy and mathematics. Clearly, then, the Jesuits provided a natural philosophical education that the states deemed satisfactory at a price they were willing to pay. The Jesuits were not failed scientific revolutionaries but educators who were phenomenally successful at dominating the universities and colleges of Catholic Germany for two centuries, who taught what they saw fit to teach, and who generally reconciled to their own satisfaction the demands of their theology, their natural philosophy, and their identities as Jesuits.

APPENDIX

Jesuit Colleges Teaching Philosophy in the German Assistancy

	College founded:	*University founded:*[a]	*Physics from:*	*Mathematics from:*
Upper Germany (total colleges: 27)[b]				
Burghausen	1627		1726/27	1771/72
Konstanz	1604		1686/87	1690/91–1697/98
Dillingen	1562	1551 (1563)	1562	1562
Freiburg im Breisgau	1620	1457 (1620)	1620/21	1620/21
Fribourg, Switzerland	1582		1700/01	1763/64
Ingolstadt	1556	1472 (1585)	1556	1556
Luzern/Lucerne	1577		1675/76	1771/72
Munich	1559		1730/31[c]	1765/66
Innsbruck	1562	1672	1670/71	1678/79
Regensburg	1589	1672	1675/76	1678/79

	College founded:	*University founded:*[a]	*Physics from:*	*Mathematics from:*
Lower Rhine (total colleges: 19)				
Aachen	1600		1687/88	1688/89
Cologne	1556	1388 (1556)	1556	1556
Hildesheim	1598		1664/65	1666/67
Münster	1625	1773	1626/27	1626/27
Osnabrück	1628		1664/65	1666/67
Paderborn	1585	1614	1615/16	1615/16
Trier	1560	1472 (1560)	1560	1560
Upper Rhine (total colleges: 16)				
Bamberg	1611	1648	1687/88	1688/89
Fulda	1574	1732	1659/60	1734/35
Heidelberg	1629	1386 (1706)	1704/05	1723/24
Würzburg	1567	1582 (1582)	1582	1589
Mainz	1561	1477 (1561)	1561	1561
Molsheim	1580	1617	1619/20	1619/20–1676/77
Austria (total colleges: 31)				
Budapest	1685		1713/14	1744/45
Kosice/Kaschau	1633	1658	1659/60	1659/60
Klagenfurt	1604		1654/55	1656/57
Klausenburg	1583		1710/11	1713/14
Gorizia	1621		1655/56	1748/49
Graz	1573	1585	1582/83	1582/83
Györ/Raab	1627		1746/47	1761/62
Ljubljana/Laibach	1597		1710/11	1710/11
Linz	1629		1670/71	1670/71
Trnava/Tyrnau	1561	1635	1636/37	1636/37
Vienna	1550	1365 (1623)	1550	1550
Theresianum, Vienna	1561		1748/49	1770/71
Zagreb	1622		1663/64	1770/71
Bohemia (total colleges: 28)				
Olomouc/Olmütz	1566	1573	1566	1566
Prague	1555	1348 (1622)	1555	1555
Wroclaw/Breslau	1646	1701	1643/44	1643/44[d]

	College founded:	*University founded:*[a]	*Physics from:*	*Mathematics from:*
Flandro-Belgica (total colleges: 20)				
Antwerp	1622		1650/51	1653/54
Louvain	1547		1632/33	1632/33–1655/56
Gallo-Belgica (total colleges: 20)				
Douai	1568	1562	1613/14	1617/18
England (total colleges: 1)				
Liège	1614		1625/26	1625/26

Source: The bulk of these data are reproduced from Karl Adolf Franz Fischer, "Jesuiten Mathematiker in der deutschen Assistenz bis 1773," *AHSI* 47 (1978): 159–224, supplemented and corrected from Karl Hengst, S.J., *Jesuiten an Universitäten und Jesuitenuniversitäten: Zur Geschichte der Universitäten in der Oberdeutschen und Rheinischen Provinz der Gesellschaft Jesu im Zeitalter der konfessionellen Auseinandersetzung* (Paderborn: Ferdinand Schöningh, 1981), and other sources.

[a] Dates in parentheses indicate the date the college was amalgamated into a preexisting university.

[b] Total number of colleges is for 1725, taken from a map by O. Werner, S.J., inserted in Pachtler, vol. 3. This includes colleges which did not teach the philosophy curriculum.

[c] Surviving theses indicate that Physics was taught at Munich well before this date.

[d] These dates appear to be erroneous, as Fischer, "Jesuiten Mathematiker," and dB/S agree on the foundation date of 1646.

NOTES

Introduction

1. Jerónimo Nadal, quoted in William V. Bangert, S.J., *Jerome Nadal, S.J., 1507–1580: Tracking the First Generation of Jesuits* (Chicago: Loyola University Press, 1992), 147.

2. For an institutional history of the founding of the colleges, see Karl Hengst, S.J., *Jesuiten an Universitäten und Jesuitenuniversitäten: Zur Geschichte der Universitäten in der Oberdeutschen und Rheinischen Provinz der Gesellschaft Jesu im Zeitalter der konfessionellen Auseinandersetzung* (Paderborn: Ferdinand Schöningh, 1981).

3. For an introduction, see John W. O'Malley, "The Historiography of the Society of Jesus: Where Does It Stand Today?" in *The Jesuits: Culture, Sciences, and the Arts, 1540–1773*, ed. John W. O'Malley et al. (Toronto: University of Toronto Press, 1999).

4. The Society of Jesus was founded by Ignatius of Loyola and nine companions in Paris on 15 August 1534. It received papal sanction from Paul III in the bull "Regimini militantis" of 27 September 1540 and again from Julius III on 21 July 1550 ("Exposuit dubitum"). It was suppressed at Pope Clement XIV's decree ("Dominus ac Redemptor Noster") on 21 July 1773.

5. Cited in Notker Hammerstein, *Aufklärung und katholisches Reich: Untersuchungen zur Universitätsreform und Politik katholischer Territorien des Heiligen Römischen Reichs deutscher Nation im 18. Jahrhundert* (Berlin: Duncker & Humboldt, 1977), 13. Actually the Jesuit administration in Paraguay was by most accounts much more humane than Spanish and Portuguese dominion in the rest of South America, but Moser's implication is clear: the Jesuits were a foreign influence that sought to convert and rule Germany like a colonial power among ignorant natives. See also Richard van Dülmen, "Antijesuitismus und katholische Aufklärung in Deutschland," *Historisches Jahrbuch* 89 (1969): 52–80.

6. The "Jesuit Law" (Jesuitengesetz) of 1872 expelled the Society from Germany. The law was softened in 1904 and repealed in 1917.

7. Carl Prantl, *Geschichte der Ludwig-Maximilians-Universität in Ingolstadt, Landshut, München* (Munich: C. Kaiser, 1872), 220. A marked contrast was Rudolf Kink's much more sympathetic portrayal of the Jesuits in his history of the University of Vienna, *Geschichte der kaiserlichen Universität zu Wien* (Vienna: C. Gerold, 1854). In 1898 Franz Romstöck challenged the anti-Jesuit prejudices that characterized Prantl's history. In the charmingly titled *Die Jesuitennullen Prantl's an der Universität Ingolstadt und ihre Leidensgenossen: Ein biobibliographische Studie* (Prantl's Jesuit zeros at the University of Ingolstadt and their companions in suffering) (Eichstatt: Brönner, 1898), Romstöck rescued the Jesuit professors from "the literary stocks" with a 459-page catalogue of manuscripts and printed sources authored by those Jesuits at Ingolstadt whom Prantl had dismissed as not having written anything worth mentioning.

8. In his *Die deutsche Schulphilosophie im Zeitalter der Aufklärung* (Hildesheim: Georg Olms, 1964), Max Wundt devoted a mere two pages to the Catholic universities, and much of that dealt with the period after 1773. Friedrich Paulsen's *Geschichte des gelehrten Unterrrichts auf Universitäten vom Ausgang des Mittelalters bis zur Gegenwart,* 2 vols. (Leipzig: Veit, 1919), was somewhat more balanced.

9. On the conflict thesis, see the editors' introduction in David C. Lindberg and Ronald L. Numbers, eds., *God and Nature: Historical Essays on the Encounter between Christianity and Science* (Berkeley: University of California Press, 1986). On the historical relationship of the two cultural systems loosely termed science and religion, see John Hedley Brooke, *Science and Religion: Some Historical Perspectives* (Cambridge: Cambridge University Press, 1991).

10. Robert Haaß, *Die geistige Haltung der katholischen Universitäten Deutschlands im 18. Jahrhundert* (Freiburg i. Br.: Herder, 1952); Hammerstein, *Aufklärung und katholisches Reich.*

11. Laetitia Boehm, "Universität in der Krise? Aus der Forschungsgeschichte zu katholischen Universitäten in der Aufklärung am Beispiel der Reformen in Ingolstadt und Dillingen," *Zeitschrift für bayerische Landesgeschichte* 54 (1991): 107–57. Much like Boehm, in his recent survey of education in early modern Germany, Anton Schindling laments the general overestimation of the University of Göttingen in the historiography and the neglect of the southern, Catholic universities. Anton Schindling, *Bildung und Wissenschaft in der frühen Neuzeit, 1650–1800* (Munich: R. Oldenbourg, 1994).

12. There is unfortunately no bibliography dedicated solely to Jesuit science. Laszlo Polgar, S.J., *Bibliographie sur l'histoire de la Compagnie de Jesus* (Rome: Institutum Historicum Societatis Jesu, 1981–90), can be used to find material on individual Jesuits and particular sciences. The *Dictionary of Scientific Biography,* ed. Charles Coulston Gillispie (New York: Scribner, 1981), has entries with useful information on many Jesuit scientists. Much material on Jesuit science can be found in the *Isis* bibliographies. The standard bibliography of primary texts is dB/S.

13. Some of the best work on Jesuit science illuminates aspects of the interaction between Galileo and the Jesuits, in particular William Wallace's *Galileo and His Sources: The Heritage of the Collegio Romano in Galileo's Science* (Princeton: Princeton University Press, 1984) and his collection of essays *Galileo, the Jesuits and the Medieval Aristotle* (Brookfield, Vt.: Ashgate, 1991). The work of

Ugo Baldini on Jesuit science and institutions is particularly important. See in particular *Legem impone subactis: Studi su filosofia e scienza dei gesuiti in Italia, 1540–1632* (Rome: Bulzoni, 1992). The essays in Luce Giard, ed., *Les jésuites à la Renaissance: Système éducatif et production du savoir* (Paris: Presses Universitaires de France, 1995), offer numerous insights into the early period of the Society.

14. On the development of the *Ratio studiorum,* see Ladislaus Lukács, S.J., "Introductio generalis," in *Mon. paed.,* 2:1*–34*, available in English as "A History of the Jesuit *Ratio Studiorum,*" in *Church, Culture, and Curriculum: Theology and Mathematics in the Ratio Studiorum,* ed. Frederick A. Homan (Philadelphia: St. Joseph's University Press, 1999); Gabriel Codina Mir, S.J., *Aux sources de la pédagogie des jésuites: Le "modus Parisiensis"* (Rome: Institutum Historicum Societatis Iesu, 1968); and Dennis A. Bartlett, "The Evolution of the Philosophical and Theological Elements of the Jesuit *Ratio Studiorum:* An Historical Survey, 1540–1599" (Ed. D. diss., University of San Francisco, 1984).

15. We should be wary of exaggerating the long-term effects of the condemnation, as John L. Russell, S.J., points out in "Catholic Astronomers and the Copernican System after the Condemnation of Galileo," *Annals of Science* 46 (1989): 365–86.

16. For example, Richard J. Blackwell, *Galileo, Bellarmine, and the Bible: Including a Translation of Foscarini's "Letter on the Motion of the Earth"* (Notre Dame: University of Notre Dame Press, 1991), 150.

17. See, for example, John L. Heilbron, *Electricity in the Seventeenth and Eighteenth Centuries: A Study in Early Modern Physics* (Berkeley: University of California Press, 1979); Peter Dear, *Discipline and Experience: The Mathematical Way in the Scientific Revolution* (Chicago: University of Chicago Press, 1995). A prominent exception to this neglect is work on the German polymath Athanasius Kircher, who continues to attract considerable scholarly (and not so scholarly) attention. There are other non-German exceptions. For example, Alfredo De Oliveira Dinis, "The Cosmology of Giovanni Battista Riccioli (1598–1671)" (Ph.D. diss, University of Cambridge, 1989), gives deserved attention to an important Jesuit astronomer. The chief exception to the rule of neglect in the eighteenth century is Roger Boscovich.

18. See Steven J. Harris, "Transposing the Merton Thesis: Apostolic Spirituality and the Establishment of the Jesuit Scientific Tradition," *Science in Context* 3 (1989): 29–65.

19. There is no systematic catalogue of this material, and the dB/S is incomplete in its coverage of university dissertations and disputations. Manfred Komorowski, "Die Hochschulschriften des 17. Jahrhunderts und ihre bibliographische Erfassung," *Wolfenbütteler Barock-Nachrichten* 24 (1997): 19–42, provides an introduction. Gunter Lind, *Physik im Lehrbuch, 1700–1850: Zur Geschichte der Physik und ihrer Didaktik in Deutschland* (Berlin: Springer, 1992), is a valuable study of physics textbooks at both Catholic and Protestant institutions.

20. Franz X. Wegele, for example, appended an entire volume of documents to his 1882 history of the University of Würzburg but did not include in it a single example of a disputation or any other text indicating what was taught in the classroom. Franz X. Wegele, *Geschichte der Universität Wirzburg* (Würzburg: Stahel, 1882).

21. Of course, not all histories of science at particular universities where the Jesuits taught do this. A selection of histories includes Josef Schaff, *Geschichte der Physik an der Universität Ingolstadt* (Erlangen: Wiedemann, 1912); Peter Stötter, "Vom Barock zur Aufklärung: Die Philosoph-

ische Fakultät der Universität Ingolstadt in der zweiten Hälfte des 17. und im 18. Jahrhundert," in *Die Ludwig-Maximilians-Universität in ihren Fakultäten,* vol. 2, ed. Laetitia Boehm and Johannes Spörl (Berlin: Duncker & Humboldt, 1980); Georg Schuppener, *Jesuitische Mathematik in Prag im 16. und 17. Jahrhundert (1556–1654)* (Leipzig: Universitätsverlag, 1999); Thomas Specht, *Geschichte der ehemaligen Universität Dillingen (1549–1804) und der mit ihr verbundenen Lehr- und Erziehungsanstalten* (Freiburg i Br.: Herder, 1902); Dana Koutná-Karg, "Experientia fuit, Mathematicum paucos discipulos habere . . . : Zu den Naturwissenschaften und der Mathematik an der Universität Dillingen zwischen 1563 und 1632," in *Das andere Wahrnehmen: Beiträge zur europäischen Geschichte,* ed. Wolfgang Stürner, Martin Kintzinger, and Johannes Zahlten (Cologne: Böhlau, 1991); Maria Reindl, *Lehre und Forschung in Mathematik und Naturwissenschaften, insbesondere Astronomie, an der Universität Würzburg von der Gründung bis zum Beginn des 20. Jahrhundert* (Neustadt: Degener, 1966).

22. See, for example, Charles B. Schmitt, "Towards a Reassessment of Renaissance Aristotelianism," *History of Science* 11 (1973): 159–93, and *Aristotle and the Renaissance* (Cambridge, Mass.: Harvard University Press, 1983).

23. James Conant's *On Understanding Science: An Historical Approach* (New Haven: Yale University Press, 1947), a key work in development of the history of science as an independent discipline after the Second World War, devoted considerable attention to the mercury tube and the air pump.

24. Steven Shapin and Simon Schaffer, *Leviathan and the Air-Pump: Hobbes, Boyle, and the Experimental Life* (Princeton: Princeton University Press, 1989). The pump also plays a reprise role in Shapin's later works, *A Social History of Truth: Civility and Science in Seventeenth-Century England* (Chicago: University of Chicago Press, 1994), and *The Scientific Revolution* (Chicago: University of Chicago Press, 1996).

25. This book will also attempt to complement and further Peter Dear's work, which has recognized the important role played by Jesuit mathematicians and natural philosophers in answering questions about what it meant to do experimental philosophy. See Dear, *Discipline and Experience.*

26. Chapter 3 will make the distinction between colleges and universities clearer.

27. The surviving archives of the Upper Rhenish province are held in the Stadtarchiv Mainz and are rather disappointing, consisting mainly of details of the Jesuits' rural properties that were taken over by the state. The library of the old University of Mainz was essentially deposited in the Mainz city library when the university was dissolved by the French during the revolutionary wars. This library has an excellent, if inadequately catalogued, collection of Jesuitica, including many dissertations. It also has an unknown, but extremely large, number of dissertations from other German universities, both Catholic and Protestant. Some materials on the Universities of Mainz and Würzburg are held in the Bayerisches Staatsarchiv, Würzburg. The Würzburg university library has another excellent collection of dissertations, catalogued in Gottfried Mälzer, *Würzburger Hochschulschriften, 1581–1803 Bestandsverzeichnis* (Würzburg: Universitäts-Bibliothek, 1992). Unfortunately, only those dissertations defended at the University of Würzburg are included in the catalogue.

28. Fortunately Ingolstadt was in the Upper German province centered on Bavaria, which was much more conscientious than any other German state in preserving the Jesuit archives after

the suppression of the Society. Today the provincial archives are stored chiefly in the Bayerisches Hauptstaatsarchiv, Munich, catalogued as "Jesuiten," with some elements scattered throughout the manuscript collection of the Bayerische Staatsbibliothek. The archives of the former University of Ingolstadt, which eventually became the University of Munich, are now housed in the archives of the Ludwig-Maximilians-Universität, although much of the material that pertains to Jesuits is in the Hauptstaatsarchiv, catalogued under "Jesuiten" or "Gerichtsliteralien," or in the manuscript collection of the Staatsbibliothek.

29. In this book, *Freiburg* refers to the southwest German city of Freiburg im Breisgau, not Freiburg in der Schweiz, where there was another Jesuit college that will be referred to as Fribourg.

30. See Andrew Cunningham, "How the *Principia* Got Its Name: Or, Taking Natural Philosophy Seriously," *History of Science* 29 (1991): 377–92, and the provocative, if not completely convincing, article by Andrew Cunningham and Perry Williams, "De-centring the 'Big Picture': The Origins of Modern Science and the Modern Origins of Science," *British Journal for the History of Science* 26 (1993): 407–32.

31. Dennis Des Chene, *Physiologia: Natural Philosophy in Late Aristotelian and Cartesian Thought* (Ithaca: Cornell University Press, 1996), 3.

1. Managing Philosophy in the Society of Jesus

1. On the development of the *Ratio studiorum*, see Lukács, "A History"; Allen P. Farrell, S.J., *The Jesuit Code of Liberal Education: Development and Scope of the Ratio Studiorum* (Milwaukee: Bruce, 1938); Bartlett, "The Evolution." None of these works is, however, by any means perfect. The most important documents regarding the development of the Society's teaching mission and the *Ratio studiorum* have been published in the seven-volume *Mon. paed.*

2. Not all Church fathers shared this positive view of philosophy. Tertullian was particularly disparaging. See G. R. Evans, Alister E. McGrath, and Allan D. Galloway, *The Science of Theology* (Grand Rapids, Mich.: Eerdmans, 1986), particularly chs. 1–5.

3. On the interaction of Catholic theology and philosophy in the medieval period, see Roger French and Andrew Cunningham, *Before Science: The Invention of the Friars' Natural Philosophy* (Aldershot: Ashgate, 1996); G. R. Evans, *Philosophy and Theology in the Middle Ages* (New York: Routledge, 1993), and *Old Arts and New Theology: The Beginnings of Theology as an Academic Discipline* (Oxford: Clarendon, 1980); John Marenbon, *Later Medieval Philosophy (1150–1350)* (New York: Routledge, 1987); and Evans et al., *Science of Theology.*

4. For a brief overview of this process, see John W. O'Malley, *The First Jesuits* (Cambridge, Mass.: Harvard University Press, 1993), ch. 6; Lukács, "A History."

5. On the influence of the University of Paris on the first Jesuits' ideas on pedagogy, see Codina Mir, *Aux sources.*

6. Jerónimo Nadal, "De artium liberalium studiis," *Mon. paed.*, 2:254.

7. Ignatius of Loyola, *Counsels for Jesuits: Selected Letters and Instructions of Saint Ignatius Loyola*, ed. Joseph N. Tylenda, S.J. (Chicago: Loyola University Press, 1985), 53–57.

8. Karl Joseph Lesch, *Neuorientierung der Theologie im 18. Jahrhundert in Würzburg und Bamberg* (Würzburg: Echter, 1978), 16–20.

9. Francisco Suárez, S.J., "Ad lectorem," *Metaphysicarum disputationum in quibus et universa naturalis theologia ordinate traditur,* 2 vols. (Venice, 1610).

10. *Mon. Ig.,* letter 1848, 3:499–503. Translation in Ignatius of Loyola, *The Letters of St. Ignatius of Loyola,* ed. and trans. William John Young, S.J. (Chicago: Loyola University Press, 1959), 236.

11. Clearly not all Reformers dismissed philosophy, as is shown by the work of Melanchthon. See Sachiko Kusukawa, *The Transformation of Natural Philosophy: The Case of Philip Melanchthon* (Cambridge: Cambridge University Press, 1995), and Charlotte Methuen, "The Role of the Heavens in the Thought of Philip Melanchthon," *Journal of the History of Ideas* 57 (1996): 385–403.

12. Jakob Gretser, S.J., *Luther Academicus,* in *Lutherus Academicus, et Waldenses,* vol. 12 of *Opera omnia* (1610; reprint, Regensburg, 1738), 123–25. Naturally one cannot rely on Gretser for an objective reporting of Luther's views, but what Luther actually said is quite irrelevant here; of interest is how Gretser defended reason and philosophy.

13. *Mon. paed.,* 4:201.

14. Wilhelm Elderen, S.J., to the Cologne Jesuits, 27 April 1561, in *PCE,* 8:720. The standard biography of Canisius is James Brodrick, S.J., *Saint Peter Canisius, S.J., 1521–1597* (London: Sheed and Ward, 1935).

15. *PCE,* 6:60–62.

16. On Averroes, see Oliver Leaman, *Averroes and His Philosophy* (Oxford: Clarendon, 1988), and Kurt Flasch, *Einführung in die Philosophie des Mittelalters* (Darmstadt: WBG, 1987), 108–16. Averroes wrote a work specifically on the proper relationship of philosophy and theology, but it was not known to the West until the middle of the nineteenth century: *On the Harmony of Religion and Philosophy,* trans. George F. Hourani (London: Luzac, 1961).

17. *PCE,* 6:61. The reading of profane authors would be limited by the *Ratio studiorum.* Racy authors such as Catullus and Ovid were banned, and even model authors such as Cicero had to be carefully vetted before being given to students. On the vetting of texts, see Pierre-Antoine Fabre, "Dépouilles d'Egypte: L'expurgation des auteurs latins dans les collèges jésuites," in Giard, *Les jésuites.*

18. *PCE,* 6:136–37.

19. Such derogatory statements about Thomas were later forbidden in the 1599 *Ratio studiorum,* which stated that Thomas's views were to be defended or simply omitted (*RS,* "Rules for the Professor of Scholastic Theology," 13).

20. *PCE,* 6:137.

21. Pisano to Borja, 10 December 1567, *PCE,* 6:136.

22. Pastelius to Borja, 11 December 1567, *PCE,* 6:142.

23. Provisions for removing those who spread discord are in *Const.,* III, 2, § 6.

24. Pisanus to Canisius, 12 December 1567, *PCE,* 6:143.

25. Peter Canisius to Borja, 26 September 1567, *PCE,* 6:67.

26. *Mon. paed.,* 3:40. On the relative functions of the general and provincial congregations, see Adrien Demoustier, "La distinction des fonctions et l'exercise du pouvoir selon les règles de la Compagnie de Jésus," in Giard, *Les jésuites.*

27. An English translation is in *ILPW,* 251–60. The original text is in *Mon Ig.* 4:669–81 (letter 3304). It was published in numerous editions and was required reading for all members. Ignatius's successors as general referred to it frequently in their own letters to the Society.

28. *ILPW,* 252.

29. *ILPW,* 254.

30. *ILPW,* 255–56.

31. Cited in Manuel María Espinosa Pólit, S.J., *Perfect Obedience: Commentary on the Letter of Obedience of Saint Ignatius of Loyola* (Westminster: Newman Bookshop, 1947), 82.

32. *ILPW,* 260.

33. *ILPW,* 257.

34. *Const.,* III, 1, § 18.

35. See Ignatius's letter of 1554, "On the Method of Dealing with Superiors," in Ignatius of Loyola, *Letters,* 390–92 (original in *Mon. Ig.,* 9:90–92, letter 5400a). Of course, if obedience to a command would put the inferior into a state of sin, no obedience was necessary, for the superior could not possibly be representing the will of God. On the exercise of authority in the Society, see Demoustier, "La distinction." The arbitrary exercise of power was checked in many ways. For example, no Jesuit was solely subordinate to the authority of a single superior. In a way, according to Demoustier, the Society represented the perfect form of government as defined by scholastic theorists because it was a mixture of monarchical, aristocratic, and democratic government.

36. "In logic, natural and moral philosophy, and metaphysics, the doctrine of Aristotle should be followed, as also the other liberal arts." *Const.,* IV, 14, § 3.

37. See the essays in Norman Kretzmann, Anthony Kenny, and Jan Pinborg, eds., *Cambridge History of Later Medieval Philosophy* (Cambridge: Cambridge University Press, 1982), and David C. Lindberg, ed., *Science in the Middle Ages* (Chicago: University of Chicago Press, 1978). For more general accounts, see Marenbon, *Later Medieval Philosophy,* and Evans, *Science of Theology.*

38. Cong. Gen. III, Decr. 8, in *ISJ,* 2:228. This decree was part of the ongoing struggle against Averroism. One should note, however, that Averroism was merely a convenient shorthand for the brand of neo-Aristotelianism taught at Italian universities such as Padua. See Charles H. Lohr, S.J., "Jesuit Aristotelianism and Sixteenth-Century Metaphysics," in *Paradosis: Studies in Memory of Edwin A. Quain,* ed. H. G. Fletcher III and M. B. Scholtel (New York: Fordham University Press, 1976).

39. Gretser, *Lutherus Academicus,* 127–28.

40. For the sixteenth century, see Lohr, "Jesuit Aristotelianism," as well as his more recent "Les jésuites et l'aristotelisme du XVIe siècle," in Giard, *Les jésuites.* There is also the somewhat dated Andreas Inauen, S.J., "Stellung der Gesellschaft Jesu zur Lehre des Aristoteles und des hl. Thomas vor 1583," *Zeitschrift für katholische Theologie* 40 (1916): 201–37.

41. *RS,* "Rules for the Professor of Philosophy," 2. The eighth session of the Fifth Lateran Council of 1513 decreed that philosophy professors must not only explain how certain philosophical errors (such as the eternity of the world and the mortality of the soul) departed from Christian truths but also demonstrate the truths of the Christian faith and their compatibility with correct philosophy.

42. *RS,* "Rules for the Professor of Philosophy," 4.

43. *RS,* "Rules for the Professor of Philosophy," 12.

44. *Mon. paed.,* 7:605.

45. "Decretum R. P. N. Generalis Praepositi Francisci Borjae in mense Novembri 1565," *Mon. paed.,* 3:382–85. The affair is treated in Lohr, "Jesuit Aristotelianism."

46. "An et quae opiniones sint professoribus prohibendae, et qua ratione?" *Mon. paed.,* 4:196–204.

47 The Collegio Romano was the Society's most prominent college, and its professors were the best in the Society. They acted as advisors to the general in educational matters. Modifications to the Society's teaching practices were often first discussed and tested at the Collegio Romano. For an introduction to the role of the Collegio Romano see Luce Giard, "Le Collège romain: La diffusion de la science (1570–1620) dans le réseau jésuite," in *XIXth International Congress of History of Science. Symposia Survey Papers,* ed. Jean Dhombres, Mariano Hormigan, and Elena Ausejo (Zarazoga, Spain: International Union of History and Philosophy of Sciences, 1993). The standard history of the Collegio Romano is Ricardo Garcia Villoslada, *Storia del Collegio romano dal suo inizio (1551) alla soppressione della Compagnia di Gesù (1773)* (Rome: Gregorian University, 1954).

48. Borja's list of specific opinions was officially abandoned by General Acquaviva in 1582. *Mon. paed.,* 6:25–26.

49. *Mon. paed.,* 6:6.

50. *Mon. paed.,* 6:14.

51. *Mon. paed.,* 6:24.

52. Bellarmine was referring specifically to theological opinions. *Mon. paed.,* 7:44.

53. The list is in *Mon. paed.,* 5:107. For examples of objections from the German provinces, see *Mon. paed.,* 6:283, 285.

54. *Const.* IV, 14, § 1B.

55. *Mon. paed.,* 6:34.

56. *Mon. paed.,* 6:35.

57. *Mon. paed.,* 6:35, paras. 5 and 10.

58. *Mon. paed.,* 6:35–36.

59. *Mon. paed.,* 6:42.

60. Some impression of the widespread use and influence of Jesuit textbooks can be gained from the essays by Eckhard Kessler, Charles Lohr, and Charles Schmitt in *The Cambridge History of Renaissance Philosophy,* ed. Charles B. Schmitt et al. (Cambridge: Cambridge University Press, 1988).

61. Francisco Toledo, S.J., *Commentaria una cum quaestionibus in universam Aristotelis logicam* (Rome, 1572), and *Commentaria una cum quaestionibus in VIII libros, De physica auscultatione* (Cologne, 1574).

62. *Mon. paed.,* 4:580.

63. *Mon. paed.,* 4:679.

64. "Instruttione per il visitatore degli studi," *Mon. paed.,* 7:481, para. 5.

65. *Mon. paed.,* 7:504. This was confirmed by the provincial, Alfonso Carrillo (*Mon. paed.,* 7:510). The textbook was Pedro da Fonseca, *Institutionum dialecticarum, libri octo: Emendatius quam antehac editi . . .* (Lisbon, 1564; reprint, Ingolstadt, 1604).

66. *Mon. paed.*, 7:518.

67. *Mon. paed.*, 7:519.

68. That is, the series of commentaries written by the Jesuit professors at the University of Coimbra in Portugal between 1591 and 1606 that eventually covered virtually every part of the Aristotelian corpus.

69. *Mon. paed.*, 7:519.

70. *Mon. paed.*, 7:520. The *Ratio studiorum* required that professors obey the prefect "in all matters that pertain to studies." *RS*, "Common Rules for All Professors of Higher Studies," 4.

71. Cong. Gen. VII, Decr. 83, in *ISJ*.

72. Borja's decision was strongly influenced by a number of documents on higher studies prepared in 1564 by Diego Ledesma: "Relatio de professorum consultationibus circa Collegii Romani studia," *Mon. paed.*, 2:464–81; "Quaedam circa studia et mores Collegii Romani data R. P. Generali," *Mon. paed.*, 2:481–90; "Quaedam quae docenda et defendenda sunt in philosophia," *Mon. paed.*, 2:496–503.

73. "Decretum R. P. N. Generalis Praepositi Francisci Borjae in mense Novembri 1565," *Mon. paed.*, 3:383.

74. So Lohr argues in "Jesuit Aristotelianism."

75. Right from the time Borja promulgated his list, there was a great deal of slippage between "most common," "more common," and "common." Naturally it made quite a difference if any common opinion was acceptable rather than just the most common one. Various decrees and letters from the generals used "more common," others used "most common." *Mon. paed.*, 4:746.

76. *Mon. paed.*, 6:25–26.

77. *Mon. paed.*, 7:563.

78. *Mon. paed.*, 3:494. The consultors were senior Jesuits in each province who acted as the provincial's advisors.

79. Hoffaeus was well known for speaking his mind openly, even to the general. This led to his dismissal from his position of German assistant and admonitor to General Acquaviva in 1591. See Burkhart Schneider, S.J., "Der Konflikt zwischen Claudius Acquaviva und Paul Hoffaeus," *AHSI* 26 (1957): 6–7.

80. *Mon. paed.*, 6:746 n. 7.

81. *PCE*, 6:309–10.

82. *Mon. paed.*, 4:703–12. In the *Ordinationes de studiis* for the Upper German province of 1594, Hoffaeus decreed that professors should make no mention of "new books" (*Mon. paed.*, 7:473).

83. *Mon. paed.*, 4:721–22.

84. "Relatio de Professorum Consultationibus circa Collegii Romani studia," *Mon. paed.*, 2:478. The comment on too much liberty in Italian academies is a reference to the secular Aristotelianism taught at universities such as Padua, which Ledesman regarded as excessively Averroist.

85. Lohr, "Jesuit Aristotelianism."

86. In particular, the *Constitutions*, the *Spiritual Exercises*, and the letters of Ignatius. For a treatment of Jesuit spirituality in this period, see the sections covering Acquaviva's generalate in "Jesuits," in *Dictionnaire de spiritualité ascétique et mystique, doctrine et histoire*, 17 vols., ed. Marcel Viller (Paris: Beauchesne, 1932–95), vol. 8, cols. 979–94.

87. Denz., para. 1997. On the dispute, see "Molinisme," in *Dictionnaire de la théologie catholique,* 16 vols. (Paris: Letouzey et Ané, 1903–72), vol. 8, cols. 2094–2187.

88. This was Leonard Lessius, *De gratis efficaci, decretis divinis libertate arbitrarii et praescientia Dei conditionata* (Antwerp, 1610). See Blackwell, *Galileo, Bellarmine, and the Bible,* 51, and Cecil H. Chamberlain, "Leonard Lessius," in *Jesuit Thinkers of the Renaissance,* ed. Gerard Smith (Milwaukee: Marquette University Press, 1939), 150–51. Acquaviva first approved Lessius's text in 1610 but then changed his mind in 1611, insisting on obedience.

89. "R. P. N. Claudii Aquavivae epistola ad provinciales Societatis de soliditate et uniformitate doctrinae (24 May 1611)," in Pachtler, 3:12–14. The numerous responses are bound together in ARSI, Inst. 213 ("De promovenda doctrinae uniformitate in Soc. Jesu 1611–1613").

90. "Ordinatio R. P. Gen Claudii Aquavivae de soliditate atque uniformitate doctrinae a nostris servandae," in Pachtler, 3:15–20.

91. "Literae P. Henr. Schereni, Provincialis Rheni, circa decretum de soliditate et uniformitate doctrinae. 16. Martii a. 1614," in Pachtler, 3:20–21.

92. "Claudius Aquaviva Generalis de opinionum delectu. A. 1613," in Pachtler, 3:21–40.

93. Consider, for example, the 1616 appeal by the mathematician Christoph Grienberger on the question of the constitution of the heavens: "[I]t seems to me that the time has now come for a greater degree of freedom of thought to be given to both mathematicians and philosophers on this matter, for the liquidity and corruptibility of the heavens are not absolutely contrary to theology or philosophy and even much less to mathematics." Reprinted in Blackwell, *Galileo, Bellarmine and the Bible,* 152.

2. Censorship and Its Limits

1. *Const.*, III, 1, § 18.

2. On the revisers general and their judgments that survive in the ARSI (FG 656–75), see Ugo Baldini, "Una fonte poco utilizzata per la storia intellettuale: Le 'censurae librorum' e 'opinionum' nell'antica Compagnia di Gesù," *Annali dell'Istituto storico italo-germanico in Trento* 9 (1985): 19–67.

3. "Instructio R. P. N. Claudii Aquavivae pro censoribus librorum per provincias (1601)," StAMz, 14/241: 14.

4. While the records of the provincial revisers of the Upper German province are particularly well preserved (BHSA, Jesuiten 710–842), those of the Upper and Lower Rhenish provinces do not appear to have survived at all. On provincial censorship, see Theodor Heigel, "Zur Geschichte des Censurwesens in der Gesellschaft Jesu," *Archiv für Geschichte des deutschen Buchhandels* 6 (1891): 162–67.

5. In 1661, for example, General Goswin Nickel instructed the German Jesuits not to respond in kind to the attacks made on the Society by the Capuchin Valeriano Magni, stating that the Society wished to show that it did not lack charity toward those who unjustly attacked it. Bernhard Duhr, S.J., *Geschichte der Jesuiten in den Ländern deutscher Zunge,* vol. 3, *In der zweiten Hälfte des XVII. Jahrhunderts* (Munich-Regensburg: Manz, 1921), 533.

6. BHSA, Jesuiten 38: 8v. This volume is a manuscript collection of various rulings on matters dealing with censorship.

7. BSB, Clm 24076, unpaginated ruling dated 18 September 1632.

8. Gregorius Biegevesen, in his evaluation of Tobius Lohner's calendars of 1607, commented that although his *Adeliger Personen Calender* used good sources in discussing "the origin, diversity and propagation of the nobility," it should all be checked "lest perhaps politicians [*Politicis*], who are very discriminating, and nobles, who are very fastidious about points of honor, are offered any grounds for affront or offense." BHSA, Jesuiten 716: 3–4. See also the anonymous evaluation of a 1650s history of Bavaria by Johannes Vervaux (BHSA, Jesuiten 724: 2r). Alois Schmid has shown that the Bavarian Jesuits' historical scholarship was a form of dynastic politics in "Geschichtsschreibung am Hofe Kurfürst Maximilians I. von Bayern," in *Um Glauben und Reich: Kurfürst Maximilian I.*, ed. Hubert Glaser (Munich: Himer, 1980).

9. "Instructio R. P. N. Claudii Aquaviva pro censoribus librorum per provincias [1601]" and "De censoribus librorum [31 December 1623]," StAMz, 14/241: 14 ff.

10. Blackwell, *Galileo, Bellarmine, and the Bible,* 149. On Biancani's criticism of Aristotle and Camerota's response, see also Michel-Pierre Lerner, "L'entrée de Tycho Brahe chez les jésuites, ou le chant du cygne de Clavius," in Giard, *Les jésuites.*

11. Cited in Blackwell, *Galileo, Bellarmine, and the Bible,* 150. See also Ugo Baldini, "Additamenta Galilaeana: I. Galileo, le nuova astronomia e la critica all'Aristotelismo nel dialogo epistolare tra Giuseppe Biancani e i revisori romani della Compagnia de Gesu," *Annali dell'Instituto e Museo di Storia della Scienza di Firenze* 9 (1984): 13–43.

12. For two very different views of the role of the Society of Jesus in the Galileo affair, see Pietro Redondi, *Galileo Heretic* (Princeton: Princeton University Press, 1987), and Rivka Feldhay, *Galileo and the Church: Political Inquisition or Critical Dialogue?* (Cambridge: Cambridge University Press, 1995). For a balanced overview of the Galileo affair, see Annibale Fantoli, *Galileo: For Copernicanism and for the Church* (Vatican City: Specola Vaticana, 1994).

13. See Michael John Gorman, "A Matter of Faith? Christoph Scheiner, Jesuit Censorship, and the Trial of Galileo," *Perspectives on Science* 4 (1996): 293–95.

14. There are numerous examples of this in the reports of the revisers general in the ARSI, particularly FG 656 AI, 656 AII, and 657. By *novelty,* the rules governing censorship did not mean everything new; since books that merely repeated what had been said on a given issue could not be published, there had to be something new about every Jesuit book. Rather, as used by Jesuit censors and in Jesuit natural philosophy as a whole, the term *novelty* came to be a code word for any opinion that did not have the authority of Aristotle or the scholastic doctors and was peddled by *neoterici* or *novatores* who spurned solid doctrines and were enamored with novelty for its own sake.

15. One such appeal was made to the general in 1621, ARSI, FG 660: 80.

16. Redondi, *Galileo Heretic,* passim, especially ch. 7. Redondi's work in drawing attention to the significance of atomism and the physics of the Eucharist in Jesuit natural philosophy is particularly important. It is now necessary, however, to examine how the issue continued to play out after the Galileo affair.

17. ARSI, Congr. 20e: 49.

18. On the background to the Ninth General Congregation and the development of the *Ordinatio pro studiis superioribus*, see Marcus Hellyer, "The Construction of the *Ordinatio pro Studiis Superioribus* of 1651," *AHSI* 72 (2003): 3–44. On the dispute between Pallavicino and the revisers general, see also Claudio Costantini, *Baliani e i gesuiti: Annotazioni in margine alla corrispondenza del Baliani con Gio Luigi Confalonieri e Orazio Grassi* (Florence: Giunti, 1969).

19. ARSI, Congr. 20e: 49r.

20. The complaints concerning studies came from Milan (ARSI, Congr. 72: 55r), Lyon (ARSI, Congr. 72: 185r), and Upper Germany (ARSI, Congr. 72: 222r–222v).

21. ARSI, Congr. 20e: 223r–233v.

22. See Hellyer, "Construction of the *Ordinatio.*"

23. The text of the Ordinatio is in Pachtler, 3:77–97. The final draft manuscript of the Ordinatio is in ARSI, FG 657: 641–67.

24. "41. Gravitas et levitas non differunt specie, sed tantum secundum magis et minus." Pachtler, 3:93.

25. Cited in Gorman, "A Matter of Faith?" 295.

26. ARSI, FG 671: 351.

27. Pierre Le Cazre, S.J., "Animadversiones quaedam circa propositiones quae propositae sunt ut censurae subjicerentur," ARSI, Congr. 20e: 234r–235r (reprinted in Hellyer, "Construction of the *Ordinatio*"). Le Cazre (also Casre or Cazraeus) occupied virtually every position below that of general; he taught humanities, philosophy, mathematics, and theology, was rector of the colleges at Metz, Dijon, and Nancy; became provincial of Champagne; and finally was appointed assistant of France in 1652.

28. Pierre Le Cazre, S.J., *Physica demonstratio qua ratio, mensura, modus ac potentia accelerationis motus in naturali decensu gravium determinatur* . . . (Paris, 1645). The letters to Gassendi are in Pierre Gassendi, *Opera omnia* (Lyon, 1658), 6:448–52 and 3:589.

29. Here Le Cazre was clearly using the term *experientia* in a sense more like the modern meaning of "experiment," rather than in the peripatetic sense of repeated, everyday experience of natural phenomena (in fact, as was typical of the time, he used the terms *experimenta* and *experientia* interchangeably). This usage became increasingly common throughout the seventeenth century in Jesuit mathematics and natural philosophy. Peter Dear's *Discipline and Experience: The Mathematical Way in the Scientific Revolution* (Chicago: University of Chicago Press, 1995) is particularly important on this issue.

30. ARSI, Congr. 20e: 234r.

31. Le Cazre, "Animadversiones," ARSI, Congr. 20e: 235r.

32. Cornaeus taught philosophy and then scholastic and polemical theology at Mainz and Würzburg in the Upper Rhenish province. Because of the Swedish occupation of Mainz and Würzburg in the Thirty Years' War, he taught philosophy for seven years in Toulouse. He rose to the position of rector of the Jesuit colleges at the Universities of Würzburg and Mainz.

33. The work was first published in Rome in 1656 under the title *Itinerarium exstaticum coeleste.* Schott's edition appeared in 1660 in Würzburg. In a letter to Kircher, Schott described his own and Melchior Cornaeus's enthusiastic reception of the work. When Rector Cornaeus had finished his duties, he would read the book night and day and discuss it with Schott (APUG, 561: 279).

34. The first two propositions dealt with the ability of God to make as many worlds as he wanted and in any way he wanted. The critic claimed that Kircher asserted that the world could not be more perfect, which would limit God's omnipotent capability to make a better one. The third concerned the number and distinctness of primary qualities. According to the critic, Kircher claimed that light, brightness, and heat not only were indistinct from each other but were not even accidents but substances and body, namely fire. Athanasius Kircher, *Iter exstaticum*, ed. Kaspar Schott (Würzburg, 1671), 488–89. Although the denial of accidents could be seen as leading to atomism and thus possibly as a danger to the orthodox view of transubstantiation, the critic did not make either of these connections. The author of the attack may well have been a Jesuit; several of the censors of Kircher's original manuscript had been very hostile toward it. Materials concerning the censorship of Kircher's *Iter exstaticum* are preserved in ARSI, FG 661: 29 ff and FG 663: 134. The criticism of Kircher is also dealt with by Carlos Ziller Camenietzki, "L'harmonie du monde au XVIIe siècle: Essai sur la pensée scientifique d'Athanasius Kircher" (Ph.D. diss, Université de Paris IV–Sorbonne, 1995), 173 ff.

35. Kaspar Schott, in Kircher, *Iter exstaticum*, 509.

36. "Apologeticon R. P. Melchior Cornaeus," in Kircher, *Iter exstaticum*, 509–12. Cornaeus concluded frankly that "this critique is too harsh and is advanced without sufficient basis. I fear that something other than a natural love of the truth impelled the mind of this severe censor into such a sinister judgment."

37. Cornaeus did not accept authority uncritically. He thought that the Roman Inquisition was too willing to put books on the Index of Prohibited Books, and he agreed with the statement of a fellow Würzburger Jesuit, Vitus Erbermann, that the Index was in danger of "increasing infinitely." Erbermann to Johann Christian von Boyneburg, 7 August 1660, in BSAW, Schönborn Archiv, Johann Philipp, 2916.

38. Unpaginated folio dated 4 July 1654 in BSB, Clm 24076. Duhr writes that in 1653 Nickel called for stricter censorship as Jesuit authors kept appearing on the Index. Duhr, *Geschichte der Jesuiten*, 3:531.

39. As soon as Schott died, the work was published again as *Joco-seriorum naturae et artis, sive Magiae naturalis centuriae tres* (Würzburg, 1666) and was followed by several further editions, including a German translation, indicating that the reading public did not share Oliva's disdain for the book. The letters from Oliva concerning that book are in the ARSI, Rhen. Sup. 2: 540, 556, 563, 565. The incident is also treated by Duhr, *Geschichte der Jesuiten*, 3:591.

40. The best-known examples are Orazio Grassi's works against Galileo, published under the name of Lothario Sarsi.

41. André Goddu, "The Dialectic of Certitude and Demonstrability According to William of Ockham and the Conceptual Relation of His Account to Later Developments," in *Studies in Medieval Natural Philosophy*, ed. Stefano Caroti, Biblioteca di Nuncius, Studi e Testi 1 (Florence: Olschki, 1989), 122.

42. The English text of Oresme's work is available as "The Compatibility of the Earth's Diurnal Rotation with Astronomical Phenomena and Terrestrial Physics," in *A Source Book in Medieval Science*, ed. Edward Grant (Cambridge, Mass.: Harvard University Press, 1974), 503–10.

43. For example, Peter Udr, S.J., discussed Copernican cosmology as early as 1635 in *Stellae vigiliis astronomicis elucubratae . . . apud Friburgenses Helvet . . .* (Bern, 1635).

44. Melchior Cornaeus, S.J., *Curriculum philosophiae peripateticae, ut hoc tempore in scholis decurri solet*... (Würzburg, 1657), 2:530 ff.

45. On the use of the term *hypothesis* in Catholic astronomy in this period, see Russell, "Catholic Astronomers," 369–70.

46. Kaspar Knittel, S.J., *Theses curiosiores ex universa Aristotelis philosophia... propugnatae*... (Prague, 1682), thesis 27.

47. There were few, if any, Jesuit supporters of a Ptolemaic world system in Germany by the middle of the seventeenth century.

48. Alfredo de Oliveira Dinis, "Cosmology of Giovanni Battista Riccioli," 235–37, concludes that the Jesuit astonomer Riccioli was sincere in his adherence to an essentially Tychonic view and his rejection of heliocentrism.

49. On Arriaga's career, see Karl Eschweiler, "Roderigo de Arriaga S.J.: Ein Beitrag zur Geschichte der Barockscholastik," in *Gesammelte Aufsätze zur Kulturgeschichte Spaniens*, ed. Johannes Vincke, Spanische Forschungen der Görresgesellschaft, 1st ser., vol. 3 (Münster: Aschendorf, 1931): 253–85.

50. "Litterae Adm. R. P. N. Vincentii (Caraffae) ad R. P. Prov. (Rh. Sup.) Nithard Biberum. 3. Mart. 1649. Prohibetur sententia Zenonis de quantitate: Constare ex meris punctis," Pachtler, 3:75.

51. For a full listing of editions, see dB/S, vol. 1, cols. 578–81.

52. Rodrigo de Arriaga, S.J., "Prologus," in *Cursus philosophicus*... (Lyon, 1653).

53. Schott described the meeting in a letter to Kircher, 17 August 1658, APUG, 561: 284.

54. The text of the ban is in Pachtler, vol. 3. Schott's encyclopedia is *Cursus mathematicus, sive Absoluta omnium mathematicum disciplinarum encyclopaedia, in libros XXVIII. digesta*... (Würzburg, 1661). Schott outlined his problem in a letter to Kircher of 18 Jan 1659, APUG, 561: 287. The role of the emperor's authority in the publication of the work is emphasized in the frontispiece, which depicts a crowned female with mathematical instruments in her left hand and a book called *Cursus mathematicus* borne by a putto in her right. While a putto bearing a laurel crown and palm frond flies over her head, she presents the book to the enthroned emperor.

55. The ban does not seem to have been rigorously enforced; Steven Harris has informed me that forty-eight texts on various aspects of military science were published by Jesuit authors, the majority of them after 1648.

56. Joseph Pfriemb, S.J., *Apologia qua errores R.P. Andreae Gordon, O.S.B. contra philosophiam scholasticam*... (Mainz, 1748).

57. Ibid., 12.

58. Ibid., 13.

59. Cornaeus to Kircher, 22 October 1653, APUG, 567: 100.

60. General Goswin Nickel to Cornaeus, 15 November 1653, ARSI, Rhen. Sup. 2: 202. As letters sent to the Society's Roman hierarchy were not preserved after the late 1500s, the text of Cornaeus's letter is not extant.

61. Cornaeus, *Curriculum philosophiae*, 2:100–101. For more on this text, see Paul Richard Blum, "Sentiendum cum paucis, loquendum cum multis: Die aristotelische Schulphilosophie und die Versuchungen der Naturwissenschaften bei Melchior Cornaeus SJ," in *Aristoteles Werk*

und Wirkung, vol. 2, *Kommentierung, Überlieferung, Nachleben*, ed. Jürgen Wiesner (Berlin: de Gruyter, 1987).

62. Cornaeus, *Curriculum philosophiae*, 2:107.

63. Ibid., 2:110.

64. Ibid., 2:110.

65. Melchior Cornaeus, S.J., *Ens rationis Luthero-Calvinicum ab Aristotele redivivo explicatum* (Würzburg, 1659).

3. The Colleges

1. The image of Ignatius as an anti-Luther was deliberately created by Jerónimo Nadal after Ignatius's death. See O'Malley, *First Jesuits*, 272–80, and "Historiography."

2. *Mon. Ig.* 3:401–2 (letter 1721). Translation in Ignatius of Loyola, *Letters*, 232.

3. See Kink, *Universität zu Wien*, esp. 1:231–308, on the decay of the university. In the spring term of 1557, each of the twenty-three professors at the University of Vienna was supposed to give forty-two lectures. One gave twenty-three, three between fifteen and eighteen, eight from one to six, and eleven none whatsoever. In 1556 the Hebrew professor Dr. Plancus had "around four or five students," and the grammar professor Master Zadesius had "around three or four students, sometimes even more" (2:168–69).

4. See Arno Seifert, "Die jesuitische Reform: Geschichte der Artistenfakultät im Zeitraum 1570 bis 1650," in Boehm and Spörl, *Die Ludwig-Maximilians-Universität*, 65–89.

5. In Würzburg, for example, the bishop himself needed an armed escort merely to ride through the city to his cathedral. Lesch, *Neuorientierung der Theologie*, 15. On the decline of the arts faculty at Mainz, see Jürgen Steiner, *Die Artistenfakultät der Universität Mainz, 1477–1562: Ein Beitrag zur vergleichenden Universitätsgeschichte* (Stuttgart: Franz Steiner, 1989).

6. The best work on the establishment of Jesuit universities in Germany is Hengst, *Jesuiten an Universitäten*. Also useful is Otto Krammer, *Bildungswesen und Gegenreformation: Die Hohen Schulen der Jesuiten im katholischen Teil Deutschlands von 16. bis zum 18. Jahrhundert* (Würzburg: Vogel, 1988). On Canisius's role in persuading princes to endow colleges, see Brodrick, *Saint Peter Canisius*.

7. The subject of Jesuit economics is certainly worthy of further study. France Martin Dolinar's examination of the finances of the Jesuit college in Lubljana concludes that the college's resources did not extend much beyond a comfortable subsistence. France Martin Dolinar, *Das Jesuitenkolleg in Laibach und die Residenz Pleterje, 1597–1704* (Lubljana: Theologische Fakultät Laibach, 1977).

8. Quoted in Ernst Schubert, *Materielle und organisatorische Grundlagen der Würzburger Universitätsentwicklung, 1582–1821* (Neustadt a. d. Aisch: Degener, 1973), 57–61. The situation at the German colleges was very similar to that in France outlined by A. Lynn Martin, *The Jesuit Mind: The Mentality of an Elite in Early Modern France* (Ithaca: Cornell University Press, 1988).

9. For more detail, see Hengst, *Jesuiten an Universitäten*, and Krammer, *Bildungswesen und Gegenreformation*. There were examples of the Jesuits taking over existing universities in other countries, but nowhere was this as common as in the Holy Roman Empire.

10. See Hengst, *Jesuiten an Universitäten*, 88–91.

11. A brief account of these tensions at Ingolstadt is provided by Seifert, "Die jesuitische Reform," 65–89. Prantl is perhaps at his most vitriolic and inaccurate in his coverage of this period in his *Geschichte der Ludwig-Maximilians-Universität.*

12. Hengst, *Jesuiten an Universitäten,* 89–97; Seifert, "Die jesuitische Reform," 71–72.

13. The case of theology is not as neat as that of philosophy/arts because in some cases at older universities non-Jesuits continued to occupy some chairs of theology.

14. See Farrell, *Jesuit Code,* for the history of the development of the Jesuit humanities curriculum. With the completion of the *Mon. paed.,* this work is now somewhat outdated.

15. On Mainz, see Leo Just and Helmut Mathy, *Die Universität Mainz: Grundzüge ihrer Geschichte* (Maulheim: Mushak, 1965); Steiner, *Artistenfakultät.*

16. Kink, *Universität zu Wien,* 332.

17. Josef Kuckhoff, *Die Geschichte des Gymnasiums Tricoronatum* (Cologne, 1931).

18. Christian Bönicke, *Grundriß einer Geschichte von der Universität zu Wirzburg* (Würzburg, 1782); Wegele, *Geschichte der Universität Wirzburg.*

19. Specht, *Universität Dillingen.*

20. For details, see Hengst, *Jesuiten an Universitäten.*

21. The Sanctio Pragmatica of 1623 at Vienna, for example, guaranteed the Society the privilege of appointing all the professors of the philosophy faculty and of teaching however and whatever it wanted. Kink, *Universität zu Wien,* 357–61.

22. BHSA, GL Fasz 1489/39, contains material concerning the ongoing dispute between the Ingolstadt faculties of philosophy and law over this matter.

23. The Society opposed the biennium, and in most colleges it was not successfully introduced until well into the eighteenth century, even at Ingolstadt, where the state first attempted to introduce it in 1642.

24. ARSI, Rhen. Sup. 26 1: 76r. From the surviving records of the Society of Jesus at the ARSI, it is relatively easy to determine the staff of any Jesuit college in Germany. StAMz has records for the Upper Rhenish province, and the BHSA has those for the Upper German province.

25. Franz Eulenberg, *Die Frequenzen der deutschen Universitäten von ihrer Gründung bis zur Gegenwart* (Leipzig: Teubner, 1904).

26. Joannes Hartung, S.J., *Iter peripateticum triennali labore feliciter confectum, festivo carmine celebratum . . .* (Würzburg, 1713).

27. Franz Huter and Anton Haidacher, eds., *Die Matrikel der Universität Innsbruck,* 3 vols. (Innsbruck: Wagner, 1952–61), 1:xli–xlii.

28. Ibid., 3:xxxvii–xxxviii.

29. *Const.,* VII, 4, titled "Ways in Which the Houses and Colleges Can Help Their Fellow Man," lists ways Jesuit communities could help "souls" in the surrounding area.

30. As the proverb goes, "Bernard loved the valleys, Benedict the mountains, Francis the towns, but Ignatius the great cities." On the urban orientation of the Jesuits, see Thomas M. Lucas, S.J., *Landmarking: City, Church, and Jesuit Urban Strategy* (Chicago: Jesuit Way, 1997).

31. Trevor Johnson, "Blood, Tears and Xavier-Water: Jesuit Missionaries and Popular Religion in the Eighteenth-Century Upper Palatinate," in *Popular Religion in Germany and Central Europe, 1400–1800,* ed. Robert Scribner and Trevor Johnson (New York: St. Martin's, 1996).

32. BHSA, Jesuiten 118.

33. On thesis sheets, see Wolfgang Seitz and Bernard Schemmel, eds., *Die graphischen Thesen- und Promotionsblätter in Bamberg* (Wiesbaden: Harrassowitz, 2001); Werner Telesko, *Thesenblätter österreichischer Universitäten* (Salzburg: Das Museum, 1996); Wolfgang Seitz, "Die graphischen Thesenblätter des 17. und 18. Jahrhunderts: Ein Forschungsvorhaben über ein Spezialgebiet barocker Graphik," *Wolfenbütteler Barock-Nachrichten* 11 (1984): 105–14.

34. Such programs, containing the names of graduating students and the titles of the graduation speeches, survive at the University of Würzburg, although the collection is far from complete (WüUB, 36/D 20). The SBD preserves a complete set of the single-page prints advertising graduations from 1555 to 1759, giving the title of the graduation speeches (XIj 133/1 for 1555–1632, XIj/133/2 for 1633–1759).

35. WüUB, 36/D 20. This question also occurred frequently as a thesis in dissertations throughout the seventeenth century and into the eighteenth. The positive answer to this question was placed on the 1651 Ordinatio's list of banned opinions: "65. Cures of wounds or diseases can occur through medicaments applied at a distance or through an unguent that is called an armarium."

36. WüUB, 36/D 20. Dillingen's speeches clearly show an increasing concern with experiment as a method of natural philosophical investigation, as well as with the issue of integrating Aristotle and the modern philosophers; see SBD, XIj/133/2.

37. Biographical information on Friedrich Spee is taken from Michael Sievernich, S.J., *Friedrich von Spee, Preister-Poet-Prophet* (Frankfurt a. M.: Josef Knecht, 1986), 15–30; Theo G.M. van Oorschot, S.J., "Die Lebensdaten," in *Friedrich Spee im Licht der Wissenschaften: Beiträge und Untersuchungen*, ed. Anton Arens (Mainz: Selbstverlag der Gesellschaft für Mittelrheinische Kirchengeschichte, 1984).

38. *Const.*, VII, 4, § 9.

39. For biographical details on Jesuit philosophy professors at the University of Würzburg, see Karl-Heinz Logermann, "Personalbibliographien von Professoren der Philosophischen Fakultät der Alma Mater Julia Wirceburgensis: Mathematik, Experimentalphysik, theoretische Physik, Botanik, Chemie, Naturgeschichte von 1582–1803" (Ph.D. diss., Universität Erlangen-Nürnberg, 1970); Winfried Stosiek, "Die Personalbibliographien der Professoren der Aristotelischen Physik in der Philosophischen Fakultät der Alma Mater Julia Wirceburgensis von 1582–1773" (Ph.D. diss., Universität Erlangen-Nürnberg, 1972); and Gudrun Uhlenbrock, "Personalbibliographien der Professoren der Philosophischen Fakultät der Alma Mater Julia Wirceburgensis von 1582–1803" (Ph.D. diss., Universität Erlangen-Nürnberg, 1973). Fritz Krafft lists all the Jesuit professors at the University of Mainz, although without biographical detail, in "Jesuiten als Lehrer an Gymnasium und Universität Mainz und ihre Lehrfächer: Eine chronologisch-synoptische Übersicht, 1561–1773," in *Tradition und Gegenwart: Studien und Quellen zur Geschichte der Universität Mainz*, vol. 1, *Aus der Zeit der Kurfürstlichen Universität*, ed. Helmut Mathy and Ludwig Petry (Wiesbaden: Franz Steiner, 1977). On Ingolstadt professors, see Laetitia Boehm, ed., *Biographisches Lexikon der Ludwig-Maximilians-Universität München* (Berlin: Duncker & Humboldt, 1998).

40. *RS*, "Rules for the Provincial," 19, §§ 4, 11.

41. The *Ratio studiorum* specified different levels of theological training for members of the Society. Those who were to be occupied solely with pastoral duties, for example, received a shorter, more practical training. The full course consisted of four years of theological classes followed by two years of independent study. The 1599 *Ratio studiorum* states: "Professors of philosophy (unless pressing need demands otherwise) should not only have finished the course of theology but also have reviewed it for two years, so that their doctrine may be the more firmly established and serve theology the better. If there are any too prone to innovations, or too liberal in their views, they shall certainly be removed from the responsibility of teaching" (*RS*, "Rules for the Provincial," 16). For an early example of this principle, see Ledesma's 1564–65 "Relatio de professorum consultationibus circa Collegii Romani studia," composed after discussion with the professors of the Collegio Romano, which recommended that philosophy professors be trained theologians. *Mon. paed.*, 2:476.

42. Duhr, in *Geschichte der Jesuiten,* 3:417, concurs that professors usually taught the triennium only once in their careers.

43. The situation was different with mathematics professors due to differing career patterns, as will be discussed later.

44. General Mutius Vitelleschi, for example, had taught Physics at the Collegio Romano. His surviving lectures are discussed by William Wallace, "Causes and Forces in Sixteenth-Century Physics," *Isis* 69 (1978): 400–412. According to dB/S, Generals Carrafa, Piccolomini, Nickel, Oliva, Gonzalez, Tamburini, Gottifredi, Noyelle, Retz, Visconti, Centurione, and Ricci had all taught philosophy.

45. Eschweiler, "Roderigo de Arriaga."

46. Admittedly Kircher was not solely a mathematician in the strict sense and published much on natural philosophical topics, but his philosophy cannot be considered scholastic or peripatetic philosophy in the conventional sense, although it contained clear scholastic elements. See Thomas Leinkauf, "Athanasius Kircher und Aristoteles: Ein Beispiel für das fortleben Aristotelischen Denkens in fremden Kontexten," in *Aristotelismus und Renaissance,* ed. Eckhard Keßler, Charles H. Lohr, and Walter Sparn (Wiesbaden: Harrassowitz, 1988), on the scholastic elements in Kircher's natural philosophy, as well as Thomas Leinkauf, *Mundus Combinatus: Studien zur Struktur der barocken Universalwissenschaft am Beispiel Athanasius Kirchers SJ (1602–1680)* (Berlin: Akademie, 1993), for an ambitious attempt to understand the foundations of Kircher's worldview.

4. The Curriculum in the Seventeenth Century

1. However, it must be emphasized that Jesuit superiors were always to use their own initiative and take local conditions into account in all their decisions if that was more conducive to the greater good. See, e.g., *Const.*, IV, 10, § 5; IV, 13, § 2.

2. Ignatius reprimanded a young Jesuit he had sent to Paris to study when, against Ignatius's advice, he tried to enter advanced classes for which he was ill prepared.

3. Codina Mir, *Aux sources.*

4. On examinations, see *RS*, "Rules for the Prefect of Studies," 20–23, and *RS*, "Rules for the Writing of Examinations." There were complaints from outside the Society that exams were purely pro forma and that everyone passed them.

5. *RS*, "Rules for the Professor of Philosophy," 20. For details on who should dispute with whom and when, see *Const.*, IV, 6, §§ 10 and 11; IV, 6, § 3, and *RS*, "Rules for the Prefect of Studies," 6–19, and "Rules for the Professor of Philosophy," 17–20.

6. The *Ratio studiorum* is a curious document in that it is structured not by topics but by roles, such as rector, prefect of studies, and professor. The structure and content of the philosophical curriculum are discussed in the section of the *Ratio studiorum* entitled "Rules for the Professor of Philosophy," although other sections have relevant material.

7. *RS*, "Rules for the Provincial," 29.

8. Suàrez, preface to the *Metaphysicarum disputationum.*

9. *RS*, "Rules for the Professor of Philosophy," 9, §§ 1–6. Toledo, *Commentaria una cum quaestionibus in universam Aristotelis logicam;* Fonseca, *Institutionum dialecticarum.*

10. *RS*, "Rules for the Professor of Philosophy," 9, § 5.

11. *RS*, "Rules for the Professor of Philosophy," 10, §§ 1–3.

12. *RS*, "Rules for the Professor of Philosophy," 11, §§ 1–2.

13. For example, Joseph Vogler's disputation, *Corpus humanum, sive Disputatio physica de fabrica, nutritione, et vita partium corporis humani . . .* (Dillingen, 1696).

14. Toledo published, in addition to his logic commentary, commentaries on the *Physics* (1573), *De anima* (1574), and *De generatione et corruptione* (1575). Fonseca also published a commentary on the *Metaphysics.* Both Toledo and Fonseca's works were published numerous times in Germany. The commentaries of the fathers of the Jesuit college at Coimbra in Portugal, published between 1591 and 1606 on the entire Aristotelian corpus, were very influential, and most were printed several times in Germany. The commentaries of Antonio Ruvio, S.J. (1548–1615), which appeared in the early seventeenth century, enjoyed in general fewer editions in northern Europe but were nonetheless often cited in Jesuit texts. See dB/S for full bibliographies.

15. Suárez, *Metaphysicarum disputationum.* It displayed the eclecticism that was to characterize Jesuit philosophy. See Martin Grabmann, "Die Disputationes metaphysicae des Franz Suárez in ihrer methodischen Eigenart und Fortwirkung," in *Mittelalterliches Geistesleben: Abhandlungen zur Geschichte der Scholastik und Mystik* (Munich: Hueber, 1926), 525–60.

16. On the textbook tradition, see Mary Richard Reif, "Natural Philosophy in Some Early Seventeenth Century Scholastic Textbooks" (Ph.D. diss., Saint Louis University, 1962), and "The Textbook Tradition in Natural Philosophy, 1600–1650," *Journal of the History of Ideas* 30 (1969): 17–32; Paul Richard Blum, "Der Standardkursus der katholischen Schulphilosophie im 17. Jahrhundert," in Keßler, *Aristotelismus und Renaissance;* Charles B. Schmitt, "The Rise of the Philosophical Textbook," in Schmitt et al., *Cambridge History of Renaissance Philosophy,* and "Galilei and the Seventeenth-Century Text-Book Tradition," in *Novità celesti e crisi del sapere: Atti del Convegno Internazionale di Studi Galileiani,* ed. Paolo Galluzzi (Florence: Barbèra, 1984).

17. Outlines of Arriaga's work are provided by Bernhard Jansen, S.J., "Die scholastische Philosophie des 17. Jahrhunderts," *Philosophisches Jahrbuch der Görres-Gesellschaft,* 50 (1937): 401–44,

and "Die Pflege der Philosophie im Jesuitenorden während des 17./18. Jahrhunderts," *Philosophisches Jahrbuch der Görres-Gesellschaft* 51 (1938): 172–215, 344–66, 436–56. Eschweiler, "Roderigo de Arriaga," gives biographical information. Recently aspects of Arriaga's natural philosophical work have been examined by Dear, *Discipline and Experience.*

18. Arriaga, "Praefatio ad lectorem," in *Cursus philosophicus.*

19. Jansen, "Die scholastische Philosophie," 431, accurately describes Carleton as a solidly scholastic philosopher who opposed the eclecticism of his fellow Jesuits contemporaries such as Arriaga.

20. Cornaeus, in "Benevole lector," in *Curriculum philosophiae,* acknowledged that there had been many examples of the *cursus philosophicus* genre but believed with considerable justification that they were too difficult for new students. As far as I am aware, this was the first philosophical compendium published by a German Jesuit. On Cornaeus's *Curriculum philosophiae,* see Blum, "Sentiendum cum paucis."

21. Other Jesuit examples include Pedro Hurtado de Mendoza, *Disputationum in universam philosophiam* (Mainz, 1619); Francisco de Oviedo, *Cursus philosophicus* (Lyon, 1640); and Francisco Soares Lusitano, *Cursus philosophicus* (Coimbra, 1651). Hurtado and Oviedo were referred to frequently in German lectures and disputations, often to provide an opposing view to Arriaga.

22. On the history of *quaestio* as a didactic method, see Brian Lawn, *The Rise and Decline of the Scholastic "Quaestio Disputata": With Special Emphasis on Its Use in the Teaching of Medicine and Science* (Leiden: E. J. Brill, 1993).

23. Some are listed in Charles Lohr's *Latin Aristotle Commentaries,* vol. 2, *Renaissance Authors* (Florence: Leo S. Olschki, 1988), but there are many others. There is no overview of surviving Jesuit manuscripts and archival holdings in Germany.

24. *Const.,* IV, 6, § 8.

25. *RS,* "Rules for the Professor of Philosophy," 16.

26. *RS,* "Rules for the Prefect of Studies," 10.

27. For an overview of the genre and of the surviving holdings from all German universities, see Komorowski, "Die Hochschulschriften." Hundreds, perhaps even thousands, of philosophical disputations and dissertations defended at Jesuit college survive. The holdings of the University of Würzburg are listed in Mälzer, *Würzburger Hochschulschriften.* The Mainzer Stadtbibliothek has an excellent collection of academic disputations from throughout Germany; the Studienbibliothek Dillingen has a large collection of Dillingen dissertations; the Universitätsbibliothek München has a large number of Ingolstadt dissertations; and the Bayerische Staatsbibliothek has a huge collection from all over Germany.

28. Friederich Staudenhecht, S.J., *Summa vitae S. Francisci Xavierii . . .* (Würzburg, 1648).

29. For example, Adam Weber, S.J., *Universa philosophia quam assertionum peripateticorum tribus centuriis complexus . . .* (Würzburg, 1653).

30. Georg Saur, S.J., *Difficultates physicae Cartesianae, thesibus inauguralibus philosophicis propositae . . .* (Würzburg, 1705); Anton Nebel, S.J., *Phaenomena barometri. Quatuor dissertationibus illustrata . . .* (Würzburg, 1748).

31. For this reason I have consistently given the *praeses* as the author.

32. Acknowledgment of this point would have greatly tempered much of the criticism that has been leveled at Jesuit natural philosophy over the past two centuries, as Laetitia Boehm rightly points out in her "Universität in der Krise?"

33. "[A]ntiquam ad nauseum usque recoquunt crambem." Wolfgang Leinberer, S.J., *Disputatio philosophica de natura et perfectione mundi* . . . (Ingolstadt, 1670), 1.

34. Ibid., 1. It was obligatory for all students to attend the disputations (*RS,* "Rules for the Rector," 4), so they were a good occasion to motivate younger students to stay at the college for philosophical studies.

35. There are such massive amounts of literature on all the topics mentioned in this section that it is impossible to cite, let alone read, all of them. A fine overview of late scholastic natural philosophy is Des Chene, *Physiologia;* also worthwhile is Reif, "Natural Philosophy." An introduction to key terms in scholastic philosophy is provided by Josef de Vries, S.J., *Grundbegriffe der Scholastik* (Darmstadt: Wissenschaftliche Buchgesellschaft, 1980). Bernard Wuellner, S.J., *A Dictionary of Scholastic Philosophy,* 2d. ed. (Milwaukee: Bruce, 1966), is helpful but suffers in that it does not historicize its definitions.

36. "12. *The Texts of Aristotle.*—Let him especially endeavor to interpret well the text of Aristotle, and let him give no less effort to this interpretation than to the questions themselves. Let him also persuade his students that their philosophy will be but very partial and mutilated unless they highly esteem this study of the text." *RS,* "Rules for the Professor of Philosophy," 12. Paras. 13, 14, and 15 then detail how Aristotle is to be taught. The 1586 and 1591 versions of the *Ratio* specified that both the professors and students had to bring the text of Aristotle to class, but this was omitted from the 1599 version.

37. *Mon. paed.,* 7:445.

38. Schmitt, *Aristotle and the Renaissance.*

39. Cong. Gen. V, Decr. 41: "Regulae pro delectu opinionum pro philosophis," 2 and 3.; *RS,* "Rules for the Professor of Philosophy," 3 and 4. The *Constitutions* (IV, 14, § 1) stated that books by suspect authors should not be read even if they "may be free of all evil teaching . . . for it rarely happens that some poison is not mixed in with that which comes from a heart filled with poison."

40. ARSI, Instit. 213: 62r.

41. Kaspar Knittel, S.J., *Aristoteles curiosus et utilis in quo centum praecipuae quaestiones peripateticae problematicè disputantur, ac utriusque partis argumenta dissolvuntur* . . . (Prague, 1682).

42. *Mon. paed.,* 6:9. See chapter 1 for more details on the critique of textbooks.

43. *RS,* "Rules for the Professor of Philosophy," 6, states that the philosophy professor may disagree with Thomas if he does so reverently. For a juxtaposition of the key points of disagreement between Suárez and Thomas, see the appendix to Guido Mattiussi, S.J., *Le XXIV Tesi della Filosofia di S. Tommaso d'Aquino* (Rome: Università Gregoriana, 1925).

44. Bernhard Jansen, S.J., particularly lamented this movement away from the scholastic doctors, which, he claimed, went hand in hand with a declining level of speculative sophistication in Jesuit philosophy. See, for example, his "Die Pflege der Philosophie."

45. See Thomas Aquinas, *The Division and Methods of the Sciences Questions V and VI of His Commentary on the "De Trinitate" of Boethius,* trans. Armand Maurer (Toronto: Pontifical Institute of Medieval Studies, 1986).

46. As one dissertation put it: "Metaphysica abstrahit ab omni materia re & ratione, Mathematica a materia universali & singulari, non tamen ab intelligibili, Physica tantum a materia singulari." Engelbert Ewick, S.J., *Oculus naturae pro discernendis erroribus, detegendis imposturibus, declinandis terroribus* (Würzburg, 1631), 1. What kind of science logic was had been contested since antiquity; it provided the tools the philosopher needed to pursue truth in the speculative sciences and so fell between the practical and the speculative.

47. Since the object of physics was natural bodies, which were transitory and mutable, there was some discussion as to whether physics could be a science—that is, demonstrable, immutable knowledge. The Jesuits unanimously agreed that physics was a science, as the principles of natural bodies were immutable even if the bodies themselves were not. The question of whether physics was a science was treated in Nicolaus Wysing, S.J., *Scientia physica actualis ad amussim Aristotelis demonstrativam exacta* . . . (Ingolstadt, 1631).

48. Arriaga, *Cursus philosophicus,* 257. Laurence W. B. Brockliss, "Pierre Gautruche et l'enseignement de la philosophie de la nature dans les colleges jésuites français vers 1650," in Giard, *Les jésuites,* 190, writes that the French Jesuit Pierre Gautruche also asserted the independent existence of matter and form, contrary to Thomas, in his *Institutio totius philosophiae.* This would indicate that there was widespread agreement within the Society on this point. Nevertheless, this opinion was placed on the list of opinions banned by the 1651 *Ordinatio pro studiis superioribus.* Gautruche also favored the Scotist view that bodies consisted of numerous forms, rather than the Thomist view of a unique substantial form in each body.

49. Georg Bernard, S.J., *Assertiones philosophicae de corpore naturali principato* . . . (Ingolstadt, 1629), 6.

50. Ibid., 7.

51. For Arriaga's discussion of this complex issue, see his *Cursus philosophicus,* 379–92.

52. Reif has identified twelve different uses of the term *nature* and nine of *natural* in seventeenth-century philosophy manuals. Reif, "Natural Philosophy," 151–69.

53. Maximilian Schmid, S.J., *Veritatum peripateticarum pentadecas physica* . . . (Würzburg, 1623).

54. Friderich Geiger, S.J., *Theses inaugurales philosophicae: Sapientia mundi* . . . (Würzburg, 1694).

55. Fritz Krafft provides an excellent treatment of this issue in "'Mechanik' und 'Physik' in Antike und beginnender Neuzeit" and "Wie die 'Physik' zur Physik wurde," in his *Das Selbstverständnis der Physik im Wandel der Zeit* (Weinheim: Chemie Verlag, 1982). See chapter 5 for a discussion of this issue.

56. "Motus est *actus entis in potentia quatenus est in potentia.*" Weber, *Universa philosophia,* 13. There were a vast range of translations of Aristotle's definition of motion, due in no small part to the fact that Aristotle gave three definitions, which, although essentially identical, differed in small ways.

57. See Kaspar Lechner, S.J., *Disputatio philosophica de proprietatibus compositi physici* . . . (Ingolstatdt, 1616), 3, for a standard schematization of the types of change.

58. There were many exceptions to this. Arriaga, for example, treated the seventh and eighth books but limited his discussion to the single question of whether any creature in general and the world in particular could exist from eternity. Arriaga, *Cursus philosophicus,* 493–95.

59. *RS*, "Rules for the Professor of Philosophy," 10, § 3.

60. Peripatetic meteorology was strikingly combined with emblematics in Franz Reinzer, S.J., *Meteorologia philosophico-politica in duodecim dissertationes per quaestiones meteorologicas et conclusiones politicas divisa, appositisque symbolis illustrata . . .* (Augsburg, 1709). The work is discussed by Christoph Meinel, "Natur als moralische Anstalt: Die 'Meteorologia philosophico-politica' des Franz Reinzer, S.J.: Ein naturwissenschaftliches Emblembuch aus dem Jahre 1698," *Nuncius* 1 (1987): 37–94.

61. Kaspar Manz, *Iudicium super illa quaestione: Utrum dari possit melior et Christianae pietati conformior modus docendi philosophiam, quam sit vulgaris?* (Ingolstadt, 1648). Reprinted in Arno Seifert, "Der jesuitische Bildungskanon im Lichte zeitgenossischer Kritik," *Zeitschrift für bayerische Landesgeschichte* 47 (1984): 43–75.

62. Denz., para. 1440.

63. The Lateran Council's decree is explicitly cited on this issue by Bernard Frey, S.J., *Controversiae physicae de corpore naturali simplici . . .* (Ingolstadt, 1644).

64. On this question, see Edward Grant, *In Defense of the Earth's Centrality and Immobility: Scholastic Reaction to Copernicanism in the Seventeenth Century* (Philadelphia: American Philosophical Society, 1984).

65. See Amos Funkenstein, *Theology and the Scientific Imagination from the Middle Ages to the Seventeenth Century* (Princeton: Princeton University Press, 1986).

66. *ST*, 3a. 75, 4.

67. See G. R. Evans, "Theology: The Vocabulary of Teaching and Research, 1300–1600: Words and Concepts," in *Vocabulary of Teaching and Research between Middle Ages and Renaissance*, ed. Olga Weijers (Turnhout: Brepols, 1995).

68. See, for example, Cornaeus, *Curriculum philosophiae*, 373.

69. Ibid., 376.

70. The questions he considered include "Posse duo corpora esse in eodem loco; et idem corpus divinitus simul in duplici loco" and "Utrum corpus existens in duplici loco debeat habere utrobique eodem accidentia, et easdem imutationes." Juan de Lugo, S.J., *Disputationes scholasticae et morales . . .* (1638; reprint, Paris: Vives, 1869), 3:565–601.

71. Martin Becanus, S.J., *Theologia scholasticae* (Mainz, 1628), pt. 3, tr. 2, 222–38.

72. Cornaeus, *Curriculum philosophiae*, 373–74.

73. Edward Grant, *Much Ado about Nothing: Theories of Space and Vacuum from the Middle Ages to the Scientific Revolution* (Cambridge: Cambridge University Press, 1981); and Cees Leijenhorst, "Jesuit Conceptions of *Spatium Imaginarium* and Thomas Hobbes' Doctrine of Space," *Early Science and Medicine* 1 (1996): 355–80.

74. On the barometer, see W. E. Knowles Middleton, *The History of the Barometer* (Baltimore: Johns Hopkins University Press, 1968). On Jesuit reactions to the barometer, see Michael John Gorman, "Jesuit Explorations of the Torricellian Space: Carp Bladders and Sulphurous Fumes," *Mélanges de l'Ecole Française de Rome* 106 (1994): 7–32.

75. Dear's important work on the shift from experience to experiment draws heavily on Jesuit sources and highlights the Jesuits' important role in this process. See Dear, *Discipline and Experience*, "Jesuit Mathematical Science and the Reconstitution of Experience in the Early Seventeenth Cen-

tury," *Studies in the History and Philosophy of Science* 18 (1987): 133–75, and "Narratives, Anecdotes, and Experiments: Turning Experience into Science in the Seventeenth Century" in *The Literary Structure of Scientific Argument,* ed. Peter Dear (Philadelphia: University of Pennsylvania Press, 1991).

5. The Physics of the Eucharist

1. "Rules for Thinking with the Church," in Ignatius of Loyola, *The Spiritual Exercises of Saint Ignatius,* trans. Louis J. Puhl (New York: Vintage Books, 2000). On the frequency of communion in the Church, see Robert Taft, "The Frequency of the Eucharist throughout History," *Concilium* 152 (1982): 13–24. On the Jesuits' advocacy of frequent communion, see O'Malley, *The First Jesuits,* 152–57, and Joseph de Guibert, S.J., *The Jesuits: Their Spiritual Doctrine and Practice. A Historical Study,* trans. William J. Young, S.J. (Chicago: Institute for Jesuit Sources, 1964), 374–85.

2. *PCE,* 7:569 (sermon in Freiburg, 30 May 1583). In 1574 Canisius wrote a lengthy response to the Franciscan Jacobus Pontanus, who had criticized the Jesuits' practice of frequent communion (*PCE,* 7:710–12).

3. *PCE,* 8:720.

4. Denz., para. 1638.

5. Peter Canisius, S.J., *A Summe of Christian Doctrine* (n.p., n.d. [1592–96]; reprint, Menston: Scholar Press, 1971), 191–92. Page citations are the same in both editions.

6. For a concise account of Protestant and Catholic eucharistic theology, see Jaroslav Pelikan, *The Christian Tradition: A History of the Development of Doctrine,* vol. 4, *Reformation of Church and Dogma (1300–1700)* (Chicago: University of Chicago Press, 1984).

7. Denz., para. 1651.

8. Ignatius of Loyola, *The Autobiography of St. Ignatius Loyola with Related Documents,* ed. John C. Olin, trans. Joseph F. O'Callaghan (New York: Harper & Row, 1974), 38.

9. *ST,* 3a. 76, 1.

10. *ST,* 3a. 76, 1, ad 3.

11. Note that for Thomas dimension is the same as quantity. Thomas says that the substance of Christ maintains its dimensions through "natural concomitance" but that the dimensions are not there locally but in the way of substance. *ST,* 3a. 76, 4.

12. This had in effect been Berengar's claim in the eleventh century, which had prompted the development of a philosophically articulated account of transubstantiation based on accidents without subject. Berengar's arguments would be raised repeatedly against the orthodox account throughout the centuries. On Berengar, see Burkhard Neunheuser, O. S. B., *Eucharistie in Mittelalter und Neuzeit* (Freiburg: Herder, 1963).

13. *ST,* 3a. 77, 1.

14. *ST,* 3a, 77, 1.

15. *ST,* 3a. 77, 2.

16. According to James F. McCue, "The Doctrine of Transubstantiation from Berengar through Trent: The Point at Issue," *Harvard Theological Review* 61 (1968): 385–430, many theologians thought both before and after the Fourth Lateran Council of 1215 that consubstantiation

(i.e., the doctrine that the substance of the bread remained along with Christ's substance) was at least philosophically more probable than transubstantiation, and only Lateran's affirmation of transubstantiation prevented orthodox theologians from supporting consubstantiation's theological correctness. See also Gary Macy, "The Dogma of Transubstantiation in the Middle Ages," *Journal of Ecclesiastical History* 45 (1994): 11–41.

17. Ockham's discussion of quantity and the Eucharist can be found in William of Ockham, *Tractatus de quantitate et Tractatus de corpore Christi,* ed. Carolus A. Grassi (St. Bonaventure, N.Y.: St. Bonaventure University, 1986), and William of Ockham, *Quodlibetal Questions* 2 vols., trans. Alfred J. Freddoso and Francis E. Kelley (New Haven: Yale University Press, 1991). For discussions of Ockham's theory of quantity and the Eucharist, see Gabriel N. Buescher, O. F. M., *The Eucharistic Teaching of William Ockham* (Washington, D.C.: Catholic University of America Press, 1950), and Gordon Leff, "Ockham and Wyclif on the Eucharist," *Reading Medieval Studies* 2 (1976): 1–13.

18. Buescher, *Eucharistic Teaching,* xix–xxv, 1. Ockham and the Franciscans defended their view as being completely compatible with the real presence.

19. *RS,* "Rules for the Professor of Scholastic Theology," 2. But the instructions moderated Thomas's authority by advising professors that "they are not confined to him so closely that they are never permitted to depart from him in any matter, since even those who especially profess to be Thomists occasionally depart from him, and it would not befit the members of our Society to be bound to St. Thomas more tightly than the Thomists themselves."

20. Suárez, *Metaphysicarum disputationum,* disp. 40, S. 2, § 8.

21. Anticipating the Nominalist objection, Suárez went on to argue that a modal distinction does not suffice: substance cannot be a mode of quantity nor quantity a mode of substance, for a mode cannot be separated from that of which it is a mode. For a more detailed consideration of Suárez's treatment of this issue, see Des Chene, *Physiologia,* 97–109.

22. Arriaga, *Cursus philosophicus,* 625–26.

23. Jakob Scholl et al., *Physica disputatio de quantitate, loco, et tempore . . .* (Würzburg, 1594).

24. Lechner, *Disputatio philosophica.*

25. There may have been more that are not preserved. Those that are preserved are in the ARSI (with Fondo Gesuitico [FG] volume and page number): 14 July 1612 (656 AII: 378); 4 April 1615 (656 AII: 456); 28 April 1620 (656 AII: 582); 10 November 1620 (660: 67); undated, but probably around early 1620s (656 AII: 623); 25 February 1621 (659: 1); 27 August 1621 (659: 3); 3 August 1623 (656 AII: 633); 1624 (656 AII: 643); 16 January 1627 (656 AII: 784); 27 June 1636 (657: 289); 26 January 1649 (657: 475); 3 February 1649 (657: 480); 12 February 1649 (657: 483–90, prop. 17); and thereafter not again until 21 November 1748 (674: 259). The lack of such condemnations after 1651 is striking.

26. ARSI, FG 656 A II: 378.

27. ARSI, FG 656 A II: 456.

28. ARSI, FG 656 A II: 643.

29. Pachtler, 3:91.

30. Cornaeus, *Curriculum philosophiae,* 142.

31. Ibid., 143.

32. Denz., paras. 1652 and 1653.

33. Edward Schillebeeckx, O.P., *The Eucharist* (New York: Sheed and Ward, 1968), 55–58.

34. Denz., paras. 690 and 700.

35. The Fourth Lateran Council in 1215 was the first council to use the term *transubstantiation,* but again it devoted little attention to the question of species. The council did employ the term *species* in a series of doctrinal definitions against the Albigensian and Cathar heretics. But the council did not specifically consider which explanation of the change was the most philosophically and theologically sensible (Denz., para. 802).

36. Under the Tridentine profession of faith (13 November 1564), the faithful had only to state that the whole Christ was under each of the species, not that the species were accidents. Denz., para. 1866.

37. F. Jansen, "Eucharistiques (accidents)," in *Dictionnaire de theologie catholique,* vol. 5, cols. 1421–20. See Darwell Stone, *A History of the Doctrine of the Holy Eucharist* (London: Longmans, 1909), 360–74, for an overview of major Jesuit theologians' thought on the Eucharist.

38. Costantini, *Baliani e i gesuiti;* Redondi, *Galileo Heretic.*

39. See Middleton, *History of the Barometer;* Lino Conti, "Galileo and the Ancient Dispute about the Weight of the Air," in *Die Schwere der Luft in der Diskussion des 17. Jahrhunderts,* ed. Wim Klever (Wiesbaden: Harrassowitz, 1997).

40. This was in his *Cursus philosophicus concinnatus ex notissimis cuique principiis . . .* (Toulouse, 1653). For a translation of the relevant passage, see Middleton, *History of the Barometer,* 10–15.

41. Paolo Casati, S.J., *Vacuum proscriptum disputatio physica . . . in qua nullum esse in rerum natura vacuum ostenditur* (Genoa, 1649), 4–7, cited in Costantini, *Baliani e i Gesuiti,* 93–94.

42. See Gorman, "Jesuit Explorations" and Redondi, *Galileo Heretic,* ch. 9. In his important experimental treatise on light, the Jesuit Francesco Maria Grimaldi proposed many arguments in favor of the substantiality of light. Francesco Maria Grimaldi, S.J., *Physico-mathesis de lumine, coloribus et irida: aliisque adnexis, libri duo . . .* (Bologna, 1665).

43. Athanasius Kircher, *Musurgia universalis, sive Ars magna consoni et dissoni in X. libros . . .* (Rome, 1650), 11.

44. On Descartes and transubstantiation, see J.-R. Armogathe, *Theologia cartesiana: L'explication physique de l'Eucharistie chez Descartes et dom Desgabets* (The Hague: Nijhoff, 1977), 29–34; Stephen Menn, "The Greatest Stumbling Block: Descartes' Denial of Real Qualities," in *Descartes and His Contemporaries: Meditations, Objections, and Replies,* ed. Roger Ariew and Marjorie Greene (Chicago: University of Chicago Press, 1995); Roger Ariew, "Descartes and the Jesuits of La Flèche: The Eucharist," in *Descartes and the Last Scholastics* (Ithaca: Cornell University Press, 1999).

45. Antoine Arnauld, *On True and False Ideas; New Objections to Descartes' Meditations; and Descartes' Replies,* trans. Elmar J. Kremer (Lewiston, N.Y.: E. Mellen, 1990), 187–88. On the *Meditations* and the objections to them, see the essays in Ariew and Greene, *Descartes and His Contemporaries.* Of greatest interest on Arnauld and his objections are the essays by Vincent Carraud, "Arnauld: From Ockhamism to Cartesianism," and Menn, "Greatest Stumbling Block."

46. Arnauld, *True and False Ideas,* 190.

47. René Descartes, *The Philosophical Works of Descartes,* ed. and trans. Elizabeth S. Haldane and G. R. T. Ross (Cambridge: Cambridge University Press, 1911–12), 2:95. The original French is in

René Descartes, *Oeuvres de Descartes,* ed. Charles Adam and Paul Tannery (Paris: Cerf, 1897–1913), 10:169–70.

48. Descartes, *Philosophical Works,* 2:118.

49. Ibid., 2:118.

50. Ibid., 2:119.

51. The second edition, published in Amsterdam, did not require an imprimatur, so Descartes restored the passages from his response to Arnauld that Mersenne feared would offend the "doctors." Ariew, "Descartes and the Jesuits," 149–50.

52. Descartes, *Philosophical Works,* 2:120. This was in effect the view of the heretic Berengar in the eleventh century.

53. Ibid., 2:121.

54. Ibid., 2:121–22. The original Latin, from which this English translation is derived, is somewhat different from the French edition of Adam and Tannery, *Oeuvres de Descartes,* 10:197.

55. Thomas Compton Carleton, S.J., *Philosophia universa* (Antwerp, 1649), preface.

56. Ibid., 238.

57. "[A]ccidentia omnia Physica, & realia." Ibid., 245.

58. Ibid., 247.

59. Denz., para. 1251.

60. Carleton, *Philosophia universa,* 253.

61. Pachtler, 3:77–97.

62. Weber, *Universa philosophia,* 1.

63. "58. Gegen die Kartesianische Philosophie. 1687," Pachtler 3:121.

64. Giovanni Baptista Tolomei, S.J., *Philosophia mentis et sensuum secundum utramque Aristotelis methodum pertracta metaphysice, et empirice . . .* (Augsburg, 1698), 371.

65. This was a reference to the eucharistic theory of the Minim Emmanuel Maignan. Geiger, *Theses inaugurales philosophicae,* 80–88.

66. Saur, *Difficultates physicae Cartesianae,* preface.

67. They were Descartes's system of the world, his hypothesis on the creation of the world, his conception of the intrinsic principles of natural bodies, his account of secondary causes and the laws of motion, his explanation of the sensible states of bodies, and finally his view of the generation, life, and souls of animals. For late scholastic responses to Descartes's philosophy, see Ariew, *Descartes and the Last Scholastics,* especially chs. 8 and 9, and "Quelques condamnations du cartèsianisme: 1662–1706," *Bulletin cartésien XXII, Archives de Philosophie* 57 (1994): 1–6.

68. Saur, *Difficultates physicae Cartesianae,* 36.

69. Georg Saur, S.J., *Relatio historica judiciorum et censurarum adversus philosophorum antiperipateticum; subjunctis thesibus ex universa philosophia peripatetica* (Würzburg, 1708).

70. Ibid., 3–4.

71. Ibid., 95.

72. See F. Jansen, "Eucharistiques (accidents)."

73. Saur, *Relatio historica,* 16.

74. Ibid., 152.

6. The Tension between Mathematics and Physics

1. See Eberhard Knobloch, "Klassifikationen," in *Maß, Zahl und Gewicht: Mathematik als Schlüssel zu Weltverständnis und Weltbeherrschung,* ed. Menso Folkerts (Weinheim: VCH, Acta Humaniora, 1989).

2. The title is somewhat misleading, as the work provides considerably more material than could be covered in the time the triennium assigned to mathematics (approximately one hour per day during the second year).

3. Plutarch, *The Lives of the Noble Grecians and Romans,* trans. John Dryden (New York, n.d.), 378.

4. See Fritz Krafft, "'Mechanik' und 'Physik' in Antike und beginnender Neuzeit," in *Das Selbstverständnis der Physik.*

5. Aristotle, "Mechanical Problems," in *Minor Works,* trans. W. S. Hett (Cambridge, Mass.: Harvard University Press, 1963). For Herodotus, *techne* was impious meddling with nature (Krafft, "'Mechanik' und 'Physik,'" 64–65). Clearly this was not the only intellectual tradition available in the Middle Ages and Renaissance: alchemists, and practitioners of natural magic in general, for example, insisted that art aped nature and could reproduce its effects.

6. Paolo Mancosu, *Philosophy of Mathematics and Mathematical Practice in the Seventeenth Century* (Oxford: Oxford University Press, 1996).

7. Aquinas, *Division and Methods,* 21.

8. Robert S. Westman, "The Astronomer's Role in the Sixteenth Century: A Preliminary Study," *History of Science* 18 (1980): 108–9.

9. Bernardino Baldi published an Italian translation of Hero's *Automata* in 1589, and Francisco Barocio published a Latin edition of Hero's work on siege engines in 1572. But the most widespread of Hero's works was the *Pneumatics,* of which there are more than a hundred extant medieval and Renaissance manuscripts (see "Hero of Alexandria" in Gillispie, *Dictionary of Scientific Biography*).

10. Westman, "Astronomer's Role," 120–23, Robert S. Westman, "The Melanchthon Circle, Rheticus, and the Wittenberg Interpretation of the Copernican Theory," *Isis* 66 (1975): 164–93.

11. Agostino Ramelli, *Schatzkammer mechanischer Künste* (n.p, 1620), preface.

12. Ibid., vi.

13. Joseph Furttenbach, *Mechanische Reißladen, das ist ein gar geschmeidigte bey sich verborgen tragende Laden . . .* (Augsburg, 1644), 19.

14. Ramelli, *Schatzkammer,* ii verso.

15. W. R. Laird, "The Scope of Renaissance Mechanics," *Osiris,* 2d ser., 2 (1986): 43–68. See also Paul Lawrence Rose and Stillman Drake, "The Pseudo-Aristotelian *Questions of Mechanics* in Renaissance Culture," *Studies in the Renaissance* 18 (1971): 65–104.

16. William Wallace, "Aristotelian Influences on Galileo's Thought," in *Galileo, the Jesuits,* 368–69.

17. *Const.,* IV, 12, 3, C.

18. Christopher Duffy, *Siege Warfare: The Fortress in the Early Modern World, 1494–1660* (London: Routledge, 1979), 235.

19. See Otto Cattenius's lectures in Albert Krayer, *Mathematik im Studienplan der Jesuiten: Die Vorlesungen von Otto Cattenius an der Universität Mainz (1610/11)* (Stuttgart: Franz Steiner, 1991).

20. A. C. Crombie, "Mathematics and Platonism in the Sixteenth-Century Italian Universities and in Jesuit Educational Policy," in *Prismata: Naturwissenschaftsgeschichtliche Studien. Festschrift für Willy Hartner,* ed. Y. Maeyama and W. G. Saltzer (Wiesbaden: Franz Steiner, 1977), 67–68.

21. Giuseppe Biancani, *De mathematicarum natura dissertatio* (an appendix to *Aristotelis loca mathematica*), trans. Gyula Klima, in Mancosu, *Philosophy of Mathematics.*

22. See his three memoranda of 1581, 1582, and 1593 on improving the teaching of mathematics in the Society in *Mon. paed.,* 7:109–21. That of 1582, namely "Modus quo disciplinae mathematicae in scholis Societatis possent promoveri," is translated in Crombie, "Mathematics and Platonism," 65–66. On Clavius's mathematics, in particular his astronomy and cosmology, see James M. Lattis, *Between Copernicus and Galileo: Christoph Clavis and the Collapse of Ptolemaic Cosmology* (Chicago: University of Chicago Press, 1994).

23. Censura of Fathers Le Röy, Dunellus, and Arbico, 4 May 1654, ARSI, FG 661: 32.

24. Brunellus to General Nickel, 7 May 1657, ARSI, FG 661: 30.

25. Censura of Kaspar Schott, 21 May 1655, ARSI, FG 661: 200.

26. Censura of Gabriel Beati, 27 April 1657, ARSI, FG 661: 201.

27. BSB, Clm 27322/II: 69–75. "Puncta ad Facultatem Artisticam inter ipsos nostros professores controversa."

28. Cited in Seifert, "Der jesuitische Bildungskanon," 53. A Spanish professor in the Jesuit college in Vienna stated that "[f]ew people are drawn to [mathematics]. As for myself, I think that it is little suited to the Institute of the Company and since I see that I am almost alone in being concerned with this science—where, in my opinion, there are many vain and useless things—I was moved by the desire to leave aside mathematics and to apply myself to ordinary studies." Cited in Westman, "Astronomers' Role," 131.

29. SBD, XV 244, contains the correspondence of several German Jesuit mathematicians.

30. Krafft, "Jesuiten als Lehrer."

31. Dates taken from Karl Adolf Franz Fischer, "Jesuiten Mathematiker in der deutschen Assistenz bis 1773," *AHSI* 47 (1978): 159–224.

32. Kochanski to Kircher, 14 April 1664, APUG, 563: 79.

33. Mathematicians were also often responsible for teaching ethics in the third year of the triennium. Others performed mathematical duties in the missions. The most prominent German example is Adam Schall, who attained high rank as the mandarin directing the Chinese imperial observatory.

34. Krayer, *Mathematik im Studienplan,* provides an excellent edition of Cattenius's lectures.

35. *RS,* "Rules for the Professor of Philosophy," 10, § 2.

36. Dear, *Discipline and Experience,* "Jesuit Mathematical Science," and "Narratives, Anecdotes, and Experiments"; see also Peter Dear, "Miracles, Experiments, and the Ordinary Course of Nature," *Isis* 81 (1990): 663–83.

37. Dear, *Discipline and Experience,* 13.

38. Ibid., 41.

39. On this debate both among Jesuits and in the wider astronomical community, see William H. Donahue, "The Solid Planetary Spheres in Post-Copernican Natural Philosophy," in *The Copernican Achievement,* ed. Robert S. Westman (Berkeley: University of California Press, 1975).

40. Aristotle, *De Caelo/On the Heavens* (Cambridge, Mass.: Harvard University Press, 1971), 293a.

41. Edward Grant, *Planets, Stars, and Orbs: The Medieval Cosmos, 1200–1687* (Cambridge: Cambridge University Press, 1994), 345. For coverage of the question among Jesuits in other countries, see W. G. L. Randles, "Le ciel chez le jésuites espagnols et portugais (1590–1651)," and Lerner, "L'entrée de Tycho," both in Giard, *Les jésuites.*

42. The text of Bellarmine's lectures is in Ugo Baldini and George V. Coyne, S.J., *The Louvain Lectures (Lectiones Louvanienses) of Bellarmine and the Autograph Copy of his 1616 Declaration to Galileo,* Studi Galileiani 1/2 (Vatican City: Specola Vaticana, 1984).

43. Censura of Joannes Camerota, 15 September 1614, ARSI, FG 662: 162–63.

44. BSB, Clm 1972: 110.

45. For example, the provincial congregation of the Upper German province ruled against the opinion in 1636. BHSA, Jesuiten 84: section dated 136.

46. BHSA, Jesuiten 651: 34.

47. Cattenius, reprinted in Krayer, *Mathematik im Studienplan,* 284.

48. Ibid.

49. ARSI, FG 656 AII: 643.

50. Johann Baptista Cysat, S.J., *Mathemata astronomica de loco, motu, magnitudine, et causis cometae qui sub finem anni 1618 et initium anni 1619 in coeli fulsit* (Ingolstadt, 1619), 34.

51. For a listing of surviving German Jesuit mathematical correspondence, including this file at the BSB, see Ernst Zinner, *Entstehung und Ausbreitung der Copernicanischen Lehre* (Erlangen: Mencke, 1943), 484–92.

52. Significantly, König repeatedly used the terms *demonstrare* and *demonstratio,* indicating the certainty of the conclusions derived from observations. König to Cysatus, 22 March 1619, BSB, Clm 1609: 8/9a. This letter is also reprinted in August Ziggelaar, S.J., "Jesuit Astronomy North of the Alps: Four Unpublished Letters, 1611–1620," in *Cristoph Clavius e l'attivita scientifica dei Gesuiti nell'eta di Galileo: Atti del Convegio Internationale, Chieti 28–30 aprile 1993,* ed. Ugo Baldini (Rome: Bulzoni, 1995), 128–30.

53. Georg Gobat, S.J., *Mundus ad disputationem publicam in florentissimo Friburgensi Helvetiorum gymnasio . . .* (Bern, 1634), 10.

54. Aristotle, *Metaphysics,* XII. 8. 1073b.

55. Udr, *Stellae vigiliis,* 21. Despite being published so shortly after the trial of Galileo, this dissertation contains many approving references toward Copernicus, although none on heliocentrism itself. Mention of Galileo is, however, scrupulously avoided, even in the sections dealing with the blemishes on the moon and the satellites of Jupiter.

56. Ibid., 18.

57. Johannes Feirabent, S.J., to Cysat, 3 Feb 1620, BSB, Clm 1609: 15.

58. Carleton, *Philosophia universa,* 400. On Clavius's view of Tycho, see Lattis, *Between Copernicus and Galileo,* and Lerner, "L'entrée de Tycho."

59. Carleton, *Philosophia universa,* 400.

60. Ibid. Bellarmine's letter to Foscarini is translated in *The Galileo Affair: A Documentary History,* ed. Maurice A. Finnochiaro (Berkeley: University of California Press, 1989), 67–69.

61. Conrad Calmelet, S.J., *Coeli coelorumque virtutes . . .* (Ingolstadt, 1642).

62. "Can you, like him [God], spread out the skies, hard as a molten mirror?" Job 37:18.

63. The author provided supporting scriptural and patristic passages in his footnotes. Calmelet, *Coeli,* 5.

64. Ibid., 6.

65. Ibid., 6.

66. Grant, *Planets, Stars, and Orbs,* 349.

67. See Dear, *Discipline and Experience,* 172–73. Grimaldi's *Physico-mathesis de lumine* is another example.

68. Dear, *Discipline and Experience,* 69.

69. Christophor Haunold, *De ortu et interitu theoremata physica mathematicis permixta . . .* (Dillingen, 1645), 1.

70. Haunold cited Blancanus's work approvingly in his *Acroamata physica exotericis medicis permixta . . .* (Dillingen, 1645).

71. Haunold, *De ortu et interitu,* 36.

72. Ibid., 38.

73. Zeno's indivisibles had many times come to the attention of the revisers general. This was a very perplexing and disputed issue, and Haunold devoted considerable attention to it in his *Acroamata physica.*

74. Cornaeus, *Curriculum philosophiae.* The term *reductis* implies that Cornaeus believed he was seeking to return to an older form of knowledge in which physics and mathematics were united.

75. "Argumenta mea statica sunt, et ex statica deprompta, neq[ue] adeo vix eorum [obscure] potest ab iis qui eam Mathesios partem plane ignorant." Cornaeus to Kircher, 22 October 1653, APUG, 567: 100. Copernicus's statement is in the dedication to *De revolutionibus orbium coelestium.*

7. The Peregrinations of the Pump

1. Kircher recounted the vision in his autobiography, reproduced in Hieronymus Ambrosius Langenmantel, *Fasciculus epistolarum Adm. R. P. Athanasii Kircheri Soc. Jesu viri in mathematicis et variorum idiomatus scientiis celebratissimi . . .* (Augsburg, 1684), 38–39, reprinted in Wegele, *Geschichte der Universität Wirzburg,* 1:328.

2. Wegele, *Geschichte der Universität Wirzburg,* 1:330–33. There is a vivid Jesuit manuscript account of the fall of Würzburg in BSAW, Kloster Himmeltal 341, "Annuae Collegij Herbipolensis. Anno 1631. 34. 35. 36." According to this account, the Jesuits offered to stay and aid in the defense of the city, but the bishop expressly ordered them to leave.

3. The chronology of Schott's career is given in ARSI, Lamalle: Schott.

4. In June 1652 Schott wrote that he would like to do editorial work for Kircher, as he had found many errors in his works (APUG, 561: 280). The earliest surviving letter between them is from 8 July 1650 (APUG, 567: 50).

5. Schott to Kircher, 18 August 1652, APUG, 561: 277.

6. 8 May 1655, ARSI, Rhen. Sup. 2: 254.

7. On 19 January 1656 General Nickel wrote to Schott regretting that Schott was cold and had no time to work but expressing confidence that Schott would win esteem for mathematics (ARSI, Rhen. Sup. 2: 274).

8. In the *Prooemium* to his *Magia universalis naturae et artis, sive Recondita naturalium & artificialum rerum scientia*..., 2 vols. (Bamberg, 1657), Schott wrote that Kircher had planned a compendium on natural magic. Since he was so busy with his other works, he had given Schott a rough outline of the work along with all the necessary books and notes and told him to finish it.

9. 28 August 1659 (APUG, 561: 281).

10. For a summary of the content of Schott's works, see M. l'Abbé M*** [Barthelemy Mercier], *Notice raisonnée des ouvrage de Gaspar Schott, jésuite; contenant des observations curieuses sur la physique expérimentale, l'histoire naturelle & les arts* (Paris, 1785). For complete bibliographies of Schott's work, see dB/S. In addition to his printed works, there is a manuscript in APUG (*Tractatus de pantometro, sive Proportionis circino*) and another in WüUB containing a series of watercolor illustrations of machines (WüUB, Delin 5). There is no trace of his personal papers, which must have been quite extensive.

11. Schott to Kircher, 15 June 1658, APUG, 567: 46.

12. Oliva to Provincial Goltgens, 3 September 1661, ARSI, Rhen. Sup. 2: 512; Oliva to Goltgens, 3 September 1661, ARSI, Rhen. Sup. 2: 519.

13. 12 July 1664, ARSI, Rhen. Sup. 2: 616.

14. ARSI, Rhen. Sup. 2: 648 and 650.

15. Schott mentions his health problems in letters to J. M. Faber ("Post dierum aliquot amarissimas potiones, ac purgationes, rectius valeo," 10 October 1665, UAE, BT, Kaspar Schott). His necrology (ARSI, Rhen. Sup. 44. I) states that "Vir erat Disciplinis Mathematicis insigniter excultus, laboris penè ferrei, simplicitatis verè Religiosae, pietatis constantis et assiduae, Catholicis et Acatholicis in amoribus et Veneratione. Exhaustus demum studijs et occupationibus, propè continuis." ARSI records state that he died on 22 October 1666 in Würzburg (ARSI, Rhen. Sup. 26. I, "Catalogi Breves 1641–1709": f. 164), whereas Sommervogel gives the date and place as 12/22 May 1666 in Augsburg. I am not aware of any surviving portrait of Schott.

16. Schott, *Magia universalis*, Prooemium.

17. On the notion of Jesuit intellectual activity as a form of "apostolic spirituality," see Rivka Feldhay, "Knowledge and Salvation in Jesuit Culture," *Science in Context* 1 (1987): 195–213; Harris, "Transposing the Merton Thesis"; and Luce Giard, "Le devoir d'intelligence, ou l'insertion des jésuites dans le monde du savoir," in Giard, *Les jésuites*.

18. Kaspar Schott, *Mechanica hydraulico-pneumatica: Qua praeterquam quod aquei elementi natura proprietas, vis motrix, atque occultas cum aere conflictus, a primis fundamentis demonstratur*... (Würzburg, 1657), 382.

19. BSAW, Schönborn Archiv, Korrespondenz Archiv, Johann Philipp: 2970a.

20. EUB, BT: Kaspar Schott. Letter to J. M. Faber, 15/5 July 1665.

21. Most modern scholarship regards Schott purely as a disseminator of knowledge. The *Dictionary of Scientific Biography,* for example, states that "his contribution was essentially that of an editor who prepared much of the researches of others for the press without adding much of consequence." Gillispie, *Dictionary of Scientific Biography,* 12:210–11.

22. Otto von Guericke, *Experimenta nova (ut vocantur) Magdeburgica de vacuo spatio* (Amsterdam, 1672), bk. 3, ch. 34; for the English translation, see *The New (So-Called) Magdeburg Experiments of Otto von Guericke,* trans. Margaret Foley Glover Ames (Dordrecht: Kluwer, 1994).

23. Torricelli did not publish an account of his invention, and Magni claimed that he had not heard of Torricelli's work but had invented the mercury tube independently. Many contemporaries found this somewhat difficult to believe. Magni acknowledged the prior claim of Torricelli after his own claim was denounced by Torricelli's French supporters but still maintained he had discovered it independently. See Middleton, *History of the Barometer,* 41–42.

24. Bernhard Erdmannsdörffer, *Deutsche Geschichte vom Westfälischen Frieden bis zum Regierungsantritt Friedrich's des Grossen, 1648–1740* (Berlin, 1892), 152–53.

25. There is a description of Royer's abilities in Schott's *Mechanica hydraulico-pneumatica,* 311. Lynn Thorndike, *A History of Magic and Experimental Science* (New York: Macmillan, 1923–58), 7:596, mentions a Jacques Le Royer who boasted to Louis XIV that he could do anything with mechanical devices and published in 1660 a text entitled *Les causes de flux et reflux de la mer.*

26. WüSA, SAKAJP: 2846.

27. Von Guericke, *Experimenta nova,* xix.

28. On Magni, the void, and atomism, see Redondi, *Galileo Heretic,* 102–3, 297–301.

29. There are several surviving, if damaged, letters from Magni to Johann Philipp in BSAW, SAKAJP: 794, 2758, 2761.

30. Schott specifically stated that the experiments were performed several times in Johann Philipp's presence in his *Mechanica hydraulico-pneumatica,* 444. Schott often mentioned the prominent persons who were present at performances of natural philosophical and magical devices.

31. Friedhelm Jürgensmeier, *Johann Philipp von Schönborn (1605–1672) und die römische Kurie: Ein Beitrag zur Geschichte des 17. Jahrhunderts* (Mainz: Selbstverlag der Gesellschaft für Mittelrheinische Kirchengeschichte, 1977), 202–4.

32. General Goswin Nickel to Johannes Kreyling, 25 October 1653, ARSI, Rhen. Sup. 2: 200.

33. Schott referred to a hydraulic device from Magdeburg in a letter to Kircher that he wrote from Mainz, even before he reached Würzburg (22 July 1655, APUG, 561: 38–39). At the same meeting Johann Philipp revealed his alchemical tendencies by advising Kircher not to divulge his secrets.

34. The meeting was held in Mainz, 15 July 1655, APUG, 567: 47.

35. For a detailed description of the pump and its operation, see Schott, *Mechanica hydraulico-pneumatica,* 444 ff., and *Experimenta nova.*

36. Schott acknowledged his debt to Kircher's work on the first page of the *Mechanica hydraulico-pneumatica.*

37. WüUB, Delin 5: "Machinae artificiales. R. P. Gasparis Schotti Soc.is Jesu mathematici herbipolensis."

38. Daniel Schwenter and Georg Philipp Harsdörffer, *Deliciae physico-mathematicae, oder Mathematische und philosophische Erquickstunden* (Nuremberg, 1636–53).

39. Schott, *Mechanica hydraulico-pneumatica,* 14.

40. See Grant, *Much Ado about Nothing;* Charles B. Schmitt, "Experimental Evidence for and against a Void: The Sixteenth-Century Arguments," *Isis* 58 (1967): 352–66.

41. When Jacques Royer, the regurgitator from the Regensburg Diet, was accused of producing his remarkable performances through witchcraft, he appealed to Kircher for an affidavit to testify that this was not the case (Schott, *Mechanica hydraulico-pneumatica,* 311). Unfortunately, there is no work that considers in detail Kircher's role as arbiter of natural philosophical and magical matters.

42. Schott, *Mechanica hydraulico-pneumatica,* 307.

43. Ibid., 452.

44. Nicolò Zucchi, *Experimenta vulgata non vacuum probare, sed plenum et antiperistasim instabilire* (Rome, 1648), and *Nova de machinis philosophia* (Rome, 1649).

45. Schott, *Mechanica hydraulico-pneumatica,* 471–73.

46. Guericke himself wrote an account of the siege and sack, stating that there was "nothing other than murder, flames, plundering, torture, and beatings" and that so many people were murdered in so many ways "that it cannot be adequately described with words and mourned with tears." Otto von Guericke, *Die Belagerung, Eroberung und Zerstörung der Stadt Magdeburg am 10/20 Mai 1631* (Leipzig: Voigtlander, 1912), 74–75.

47. On Guericke's personal fortune, see Fritz Krafft, *Otto von Guericke* (Darmstadt: Wissenschaftliche Buchgesellschaft, 1978). It was this wealth that enabled him to undertake the considerable expense of his experimental program.

48. Kircher and his museum were among the greatest attractions of mid-seventeenth-century Rome. Guericke seems to have enjoyed similar standing in Magdeburg. See the visit of the duke of Chevreuse and M. de Monconys to Magdeburg, October 1663, in *GA,* 46–48. Guericke remarked to Schott that prominent people frequently visited him to see his experiments (16/26 November 1661, in *GA*).

49. Vast amounts have been written on Guericke in German. The best is probably Krafft, *Otto von Guericke.* The Magdeburg city archives where the bulk of the surviving sources on Guericke were housed was destroyed along with the city itself by Allied bombing in the Second World War, but *GA* includes a German translation of the *Experimenta nova* and Guericke's surviving correspondence. There is an English translation of the *Experimenta nova,* but no detailed treatment of Guericke's life and work in English.

50. *GA,* 293–300, contains a list of all the sources cited in the *Experimenta nova,* although it is possible that Guericke first consulted some of these works after he began corresponding with Schott in 1656 but before the publication of the *Experimenta nova* in 1672.

51. Guericke to Schott, 22 July 1656, in *GA;* Schott, *Mechanica Hydraulico-pneumatica;* and Kaspar Schott, *Technica curiosa, sive Mirabilia artis* (Würzburg, 1664). Guericke did qualify this insistence on the vacuum somewhat by asking whether one would deny that a wheat sack was empty merely because it held one grain, indicating that he acknowledged that a very small amount of air was left in the receiver. For Guericke, the measure of a good air pump was how much air it could

evacuate rather than its ease of use. Thus he thought that Boyle's air pump, which had a receiver that could be opened to allow objects to be placed inside it, must have major leaks and therefore was inferior to his own (Guericke to Schott, 15/25 April 1662, *GA*). Soon afterwards he wrote that Boyle's pump was unsuitable for creating a vacuum, as Boyle himself had to admit. Guericke's receiver, on the other hand, could remain empty of air for months (Guericke to Schott 10 May 1662, *GA*).

52. Guericke to Schott, 22 July 1656, *GA*, 20.

53. Guericke to Schott, 22 July 1656, *GA*, 19.

54. Guericke to Schott, 22 July 1656, *GA*, 19.

55. "Vim attractivam metu vacui esse principium machinarum hydro-pneumaticarum," in Schott, *Magia universalis*, 525.

56. Guericke to Schott, 16/26 November 1661 and 30 December 1661, *GA*, 22–30.

57. Guericke to Schott, 28 February 1662, *GA*, 30–31.

58. The title is another example, but in this case by a non-Jesuit, of the extent to which traditionally separate disciplines were being conflated: physical phenomena were being examined mechanically with an engine. On Boyle's air pump experiments, see the important work of Steven Shapin and Simon Schaffer, *Leviathan and the Air-Pump: Hobbes, Boyle, and the Experimental Life* (Princeton: Princeton University Press, 1989). See also Antonio Clericuzio, "Notes on Corpuscular Philosophy and Pneumatical Experiments in Robert Boyle's *New Experiments Physico-mechanical, touching the Spring of the Air*," and Michela Fazzari, "Robert Boyle: Elasticity of Air and Philosophical Polemics," both in *Die Schwere der Luft in der Diskussion des 17. Jahrhunderts*, ed. Wim Klever (Wiesbaden: Harrassowitz, 1997), 109–16 and 117–34, respectively.

59. From the Latin edition, Robert Boyle, *Noua experimenta physico-mechanica de vi aeris elastica, & ejusdem effectibus: Facta maximam partem in nova machina pneumatica* . . . (Oxford, 1661), 10.

60. Schott, *Technica curiosa*, 3. Guericke cited this passage in his *Experimenta nova*, xx.

61. Schott, *Technica curiosa*, 5.

62. Guericke articulated this in the preface to the *Experimenta nova*.

63. Schott, *Technica curiosa*, 89.

64. On virtual witnessing, see Steven Shapin, "Pump and Circumstance: Robert Boyle's Literary Technology," *Social Studies of Science* 14 (1984): 481–520, and Shapin and Schaffer, *Leviathan and the Air-Pump*. Shapin provides a short definition of virtual witnessing in *The Scientific Revolution*, 108: "Virtual witnessing involved producing in the reader's mind such an image of an experimental scene as obviated the necessity for either its direct witness or its replication. In Boyle's experimental writing this meant a highly *circumstantial* style, often specifying in excruciating detail when, how and where experiments were done; who was present; how many times they were reiterated; and with exactly what results. Experiments were to be detailed in great numbers, and failures were to be reported as well as successes."

65. Schott, *Technica curiosa*, 98; Robert Boyle, *New Experiments Physico-mechanical, touching the Spring of the Air, and Its Effects* . . . (Oxford, 1660), 10.

66. Boyle, *New Experiments*, 33 ff; Schott, *Technica curiosa*, 122 ff.

67. Schott, *Technica curiosa*, 248.

68. Ibid., 261.

69. Ibid., 275.

70. Ibid., 284–85.

71. But the states were making attempts to replace the triennium with a biennium with varying degrees of success in the face of Jesuit opposition.

72. Anne C. van Helden, "The Age of the Air-Pump," *Tractix* 3 (1991): 149–72, counts only fifteen air pumps in all of Europe between 1647 and 1670.

73. Cornaeus, *Curriculum philosophiae.*

74. Although Schott is given credit for publishing the first account of the air pump, Cornaeus's work appeared at the same time. His book was largely ignored by later scholars, even Jesuits, and is still by modern historians of science. Ironically, Cornaeus acknowledged from the outset, and well before Schott, that the surrounding air did play some role in holding up the column of mercury in the mercury, while insisting the fear of the vacuum was the primary cause (Cornaeus, *Curriculum philosophiae,* 388).

75. MzStB, HS I 545. The manuscript is untitled but covers the physics curriculum and is identified by the following inscription: "Dictatus a Reverendo Domino Patre Joanne Serrario, conscriptus a Lamberto a Campo Lingense. Anno 1659, die 6. Novembris."

76. MzStB, HS II 386: 151r.

8. Censorship and *Libertas Philosophandi* in the Eighteenth Century

1. 1 February 1707, *PSGWL,* 2:328.

2. 25 June 1707, *PSGWL,* 2:330.

3. Although this work is rare today, a substantial part of it has been reprinted, along with Leibniz's marginalia to it, in Massimo Mugnai, *Leibniz' Theory of Relations* (Stuttgart: Franz Steiner, 1992).

4. 8 May 1708, *PSGWL,* 2:349.

5. 25 June 1707, *PSGWL,* 2:332.

6. 21 July 1707, *PSGWL,* 2:335.

7. 21 July 1707, *PSGWL,* 2:337.

8. 20 August 1706, *PSGWL,* 2:312. The list is in Pachtler, vol. 3, doc. 59. Ironically, some Jesuits thought the list was aimed at Leibniz.

9. 1 September 1706, *PSGWL,* 2:313. Leibniz erroneously attributed the list to General Vitelleschi rather than Piccolomini.

10. H. G. Alexander, ed., *The Leibniz-Clarke Correspondence, Together with Extracts from Newton's Principia and Opticks* (Manchester: Manchester University Press, 1956).

11. 21 July 1707, *PSGWL,* 2:338.

12. Christian Wolff, *Preliminary Discourse on Philosophy in General,* trans. Richard J. Blackwell (Indianapolis: Bobbs-Merrill, 1963), 103–5.

13. Ibid., 108.

14. Ibid., 106.

15. Ibid., 109.

16. Josef Braun, S.J., *Theses ex universa philosophia* . . . (Würzburg, 1737), A3.

17. Michael Froehling, S.J., *Meditatio philosophica de eo, quod licitum est philosopho circa usum S. Scripturae* (Würzburg, 1753).

18. Ibid., 7.

19. An example of an appeal from the provinces is ARSI, Congr. 20h: 105. The general congregation's ruling is in *ISJ,* Cong. Gen.14, Decr. 5.

20. ARSI, Congr. 20h.

21. "Verbot einiger Kartesianischer und Leibniz'scher Lehrsätze durch den P. General Tamburini. 1706," in Pachtler, vol. 3, doc. 59.

22. Ibid.

23. *ISJ,* Cong. Gen. 16, Decr. 36.

24. For example,

> 1. Non aliter corpus concipi aut explicari potest, quam ut substantia actu extensa.
> 2. Idea materiae primae Aristotelica convenit etiam atomis Epicuri aut materiae Cartesianae.
> 3. Materia prima non est entitas substantialis incompleta, et ordinata ad aliam compartem, tanquam potentia ad actum, seu tanquam subiectum ad formam, cum qua totum substantiale componat. (Pachtler, vol. 3, doc. 61)

25. See, for example, ARSI, FG 673 I and II, as well as an interesting 1748 discussion of discrepancies between the original manuscript and printed versions of the Ordinatio in ARSI, FG 674: 259r and v.

26. StBMz, HS II 378. This is the first half of the Physics course. The second half is now in Würzburg (BSAW, HV MS q. 97).

27. StBMz, HS II 387:13.

28. Carpophorus dei Giudice, *Tractatus de accidentibus absolutis, sive Sacro-Sanctum Eucharistiae Sacramentum in principiis philosophiae peripateticae impugnatum, et ex placitis philosophiae corpusculariae propugnatum* (Paderborn, 1718).

29. Johann Adam Morasch, *Philosophica atomistica in alma electorali Universitate Ingolstadiensi disputationi subjecta* . . . (Ingolstadt, 1727).

30. Georg Hermann, S.J., *Lapis offensionis atomisticae a peripatetico motus, sive Dialogi inter peripateticum, & atomistam veritatis studiosum circa generationem formarum peripateticarum* . . . (Ingolstadt, 1730).

31. Johann Adam Morasch, *Atomismus ad injustis peripateticorum censuris et imposturis vindicatus* . . . (n.p., 1733).

32. Franz Anton Ferdinand Stebler, *Aristoteles atomista, seu Explicari solitis apud atomistas principiis non difformis, sed conformis omnino Aristotelis doctrina* (Ingolstadt, 1740). As a student Stebler had defended Morasch's *Philosophia atomistica* in disputation in 1727.

33. *Etwelche Meistens bayrische Denck- und Leß-Würdigkeiten zur Fortführung des so genannten Parnassi Bojci* . . . [Stuck] 4 (1739): 269–70.

34. Bernard Duhr, *Geschichte der Jesuiten in den Ländern deutscher Zunge,* vol. 4, *Im 18. Jahrhundert* (Munich-Regensburg: Manz, 1928), pt. 2, 45. Schaff, *Geschichte der Physik,* calls Kleinbrodt

a Cartesian. See also Boehm, "Universität in der Krise?" 127, and the entry for Kleinbrodt in Boehm, *Biographisches Lexikon.*

35. MüUB, 4° 678. They were recorded by Johann Adam Morasch.

36. Ibid., paras. 114–38.

37. Anton Kleinbrodt, S.J., *Mundus elementaris disputationi subjectus* . . . , 3 vols. (Ingolstadt, 1704). The second of these was defended by Morasch, indicating that he studied particularly closely with Kleinbrodt.

38. Ibid., 1:101 ff.

39. There is certainly nothing in the correspondence from the Jesuit curia in Rome to the Upper German province, in the university archives, or in the surviving provincial archives to indicate that Kleinbrodt was removed for adhering to Cartesian opinions. To the contrary, there is a letter dated 31 October 1705 from the general to Josef Preiss, the provincial of Upper Germany, wholeheartedly agreeing to the latter's request that Kleinbrodt be awarded a doctorate in theology, indicating that he was not suspected of any kind of heterodoxy (ARSI, Germ. Sup. 12: 105v).

40. UALMU, O-I-5 (Dekansakten 1690–1773): 28 September 1704.

41. Joseph Falck, S.J., *Mundus aspectabilis philosophice consideratus* . . . (Augsburg, 1737).

42. So Falck's necrology claimed in the *Parnassus Boicus* 1 (1737): 74–75.

43. They are in the BHSA in two folders, Jesuiten 736: "Censurae librorum autorum incertorum 1730–33" and Jesuiten 785: "Censurae philosophiae P. Jos. Falck SJ 1724–1736." The folios are out of order and the pagination is missing or incorrect. The reports of the Roman revisers general concerning the work have not survived.

44. "Judicium PP. Revisorum de Opere Philosophico P. Josephi Falck," BHSA, Jesuiten 736: no pagination. A year after writing this censura, Tschiderer became a reviser general in Rome.

45. BHSA, Jesuiten 785: 24–25. Note the appeal to the local, always an element in Jesuit practice but one that grew stronger throughout the eighteenth century. Falck had written a very positive report on one of Grammatici's works in 1720 (BHSA, Jesuiten 784: folios unnumbered), and the two conducted a long-standing correspondence. There are several letters from Falck to Nicasius Grammatici in the BSB (Clm 1609 ff.44–47, 57–58, 61–62, 65–71, 74–79), almost exclusively concerned with calendrical problems.

46. BHSA, Jesuiten 785: 23r and v.

47. One of them, Ludwig Simonstin, felt that because of this the work should not be published at all. BHSA, Jesuiten 785: 3 ff., 9.

48. Joseph Mayr occupied virtually every position of responsibility in the Society short of general: rector, provincial, and assistant. Since he was in Rome as assistant for Germany in 1736, this censura must date from the first round of reports in 1725.

49. BHSA, Jesuiten 785: 15v–16r. In his earlier dissertation *Liquidorum gravitas et aequilibrium* (Freiburg i. Br., 1713), Falck had been permitted to state openly that lightness was merely a relative concept.

50. Falck, *Mundus aspectabilis,* 19.

51. Ibid., 182, 213.

52. Franz Neuff, S.J., *Authoritatis in philosophia pondus aequa examinatum trutina, ac philosophicis praefixum thesibus* . . . (Mainz, 1763), 6.

53. Ibid., 11.

54. Roger Joseph Boscovich, S.J., *A Theory of Natural Philosophy,* trans. J. M. Child (Cambridge, Mass.: M. I. T. Press, 1966), 6. I have used the Latinized form that Boscovich himself and his contemporaries used rather than his native Rudjer Josif Boskovic.

55. See Brooke, *Science and Religion,* esp. chs. 5 and 6.

56. Zarko Dadic, "Boskovic and the Question of the Earth's Motion," in *The Philosophy of Science of Ruder Boskovic: Proceedings of the Symposium of the Institute of Philosophy and Theology,* ed. Ivan Macan and Valentin Pozaic (Zagreb: Institute of Philosophy and Theology, 1987).

57. Cited in ibid., 136.

58. ARSI, Congr. 20h: 250–51, "Postulatum ad Deputationem pro Studiis remissum."

59. Joannes Nepomuk Mederer, S.J., *Annales Ingolstadiensis Academiae,* 3:208.

60. BHSA, Jesuiten 85: 59–61 (other copies of the same document ff. 62–65, 54–57, and 66–72).

61. Ibid.

62. Joseph Mangold, S.J., *Philosophia rationalis et experimentalis hodiernis discentium studiis accomodata auctore . . . ,* 3 vols. (Ingolstadt, 1755).

63. Benedikt Stattler, S.J., *Philosophia methodo scientiis propria explanata . . .* (Augsburg, 1769–72), preface.

9. The Spread of Experiment

1. The theses are reproduced in their entirety in Anna Fechtnerová, ed., *Katalog grafickich listu univerzitních tezí ulozenich ve Státní knihovne CSR v Praze* (Prague: Státní Knihovna CSR, 1984).

2. DStB, XIj 133/2.

3. DStB, XIj 133/2: 192.

4. DStB, XIj 133/2: 214.

5. DStB, XIj 133/2: 218.

6. DStB, XIj 133/2: 281v.

7. DStB, XIj 133/2: 219.

8. DStB, XIj 133/2: 220v.

9. DStB, XIj 133/2: 221.

10. MüUB, 4° Cod. ms. 678: 29 ff. For these lectures in published form, see Anton Kleinbrodt, S.J., *Physica particularis . . . ,* 3 vols. (Ingolstadt, 1704).

11. Johannes Seyfrid, S.J., *Theses ex universa philosophia . . .* (Würzburg, 1712), 4.

12. Franz Ellspacher, S.J., *Barometron Torricellianum quaestionibus philosophicis subjectum . . .* (Dillingen, 1714).

13. Paul Zetl, S.J., *Praecipua barometri phaenomena rationibus physicis illustrata . . .* (Ingolstadt, 1718). One should also mention Joseph Falck, S.J., *Liquidorum gravitas et aequilibrium . . .* (Freiburg i. Br., 1713).

14. But as late as 1740 Edmund Voit suggested that treating the simple machines in philosophy as opposed to the "artes machinales" would create chaos among the sciences. Edmund Voit, S.J.,

Veritas philosophiae peripatetico-Christianae ex contrariis doctrinis breviter appositis magis elucescens . . . (Würzburg, 1740).

15. On collections, see Paula Findlen, *Possessing Nature: Museums, Collecting, and Scientific Culture in Early Modern Italy* (Berkeley: University of California Press, 1994).

16. The inventory is reproduced in Reindl, *Lehre und Forschung*, 176–85. There was also a mathematical cabinet at the Jesuit college at the University of Freiburg from at least as early as 1626.

17. Mederer, *Annales*, 3:176.

18. On the development of air pumps and the concerns of their makers, see van Helden, "Age of the Air-Pump," 149–72.

19. Alto Brachner, ed., *G. F. Brander, 1713–1783, wissenschaftliche Instrumente aus seiner Werkstatt* (Munich: Deutsches Museum, 1983), 342. In his *Annales* of the University of Ingolstadt, Mederer reported that the Jesuit observatory there was equipped with instruments made by Brander (3:297–303), so it is quite probable that his experimental instruments were also used at the university in Physics. It would be quite curious if the University of Dillingen, which was only a few miles from Augsburg, did not also use his instruments. A catalogue of scientific instruments at the University of Innsbruck, compiled in 1809 by the ex-Jesuit Franz von Zallinger, also lists instruments by Brander, although it does not identify the maker of the two air pumps in the collection. See Rudolf Steinmaurer, "Die Lehrkanzel für Experimentalphysik," in *Die Fächer Mathematik, Physik, Chemie an der Philosophischen Fakultät zu Innsbruck bis 1945*, ed. Franz Huter (Innsbruck: Kommissionsbuchh. in Komm., 1971), 105–14.

20. Maurice Daumas, *Scientific Instruments of the Seventeenth and Eighteenth Centuries and Their Makers* (London: Batsford, 1972), 117, 146.

21. BHSA, GL Fasz. 1489/20: no pagination.

22. On the reform of the Catholic universities to promote better legal training and the resulting establishment of chairs of history, see Hammerstein, *Aufklärung und katholisches Reich.*

23. BHSA, GL Fasz. 1489/20: no pagination (dated 5 October 1746).

24. The medical faculty was also interested in certain aspects of natural philosophy and experiment. Since, however, the Jesuits played no role in the medical faculty, such questions fall outside the scope of this study.

25. UALMU, O-I-5: 15 May 1748.

26. BHSA, GL Fasz. 1484/III/6: 1r and v.

27. BHSA, GL Fasz. 1484/III/6: 6v.

28. BHSA, GL Fasz 1484/III/6: 3r, dated 23 February 1755.

29. Electoral decree dated 14 April 1755, BHSA, GL Fasz 1484/III/6: 5r.

30. Steinmaurer, "Die Lehrkanzel," 56–60.

31. Ernst Walter Zeeden, "Die Freiburger Philosophische Fakultät im Umbruch des 18. Jahrhunderts," in *Beiträge zur Geschichte der Freiburger Philosophischen Fakultät*, by Clemens Bauer, Ernst Walter Zeeden, and Hans-Günter Zmarzlik (Freiburg i. Br.: Albert, 1957), 50–52; Theodor Kurrus, "Zur Einführung der Experimentalphysik durch die Jesuiten an der Universität Freiburg i. Br.," in *Zur Geschichte der Universität Freiburg i. Br.*, ed. Johannes Vincke (Freiburg i. Br.: Albert, 1966).

32. Heilbron, *Electricity*, 150.

33. BSAW, MRA 40/1273: 371. It is difficult to date the list exactly; from the instruments listed and the dates of the surrounding documents, it is probably from the later 1740s.

34. BSAW, MRA 40/L 1274 II: 204. The biennium had been introduced at Mainz in 1745.

35. The *Studienordnungen* are reprinted in full in Wegele, *Geschichte der Universität Wirzburg,* 2:330.

36. WüUB, Reuß IV: 76. Also reprinted in Wegele, *Geschichte der Universität Wirzburg,* 2:404.

37. Reindl, *Lehre und Forschung,* 22, 98. Henner's textbook was *Conatus physico-experimentales de corporum affectionibus tum generalibus tum specialibus, ad usum philosophiae candidatorum suscepti . . . ,* 2 vols. (Würzburg, 1756–69). The second volume was completed by his successor to the chair, his brother Georg.

38. Karl Scherffer, S.J., *Institutionum physicae . . .* (Vienna: Trattner, 1752–53); Josef Redlhamer, S.J., *Philosophiae naturalis. Pars prima, seu Physica generalis ad praefixam in scholis nostris normam concinnata . . .* (Vienna, 1755).

39. "[O]mnigenis rerum naturalium experimentis." Tolomei, *Philosophia mentis,* 374.

40. Ibid., 374.

41. Cong. Gen. 16, Decr. 36, in John W. Padberg, S.J., Martin D. O'Keefe, S.J., and John L. McCarthy, S.J., *For Matters of Greater Moment: The First Thirty Jesuit Congregations* (St. Louis: Institute of Jesuit Sources, 1994), 384–85.

42. Cong. Gen. 17, Decr. 13, in Padberg, *For Matters of Greater Moment,* 390–91. Visconti to Lower Rhenish province, 22 July 1752 (EABP, I. 12).

43. StBMz, HSS II 378, f. 205r.

44. BHSA, GL Fasz. 1489/20: no pagination.

45. Anton Mayr, S.J., *Philosophia peripatetica antiquorum principiis, et recentiorum experimentis conformata . . .* (Ingolstadt, 1739).

46. Knittel, *Aristoteles curiosus.*

47. ARSI, FG 675: 348–53. "Synopsis, seu conspectus sententiarum, quae in philosophia per Germaniam superiorem a professoribus Societatis IESU sunt docendae."

48. On Gautruche, see Brockliss, "Pierre Gautruche."

49. Redlhamer, *Philosophiae naturalis.* His phrase is "scientiae naturalis historia" (9).

50. Ibid., 7.

51. Ibid., 13.

52. Shapin, *The Scientific Revolution,* 96.

53. The Deutsches Musuem in Munich has an example manufactured by Brander in 1777. See Brachner, *G. F. Brander,* 296.

54. For a catalogue of German (including Jesuit) emblem books, see John Landwehr, *German Emblem Books, 1531–1888* (Utrecht: Haentjens Dekker & Gumbert, 1972). See also G. Richard Dimler, S.J., "A Bibliographical Survey of Jesuit Emblem Authors in German-Speaking Territories: Topography and Themes," *AHSI* 89 (1976): 129–37, and "The *Imago Primi Saeculi:* Jesuit Emblem Books and the Secular Tradition," *Thought* 56 (1981): 433–48.

55. StAK, Jes. Abt. 29a (*Studienheft* of Hubert Houben, 1722): 72v ff.

56. This emblem was the central motif of the inspirational tract by the German Jesuit Jeremius Drexelius, S.J., entitled *The Heliotropium ("Turning to HIM") or Conformity of the Human Will to the Divine* (New York: Devin-Adair, 1912).

57. Philippo Picinelli, *Mundus symbolicus, in emblematum universitate formatus, explicatus, et tam sacris, quam profanis eruditionibus ac sententiis illustratus*... (Cologne, 1694). The pulley is in 2:195, with the motto "Sine pondere pondis" (you may weigh without burden). Emblems relying on machines are also extremely scarce in the foremost modern catalogue of emblems, Arthur Henckel and Albrecht Schöne, *Emblemata: Handbuch zur Sinnbildkunst des XVI. und XVII. Jahrhunderts* (Stuttgart: Metzler, 1967).

58. The medal is at the Herzog-Anton-Ulrich Museum in Braunschweig, Germany.

59. My numbering is arbitrary; there does not seem to be any starting point or a progression through the emblems.

60. Berthold Hauser, S.J., *Elementa philosophiae ad rationis et experientiae ductum conscripta, atque usibus scholasticis accomodata*... (Augsburg, 1755–64).

61. See the appendix to Schaff, *Geschichte der Physik*, 194–220, on the history of Ingolstadt's "Armarium."

10. The Jesuits and Their Contemporaries in the *Aufklärung*

1. Jakob Brucker, *Historia critica philosophiae a mundi incunabulis ad nostram usque aetatem deducta* (Leipzig, 1742–43), 4:117.

2. Ibid., 4:121.

3. Ibid., 4:139–40. From his references to his sources, it is clear that Brucker had not read Arriaga or, most probably, the other Jesuit neoscholastics to whom he refers.

4. Ibid., 4:141 (comment on Guericke, 4:659). Franz Rassler's text is *Problema physicum inter Aristotelem et Empedoclem controversum de primis initiis corporis generalis* (Dillingen, 1685). The original manuscript of it is MüUB, 4° Cod. ms. 677. Rassler taught Anton Kleinbrodt.

5. On the *Acta Eruditorum*, see A. H. Laeven, *The "Acta Eruditorum" under the Editorship of Otto Mencke (1644–1707): The History of an International Learned Journal between 1682 and 1707* (Amsterdam: APA-Holland University Press, 1990).

6. Henri Brunschwig's important study of the Enlightenment in Prussia opens by stating that "the *Aufklärung* prevails in Prussia rather than in Germany as a whole.... Prussia is the true fatherland of the *Aufklärung*." Henri Brunschwig, *Enlightenment and Romanticism in Eighteenth-Century Prussia* (1947; reprint, Chicago: University of Chicago Press, 1974), 1.

7. For recent essays and bibliography on the Catholic Enlightenment in Germany, see Harm Klueting, ed., *Katholische Aufklärung: Aufklärung im katholischen Deutschland* (Hamburg: Felix Meiner, 1993).

8. Van Dulmen, "Antijesuitismus."

9. Cited in T. C. W. Blanning, *Reform and Revolution in Mainz, 1743–1803* (Cambridge: Cambridge University Press, 1974), 103.

10. This attitude was often nothing short of hatred. One reformer hoped that "with God's help [the Jesuits] will meet the fate of the Templers." Van Dülmen, "Antijesuitismus," 58.

11. Haaß, *Die geistige Haltung,* 123.

12. On Ickstatt's reforms at Würzburg and Ingolstadt, see Hammerstein, *Aufklärung und katholisches Reich,* 36–131.

13. See Ludwig Hammermayer, "Aufklärung im katholischen Deutschland des 18. Jahrhunderts: Werk und Wirkung von Andreas Gordon O. S. B. (1712–1751), Professor der Philosophie an der Universität Erfurt," *Jahrbuch des Instituts für deutsche Geschichte* 4 (1975): 53–109; and Michael Klein, "Das Schottenkloster St. Jakob," in *Im Turm, im Kabinett, im Labor: Streifzüge durch die Regensburger Wissenschaftsgeschichte,* ed. Martina Lorenz (Regensburg: Universitätsverlag, 1995).

14. Andreas Gordon, O. S. B., *P.S. de scripto R. P. Opffermann S.J. nuperrime hic edito* (Erfurt, 1749), 85.

15. Andreas Gordon, O. S. B., "Oratio philosophiam novam veteri praeferendam suadens," reprinted in his *Varia philosophiae mutationem spectantia* (Erfurt, 1749), 13–22. Six texts by Gordon and his opponents relating to the dispute were reprinted in the *Varia philosophiae.* For the Erfurt officials' report on the dispute, see LSA, Rep. A 37 b I, II. Tit. XVI Nr. 50.

16. Hammermayer, "Aufklärung," 84–85.

17. Andreas Gordon, O. S. B., "Oratio, philosophiam novam utilitatis ergo complectandam et scholasticam philosophiam futilitatis causa eliminandam," in his *Varia philosophiae,* 13–22.

18. Pfriemb, *Apologia,* and Joseph Pfriemb, S.J., *Connubium felix philosophiae theoreticae cum experimentali . . .* (Mainz, 1749).

19. LSA, Rep. A 37 b I, II. Tit. XVI Nr. 50.

20. Pfriemb, *Apologia,* 38.

21. Andreas Gordon, O. S. B., "Oratio philosophiam novam veteri praeferendam," in his *Varia philosophiae,* 19.

22. Andreas Gordon, O. S. B., "Oratio philosophiam novam utilitatis ergo amplectandam," in his *Varia philosophiae,* 26. This was Gordon's second oration on the subject.

23. Pfriemb, *Apologia,* 39–40.

24. Ludwig Hammermayer, *Geschichte der Bayerischen Akademie der Wissenschaften 1759–1807* (Munich: C. H. Beck, 1983), 1:37–40. Amort wrote many laudatory pieces about Jesuits or their writings in the journal *Parnassus Boicus.*

25. Richard van Dülmen, *The Society of the Enlightenment: The Rise of the Middle Class and Enlightenment Culture in Germany* (Cambridge: Polity Press, 1992), 32–33. On the relationship between the Bavarian Academy and the Jesuits, see Wilhelm Kratz, S.J., "Aus den Frühtagen der bayerischen Akademie der Wissenschaften: Zur Vorgeschichte der Aufhebung des Jesuitenordens," *AHSI* 7 (1938): 181–219.

26. For example, Lori to Töpsl, 12 May 1759, in Spindler, 33–36.

27. On the intellectual activities of the Scottish Benedictines in Germany, see Klein, "Das Schottenkloster," and Hammermayer, "Aufklärung."

28. Lori to Ulrich Weiß, O. S. B., 6 December 1759, in Spindler, 231–33. The Accademia dei Lincei, perhaps the prototypical scientific society, explicitly excluded regular clergy.

29. Anton Roschmann to Lori, 5 July 1759, in Spindler, 90–93.

30. Johann Georg von Stengel to Lori, 7 Aug 1759, in Spindler, 129–31.

31. Ickstatt to Lori, 2 July 1759, in Spindler, 84–85.

32. Lori to Ickstatt, 5 July 1759, in Spindler, 89–90.

33. Lori to Kolb, O. S. B., 2 July 1759, in Spindler, 464.

34. Andreas Kraus writes that the Jesuits stood out among the various Catholic orders for being the most receptive toward modern science. Although the Benedictines were consciously improving their cabinets of instruments, at the time of the founding of the Bavarian Academy they were unable to provide the academy with much expertise. Andreas Kraus, *Die naturwissenschaftliche Forschung an der Bayerischen Akademie der Wissenschaften im Zeitalter der Aufklärung* (Munich: C. H. Beck, 1978), 10.

35. Hammermayer, *Geschichte der Bayerischen Akademie,* 1:377, 2:399. The Jesuit Heinrich Schutz, professor of history at Ingolstadt, also won the academy's history prize in 1761/62.

36. Lorenz Westenrider, *Geschichte der baierischen Akademie der Wissenschaften, auf Verlangen derselben verfertigte,* vol. 1, *1759–1777* (Munich, 1784).

37. G. F. Brander to Dominikus von Linprun, 28 July 1759, in Spindler, 118–21, and Brander to Lori, 2 December 1759, in Spindler, 224–26. A photograph of this particular pump modeled on 'sGravesende's design, complete with its admittedly elegant legs, can be found in Brachner, *G. F. Brander,* 298.

38. Adolf Kistner, *Die Pflege der Naturwissenschaften in Mannheim zur Zeit Karl Theodors* (Mannheim: Selbstverlag des Mannheimer Altertumsvereins, 1930), 5–40.

39. Theodor Kurrus, *Die Jesuiten an der Universität Freiburg i. Br* (Freiburg i. Br.: Albert, 1963), 2:157–58.

40. On the *Studienordnung* at Vienna, see Kink, *Universität zu Wien,* 1:458–63. For a biography of Van Swieten, see Frank T. Brechka, *Gerard van Swieten and His World, 1700–1772* (The Hague: Martinus Nijhoff, 1970).

41. Cited in Helmuth Gericke, *Zur Geschichte der Mathematik an der Universität Freiburg i. Br.* (Freiburg i. Br.: Albert, 1955), 43.

42. Cited in ibid., 44.

43. Kurrus, *Die Jesuiten,* 2:163 ff.

44. On Freiburg, see Theodor Kurrus, "Zur Einführung der Experimentalphysik," and Gericke, *Zur Geschichte der Mathematik.*

45. Zeeden, "Die Freiburger Philosophische Fakultät," 47–51.

46. Kurrus, *Die Jesuiten,* 2:165.

47. See Zeeden, "Die Freiburger Philosophische Fakultät," 45–49, 56–60, on the attempts of the University of Freiburg to avoid implementing the new *Studienordnung.*

48. Cited in Kink, *Universität zu Wien,* 490–91.

49. Zeeden, "Die Freiburger Philosophische Fakultät," 55.

50. Grete Klingenstein, "Despotismus und Wissenschaft: Zur Kritik norddeutscher Aufklärer an der österreichischen Universität 1750–1790," in *Formen der europäischen Aufklärung: Untersuchungen zur Situation von Christentum, Bildung und Wissenschaft im 18. Jahrhundert,* ed. Friedrich Engel-Janosi, Grete Klingenstein, and Heinrich Lutz (Munich: R. Oldenbourg, 1976).

51. For an overview of criticisms leveled at German universities, see R. Steven Turner, "University Reformers and Professorial Scholarship in Germany 1760–1806," in *The University in Society,* vol. 2, *Europe, Scotland and the United States from the Sixteenth to the Twentieth Century,* ed.

Lawrence Stone (Princeton: Princeton University Press, 1974). On the problem of anachronistic critiques of German universities, see Boehm, "Universität in der Krise?"

52. See Eulenburg, *Die Frequenzen der deutschen Universitäten.*

53. Virtually all institutional histories note the students' appalling misbehavior.

54. See Turner, "University Reformers," 515–29, although he accepts Prantl's dismissal of the Ingolstadt Jesuits at face value.

55. On the political and legal sciences, see Hammerstein, *Aufklärung und katholisches Reich.*

56. Johann Christoph Sturm, *Collegium experimentale, sive Curiosum* . . . (Nuremberg, 1676).

57. Walter Jaenicke, "Naturwissenschaften und Naturwissenschaftler in Erlangen 1743–1993," in *250 Jahre Friedrich-Alexander-Universität Erlangen-Nürnberg. Festschrift,* ed. Henning Kössler (Erlangen: Universitätsbund Erlangen-Nürnberg, 1993), 629–34.

58. Herman Haupt, ed., "Chronik der Universität Gießen von 1607 bis 1907," in *Die Universität Gießen von 1607 bis 1907* (Gießen: Töpelmann, 1907), 380.

59. Gunther von Minnigerode, "250 Jahre Demonstrationsversuche in der Physik," in *Naturwissenschaften in Göttingen: Eine Vortragsreihe,* ed. Hans-Heinrich Vogt (Göttingen: Vandenhock & Rupprechtt, 1988), 37. On science at the University of Göttingen, see Hans Schimank, "Zur Geschichte der Physik an der Universität Göttingen vor Wilhelm Weber (1734–1830)," *Rete: Strukturgeschichte der Naturwissenschaften* 2 (1974): 207–52.

60. Cited in Carl Brinitzer, *A Reasonable Rebel: Georg Christoph Lichtenberg,* trans. Bernard Smith (London: Allen & Unwin, 1960), 120.

61. See John Edwin Gurr, *The Principle of Sufficient Reason in Some Late Scholastic Systems, 1750–1900* (Milwaukee: Marquette University Press, 1959).

62. Lind, *Physik im Lehrbuch;* on Crusius, see Georgio Tonelli, "Einleitung," in *Die philosophischen Hauptwerke,* by Christian August Crusius, ed. Georgio Tonelli (Hildesheim: Georg Olms, 1969).

11. The Transubstantiation of Physics

1. Mayr, *Philosophia peripatetica.*

2. This shift has been identified by historians such as Grabmann, "Die Disputationes Metaphysicae." Wolff outlined the structure of his philosophy in his *Preliminary Discourse.* See also Richard J. Blackwell, "The Structure of Wolffian Philosophy," *Modern Schoolman* 38 (1961): 203–18.

3. Anton Erber, S.J., *Cursus philosophicus methodo scholastica elucubrata,* 3 vols. (Vienna, 1750).

4. Hauser, *Elementa philosophiae.*

5. Stattler, *Philosophia methodo scientiis.*

6. ARSI, FG 671: 351.

7. Leibniz's admiration was perhaps due to his belief that Tolomei's philosophy closely followed his own. Consequently he felt that it would be a very good thing if Tolomei were elected general of the Society. *PSGWL,* 2:294.

8. Tolomei, *Philosophia mentis,* 373.

9. According to Krafft, "Jesuiten als Lehrer," in 1744/45 Friderich occupied the chair of physics and ethics and, unusually, the chair of mathematics as well.

10. Philipp Friderich, S.J., *Biennium philosophicum inauguratione solenni coronatum. . . . Quid sentiendum sit de methodo ac libertate philosophandi a RR. quibusdam asserta?* . . . (Mainz, 1745), 4.

11. Philipp Friderich, S.J., *Selectae ex philosophia peripatetica universa theses et antitheses parallelae, seu hypotheses Cartesii cum systemate peripatetico dissertatio irenica partim conciliatae, partim excussis illarum fundamentis ab erroribus assertae* . . . (Mainz, 1745), 12.

12. Ibid., 89.

13. Maximus Mangold, S.J., *Philosophia recentior praelectionibus publicis accomodata* (Mainz, 1763–64), 1:234.

14. Anton Nebel, S.J., *Disputatio I. Menstrua de principiis intrinsecis corporis naturalis in genere et in specie, cum resolutione quaestionis: An dentur accidentia absoluta?* (Würzburg, 1747).

15. J. Mangold, *Philosophia rationalis,* 1:211–12.

16. Ibid., 1:213.

17. Ibid., 1:219–21.

18. M. Mangold, *Philosophia recentior,* 1:67.

19. Schillebeeckx, *The Eucharist,* 62.

20. Wirceburgenses, *Theologia Wirceburgensis: Theologia dogmatica, polemica, scholastica et moralis praelectionibus publicis in alma universitate Wirceburgensi accommodata* (Paris, 1853), 5:353–56. Originally published in the late 1760s, it was compiled from various treatises and courses produced over the preceding years. See Klaus Schilling, *Die Kirchenlehre der Theologia Wirceburgensis* (Paderborn: Ferdinand Schöningh, 1969).

21. On the place of Maignan and his Minim brothers in the development of theories of eucharistic accidents, see F. Jansen, "Eucharistiques (accidents)," cols. 1431–34.

22. Carlo Sardagna, S.J., *Theologia dogmatico-polemica, qua adversus veteres novasque haereses ex Scripturis, patribus, atque ecclesiastica historia catholica veritas propugnatur* (Regensburg: 1771), 7:208–48.

23. Georg Vaeth, S.J., *Concordia veritatis ontologicae de accidentibus cum veritate theologica de speciebus Eucharisticis brevi dissertatione stabilita* . . . (Würzburg, 1772), 8.

24. Ibid., 20.

25. Ibid., 29.

26. Ibid., 38–40.

27. Ibid., 26–28.

28. Ibid., 50.

29. On Wolff's role in disseminating Newton in Germany, see Martina Lorenz, "Der Beitrag Christian Wolffs zur Rezeption von Grundprinzipien der Mechanik Newtons in Deutschland zu Begin des 18. Jahrhunderts," *Historia Scientiarum* 31 (1986): 87–100.

30. Hauser, *Elementa philosophiae,* 6:434.

31. Paul Offermann, *Theses selectae ex universa philosophia ad majorem dei optimi maximi gloriam, virginis sine labe conceptae honorem* . . . (Mainz, 1761), 74–75.

32. Hauser knew that Boscovich had attempted to reconcile Newtonian mechanics with a stationary earth, but he and the other Germans were not yet prepared to follow him.

33. Hauser, *Elementa philosophiae,* 6:448.

34. Ibid., 6:461–67.

35. M. Mangold, *Philosophia recentior,* 2:364–80.

36. Voit, *Veritas philosophiae,* 16.

37. Boscovich, *Theory of Natural Philosophy.* See J. M. Child's introduction.

38. On Boscovich's theory of matter, see Lancelot Law Whyte, "Boscovich's Atomism," and Zeljko Markovic, "Boscovich's *Theoria,*" both in *Roger Joseph Boscovich, S.J., F.R.S., 1711–1787,* ed. Lancelot Law Whyte (New York: Fordham University Press, 1961).

39. On opposition to Boscovich in France, see Markovic, "Boscovich's *Theoria,*" 147. On the reception of Boscovich in the German assistancy, see Steven J. Harris, "Boscovich, the 'Boscovich Circle' and the Revival of the [sic] Jesuit Science," in *R. J. Boscovich: Vita e Attività Scientifica. His Life and Work,* ed. Piers Bursill-Hall (Rome: Istituto della Enciclopedia italiana, 1993).

40. Stattler, *Philosophia methodo scientiis,* Proemium, § II.

41. Nicolaus Burkhaeuser, S.J., *Theoria corporis naturalis principiis Boscovichii conformata, quam una cum thesibus ex philosophica universa . . .* (Würzburg, 1770).

42. This can be seen in Christian Appel's *Theses ex universa philosophia . . .* (Würzburg, 1769), particularly 22–26.

43. Burkhäuser, *Theoria corporis naturalis,* 2–4.

44. Ernst Attlmayr, "Die fünf gelehrten Naturwissenschaftler von Zallinger aus Bosen," *Beiträge zur Technikgeschichte Tirols* 2 (1970): 36–43.

45. Jacob von Zallinger, S.J., *Lex gravitatis universalis ac mutuae cum theoria de sectione coni, potissimum elliptica proposita* (Munich, 1769), *De expositione physica demonstrationum mathematicarum in philosophia naturali dissertatio* (Dillingen, 1772), *Interpretatio naturae, seu Philosophia Newtoniana methodo exposita, atque academicis usibus adcommodata,* 3 vols. (Augsburg, 1773–75), and *Anmerkungen über den Auszug, und die Kritik eines berlinischen Herrn Recensenten des Boscovichische System betreffend* (Freiburg im Br., 1772).

46. Zallinger, *Interpretatio naturae,* 1:7v–8v, 2:2r–3v.

47. E. A. [Eusebius Amort], "Von dem Leben und gelehrten Schrifften P. Nicasii Grammatici S.J.," *Parnassus Boicus* 1 (1737): 46–53. Schaff, *Geschichte der Physik,* 5, 160, claims that Grammatici was a Copernican, but in light of Amort's eulogy this seems highly unlikely.

48. J. Mangold, *Philosophia rationalis,* 3:425.

49. Finnochiaro, *The Galileo Affair,* 67–69.

50. J. Mangold, *Philosophia rationalis,* 3:434–36. Boscovich had already claimed in a 1742 dissertation, *De annuis aberrationibus,* that the aberration of fixed stars was not in itself proof of the earth's motion (Dadic, "Boskovic and the Question," 132).

51. On Boscovich and heliocentrism, see Dadic, "Boskovic and the Question"; J. Casanovas, "Boscovich as an Astronomer," in *Bicentennial Commemoration of R. G. Boscovich,* ed. M. Bossi & P. Tucci (Milan: Edizioni Unicopli, 1988), 57–70.

52. M. Mangold, *Philosophia recentior,* 2:462.

53. Ibid., 2:453.

54. Benedikt Stattler, S.J., *Theses ex universa philosophia . . .* (Munich, 1763), theses LIX and LXII.

55. Franz Sales Stadler, S.J., *Theses ex universa philosophia* (Munich, 1768), thesis XC.

56. For example, Aegidius Frey, S.J., and Franz Zallinger, S.J., *Positiones ex astronomia physico-geometrica*... (Ingolstadt, 1766); Johann Evangelist Cronthaler, S.J., *Theses ex institutionibus philosophicis biennio explicatis selectae*... (Ingolstadt, 1766); [Sex Religiosi Societatis Jesu], *Materia tentaminis publici quod ex praelectionibus philosophicis*... (Ingolstadt, 1770); Franz Xavier Epp, S.J., *Positiones ex universa philosophia selectae*... (Munich, 1771).

57. Appel, *Theses ex universa philosophia*, 34.

Epilogue: After the Suppression

1. The unprecedented nature of the act was noted by the (by then ex-) Jesuit philosopher Benedikt Stattler in his *Freudschaftliche Verteidigung der Gesellschaft Jesu* (Berlin, 1773).

2. Ferdinand Maaß, *Der Josephinismus: Quellen zu seiner Geschichte in Österreich, 1760–1790*... (Vienna: Herold, 1951–61), vol. 1, doc. 114, draft of letter from Kaunitz to Maria Theresa, Vienna, 25 January 1768.

3. On the Suppression and its aftermath in Germany and Austria, see Winfried Müller, "Die Aufhebung des Jesuitenordens in Bayern: Vorgeschichte, Durchführung, Administrative Bewältigung," *Zeitschrift für bayerische Landesgeschichte* 48 (1985): 285–352, and *Universität und Orden: Die bayerische Landesuniversität Ingolstadt zwischen der Aufhebung des Jesuitenordens und der Säkularisation 1773–1803* (Berlin: Duncker & Humboldt, 1986); Hermann Haberzettl, *Die Stellung der Exjesuiten in Politik und Kulturleben Österreichs zu Ende des 18. Jahrhunderts* (Vienna: Verband der wissenschaftlichen Gesellschaften Österreichs Verlag, 1973); Helmut Kröll, "Die Auswirkungen der Aufhebung des Jesuitenordens in Wien und Niederösterreich," *Zeitschrift für bayerische Landesgeschichte* 34 (1971): 567–617.

4. Maaß, *Der Josephinismus*, vol. 2, doc. 20, Kaunitz to Maria Theresa, Vienna, 8 April 1773.

5. Hermann Hoffmann, *Friedrich II. von Preussen und die Aufhebung der Gesellschaft Jesu* (Rome: Institutum Historicum Societatis Iesu, 1969). Also ironic was that Catherine the Great did not promulgate the bull, so that the Society continued to exist in Russian Poland.

6. Haaß, *Die geistige Haltung*, 23.

7. Joseph Freisen, *Die matrikel der Universität Paderborn: Matricula Universitatis Theodorianae Padibornae, 1614–1844* (Würzburg, 1931), 1:16.

8. Reindl, *Lehre und Forschung*, 27, 69–71, 99–100, 107.

9. Steinmaurer, "Die Lehrkanzel für Experimentalphysik," 56–62.

10. Hammermayer, *Geschichte der Bayerischen Akademie*, 2:382–87.

Conclusion

1. On the history of the term *Scientific Revolution*, see H. Floris Cohen, *The Scientific Revolution: A Historiographical Inquiry* (Chicago: University of Chicago Press, 1994), ch. 2, and David C. Lindberg, "Conceptions of the Scientific Revolution from Bacon to Butterfield," in *Reappraisals of*

the Scientific Revolution, ed. David C. Lindberg and Robert S. Westman (Cambridge: Cambridge University Press, 1990). See also I. Bernard Cohen, *Revolution in Science* (Cambridge, Mass.: Harvard University Press, 1985), on the development of the concept of revolution in science. Regardless of the problems with the term, it remains a useful pedagogical and organizational concept that historians have yet to replace adequately. I have not been able to replace it either.

2. For example, "It now appears unlikely that the 'high theory' of the Scientific Revolution had any substantial direct effect on economically useful technology in either the seventeenth century or the eighteenth." Shapin, *The Scientific Revolution,* 140.

3. Margaret C. Jacob, *Scientific Culture and the Making of the Industrial West* (Oxford: Oxford University Press, 1997).

BIBLIOGRAPHY

Archival Material

Bancroft Library, University of California Berkeley (BL/UCB)
Boscovich Collection

Studienbibliothek Dillingen (SBD)
XV 134, 244, 247.
XVj 133/1, 133/2.

Universitätsarchiv Erlangen (UAE)
Briefsammlung Trew

Stadtarchiv Köln (StAK)
Jes. Abt. 21, 38, 717/3, 716, 727, 990, 1047, 1048, 1069.

Stadtarchiv Mainz (StAMz)
14/241, 250, 1090.
15/110, 130, 340, 432, 454, 456.

Stadtbibliothek Mainz (StBMz)
HS III/69.

Bayerisches Hauptstaatsarchiv, München (BHSA)
Jes. 10, 38, 57, 63, 85, 256, 258, 495, 651, 689, 710–842, 1766.
GL Fasz. 1479: 33, 44; 1479: 75; 1484/II: 37, 40; 1484/III: 6, 8, 11; 1489: 20, 39.

Universitätsarchiv, Ludwig-Maximilians-Universität, München (UALMU)
O-I-4, O-I-5.

Bayerische Staatsbibliothek, München (BSB)

Clm 1609, 1972, 9224, 24076, 24077, 26481a & b, 26496, 27322/I, 27322/II.

Erzbischöfliche Akademische Bibliothek, Paderborn (EABP)

Archiv des Paderborner Studienfonds:

Pa 125, 128 I & II.

I. 4, 12.

Pad 259.

Archivum Romanum Societatis Jesu, Rome (ARSI)

Rhen. Sup. 2, 3, 44. I, 26 I.

FG 656–675.

Cong. 2a, 20c I, 20, 20 f, 20h, 60, 66, 69, 70, 72, 73, 82, 83, 84, 86, 91.

Germ. Sup. 5–17.

Inst. 213.

Archive of the Pontifical University Gregoriana, Rome (APUG)

Kircher Correspondence. Now online at:

www.stanford.edu/group/STS/gorman/nuovepaginekircher/

Landeshauptarchiv Sachsen-Anhalt, Aussenstelle Wernigerode (LSA)

Rep A 37b I, II, Tit. XVI Nr. 50

Bayerisches Staatsarchiv Würzburg (BSAW)

Schönborn Archiv, Korrespondenz Archiv, Johann Philipp.

MRA 40/1273; 40/L 1274 I, II; 40/L 1275a.

Kloster Himmeltal 178, 341.

Historischer Verein M. S. f. 1208.

Manuscripts

Bayerische Staatsbibliothek München (BSB)

Clm 9423, 9424, 12409, 24994, 26092, 26525.

Stadtarchiv Köln (KStA)

Jes. Abt. 29a.

Stadtbibliothek Mainz (MzStB)

HS I 545, 612, 622; II 353, 378, 382, 386.

Universitätsbibliothek München (MüUB)

4° Cod. ms. 672, 673, 678, 722.

Bayerisches Staatsarchiv Würzburg (BSAWü)

HV MS q. 97.

Universitätsbibliothek Würzburg (WüUB)

M. ch. f. 660

WüUB Delin 5.

Published Primary Sources

Acta Eruditorum. Leipzig, 1682–1776.

Alexander, H. G., ed. *The Leibniz-Clarke Correspondence Together with Extracts from Newton's "Principia" and "Opticks."* Manchester: Manchester University Press, 1956.

Appel, Christian, S.J. *Theses ex universa philosophia....* Würzburg, 1769.

Aquinas, Thomas. *Summa theologiae.* Vol. 58. *The Eucharistic Presence (3a. 73–78).* Translated by William Barden, O.P. New York: Blackfriars, 1965.

———. *The Division and Methods of the Sciences: Questions V and VI of his Commentary on the "De Trinitate" of Boethius.* Translated with introduction and notes by Armand Maurer. Toronto: Pontifical Institute of Medieval Studies, 1986.

Arakielowicz, Gregor, S.J. *De mundi systemate dissertatio cosmologica. In qua de Copernicani systematis cum philosophia, sacrisque praesertim litteris congruentia quaestio discutitur.* Premisla, 1768.

Aristotle. *The Physics.* Cambridge, Mass.: Harvard University Press, 1963.

———. "Mechanical Problems." In *Minor Works,* translated by W. S. Hett. Cambridge, Mass.: Harvard University Press, 1963.

———. *De Caelo/On the Heavens.* Cambridge, Mass.: Harvard University Press, 1971.

Arnauld, Antoine. *On True and False Ideas; New Objections to Descartes' Meditations; and Descartes' Replies.* Translated by Elmar J. Kremer. Lewiston, N.Y.: E. Mellen, 1990.

Arriaga, Rodrigo de, S.J. *Cursus philosophicus....* 1632. Lyon, 1653.

———. *Cursus philosophicus, iam noviter maxima ex parte auctus, & illustratus, & a variis objectionibus liberatus, necnon a mendis expurgatus.* Leiden, 1669.

Averroes. *On the Harmony of Religion and Philosophy.* Translated by George F. Hourani. London: Luzac, 1961.

Banholzer, Johannes, S.J. *Philosophia legalis, seu Quaestiones dialecticae physicae et metaphysicae ad scientiam juris accomodatae.* Dillingen, 1682.

———. *Philosophia naturalis ad obsequium orthodoxae fidei circa grande Eucharistiae mysterium....* Dillingen, 1682.

Becanus, Martin, S.J. *A Controversy, in which the Communion of Calvin in whole overthrowne, and the Real Presence of Christs body in the Eucharist confirmed.* N.p., 1614. Reprinted as vol. 46 of *English Recusant Literature 1558–1640.* Menston: Scholar Press, 1970.

———. *Theologiae scholasticae.* Pt. 3, tr. 2. Mainz, 1628.

Bellum literarium in sophiae stadio equestri praelio et eruditae pacis triumpho coronatum. Mainz, 1745.

Bernard, Georg, S.J. *Assertiones philosophicae de corpore naturali principato....* Ingolstadt, 1629.

Blazicek, Oldrich J., ed. *Theses in Universitate Carolina Pragensi disputatae (saec. XVII et XVIII).* 7 vols. Prague: Pragopress, 1967–70.

Boscovich, Roger Joseph, S.J. *A Theory of Natural Philosophy.* Cambridge, Mass.: M. I. T. Press, 1966.

Boyle, Robert. *New Experiments Physico-Mechanical, touching the Spring of the Air, and its Effects; Made for the most Part, in a New Pneumatical Engine....* Oxford, 1660. Latin edition, *Noua experimenta physico-mechanica de vi aeris elastica, & ejusdem effectibus: facta maximam partem in nova machina pneumatica....* Oxford, 1661.

Branca, Giovanni. *Le macchine.* Rome, 1629. Facsimile ed., Turin, 1977.

Braun, Joseph, S.J. *Dissertatio philosophica hermeneutica scientiam interpretandi, sive Elementa & principia artis criticiae exhibens.* . . . Würzburg, 1737.

———. *Theses ex universa philosophia, quas cum adjuncta dissertatione, praecipua Cartesianae doctrinae capita examinante theses ex universa philosophia.* . . . Würzburg, 1737.

Brucker, Jakob. *Historia critica philosophiae a mundi incunabulis ad nostram usque aetatem deducta.* Leipzig, 1742–43.

Burkhaeuser, Nicolas, S.J. *Theoria corporis naturalis principiis Boscovichii conformata, quam una cum thesibus ex philosophia universa.* . . . Würzburg, 1770.

———. *Theses selectae ex logica, metaphysica et mathesi.* . . . Würzburg, 1773.

Busenbaum, Hermann, S.J. *Medulla theologiae moralis, facili ac perspectiva methodo resolvens casus conscientiae.* Leiden, 1660.

Calmelet, Conrad, S.J. *Coeli coelorumque virtutes.* . . . Ingolstadt, 1642.

———. *Philosophia per SS. Eucharistiam illustrata.* Ingolstadt, 1642.

Canisius, Peter. *A Summe of Christian Doctrine.* N.p., n.d. [1592–96]. Reprinted as vol. 35 of *English Recusant Literature 1558–1640.* Menston: Scholar Press, 1971.

———. *Beati Petri Canisii Societatis Iesu epistolae et acta.* Edited by Otto Braunsberger, S.J. Freiburg i. Br.: Herder, 1913.

Carleton, Thomas Compton, S.J. *Philosophia universa.* Antwerp, 1649.

Carpentarius, Fiacrius. *Publicae exercitationis ergo, varia logica, physica ethicaque asserta.* . . . Würzburg, 1597.

Coimbran Commentators [Conimbricenses]. *Commentarii . . . in universam dialecticam Aristotelis Stagiritae.* . . . Lyon, 1607.

———. *Commentarium . . . super octo libros physicorum Aristotelis Stagiritae.* . . . Venice, 1616.

———. *Commentarii . . . in quatuor libros. De caelo, meteorologicos & parva naturalia, Aristotelis Stagiritae.* . . . Cologne, 1631.

Cordier, Philipp, S.J. *Theses ex universa physica.* . . . Cologne, 1752.

Cornaeus, Melchior, S.J. *Curriculum philosophiae peripateticae, uti hoc tempore in scholis decurri solet.* . . . Würzburg, 1657.

———. *Ens rationis Luthero-Calvinicum ab Aristotele redivivo explicatum.* Würzburg, 1659.

Cronthaler, Johann Evangelist, S.J. *Theses ex institutionibus philosophicis biennio explicatis selectae.* . . . Ingolstadt, 1766.

Crusius, Christian August. *Die philosophischen Hauptwerke.* Edited by Giorgio Tonelli. Hildesheim: Georg Olms, 1969.

Cysat, Johann Baptista, S.J. *Mathemata astronomica de loco, motu, magnitudine, et causis cometae qui sub finem anni 1618 et initium anni 1619 in coeli fulsit.* Ingolstadt, 1619.

Dabutz, Florian, S.J. *Hydraulica. Quam una cum selectis ex mathesi corollaris exponent et demonstrabunt.* . . . Mainz, 1768.

[Daniel, Gabriel, S.J.]. *A Voyage to the World of Cartesius.* London, 1694.

De Lugo, Juan, S.J. *Disputationes scholasticae et morales.* 1638. Reprint, Paris: Vivès, 1869.

Denzinger, Heinrich, ed. *Kompendium der Glaubensbekenntnisse und kirchlichen Lehrentscheidungen.* 37th ed. Freiburg i. Br.: Herder, 1991.

Descartes, René. *Oeuvres de Descartes.* Edited by Charles Adam and Paul Tannery. Paris: Cerf, 1897–1913.

———. *The Philosophical Works of Descartes.* Translated by Elizabeth S. Haldane and G. R. T. Ross. Cambridge: Cambridge University Press, 1911–12.

———. *The World, ou Traite de la Lumiere.* Translated by Michael Sean Mahoney. New York: Aberis, 1979.

———. *A Discourse on Method; Meditations on the First Philosophy; Principles of Philosophy.* Translated by John Veitch. London: Everyman, 1994.

Dessloch, Emmanuel, S.J. *Disquisitio physico-mathematica, gemina, de vacuo itemque de attractione. . . .* Amsterdam, 1661.

———. *Dissertationes ethicae ex principiis philosophiae moralis . . . una cum thesibus ex universa philosophia. . . .* Würzburg, 1736.

Dobrynicki, Caspar Maximilian/Godefried Roos, S.J. *Armamentarium sapientiae humanae, quod oppugnandis peripateticae doctrinae hostibus. . . .* Mainz, 1620.

Drexelius, Jeremius, S.J. *The Heliotropium ("Turning to HIM") or Conformity of the Human Will to the Divine.* New York: Devin-Adair, 1912.

E. A. [Eusebius Amort]. "Von dem Leben und gelehrten Schrifften P. Nicasii Grammatici S.J." *Parnassus Boicus* 1 (1737): 46–53.

Ecker, Josef. *Positiones ex philosophia selectae. . . .* Ingolstadt, 1770.

Eimer, Ludwig, S.J. *Dissertatio inauguralis de primo philosophiae Cartesianae fundamento, seu generali de rebus omnibus dubitatione in fide periculosa. . . .* Bamberg, 1749.

Eisentraut, Peter, S.J. *Dissertatio de corporum electricorum vi attractiva & repulsiva, una cum thesibus ex universa philosophia. . . .* Würzburg, 1748.

Ellspacher, Franz, S.J. *Barometron Torricellianum quaestionibus philosophicis subjectum. . . .* Dillingen, 1714.

Epistolae praepositorum generalium ad patres et fratres Societatis Jesu. Prague, 1711.

Epp, Franz Xavier, S.J. *Positiones ex universa philosophia selectae. . . .* Munich, 1771.

Erber, Anton, S.J. *Cursus philosophicus methodo scholastica elucubrata.* 3 vols. Vienna, 1750.

Etwelche Meistens bayrische Denck- und Leß-Würdigkeiten zur Fortführung des so genannten Parnassi Bojci. . . . Ingolstadt, 1737–40.

Ewick, Engelbert, S.J. *Oculus naturae pro discernendis erroribus, detegendis imposturibus, declinandis terroribus. . . .* Würzburg, 1631.

Falck, Joseph, S.J. *Liquidorum gravitas et aequilibrium. . . .* Freiburg i. Br., 1713.

———. *Mundus aspectabilis philosophice consideratus. . . .* Augsburg, 1740.

Fechtnerová, Anna. *Katalog grafickich listu univerzitních tezí ulozenich ve Státní knihovne CSR v Praze.* Prague: Státní knihovna CSR, 1984.

Fitzpatrick, Edward A., ed. *St. Ignatius and the Ratio Studiorum.* New York: McGraw-Hill, 1933.

Fonseca, Pedro da, S.J. *Institutionum dialecticarum, libri octo. Emendatius quam antehac editi. . . .* Lisbon, 1564. Reprint, Ingolstadt, 1604.

Freudenberger, Theobaldus, ed. *Concilii Tridentini actorum,* pt. 3, vol. 3, *Summaria sententiarum theologorum super articulis Lutheranorum de sacramentis purgatorio indulgentis sacrificio missae in concilio Bononiensi disputatis. . . .* Freiburg i. Br.: Herder, 1974.

Frey, Aegidius, S.J., and Franz Zallinger, S.J. *Positiones ex astronomia physico-geometrica.* . . . Ingolstadt, 1766.

Frey, Bernard, S.J. *Controversiae physicae de corpore naturali causato.* . . . Ingolstadt, 1644.

———. *Controversiae physicae de corpore naturali principiato.* . . . Ingolstadt, 1644.

———. *Controversiae physicae de corpore naturali simplici.* . . . Ingolstadt, 1644.

———. *Controversiae physicae de proprietatibus corporis naturalis.* . . . Ingolstadt, 1644.

Friderich, Philipp, S.J. *Biennium philosophicum inauguratione solenni coronatum . . . Quid sentiendum sit de methodo ac libertate philosophandi a RR. quibusdam asserta ?* . . . Mainz, 1745.

———. *Selectae ex philosophia peripatetica universa theses et antitheses parallelae, seu hypotheses Cartesii cum systemate peripatetico dissertatio irenica partim conciliatae, partim excussis illarum fundamentis ab erroribus assertae.* . . . Mainz, 1745.

Froehling, Michael, S.J. *Meditatio philosophica de eo, quod licitum est philosopho circa usum S. Scripturae.* . . . Würzburg, 1753.

Furttenbach, Joseph. *Mechanische Reißladen, das ist ein gar geschmeidigte bey sich verborgen tragende Laden.* . . . Augsburg, 1644.

Gassendi, Pierre. *Opera omnia.* Lyon, 1658.

Gautruche, Pierre, S.J. *Philosophiae ac mathematicae totius insitutio.* . . . Vienna, 1661.

Geiger, Friderich, S.J. *Theses inaugurales philosophicae: Sapientia mundi.* Würzburg, 1694.

Gerl, Herbert, S.J. *Catalogus generalis Provinciae Germaniae Superioris et Bavariae Societatis Iesu, 1556–1773.* N.p., 1968.

Giudice, Carpophorus dei. *Tractatus de accidentibus absolutis, sive Sacro-Sanctum Eucharistiae Sacramentum in principiis philosophiae peripateticae impugnatum, et ex placitis philosophiae corpuscularíae propugnatum.* Paderborn, 1718.

Gobat, Georg, S.J. *Mundus ad disputationem publicam in florentissimo Friburgensi Helvetiorum gymnasio.* . . . Bern, 1635.

Gordon, Andreas, O. S. B. *Varia philosophiae mutationem spectantia.* . . . Erfurt, 1749.

———. *P.S. de scripto R.P. Opffermann S.J. nuperrime hic edito.* Erfurt, 1749.

Grant, Edward, ed. *A Source Book in Medieval Science.* Cambridge, Mass.: Harvard University Press, 1974.

Gretser, Jacob. *Lutherus Academicus et Waldenses.* Vol. 12 of *Opera omnia.* 1610. Reprint, Regensburg, 1738.

Grimaldi, Francesco Maria, S.J. *Physico-mathesis de lumine, coloribus et irida: Aliisque adnexis, libri duo.* . . . Bologna, 1665.

Gropp, Ignatii. *Collectio novissima scriptorum et rerum Wirceburgensium a saeculo XVI. XVII. et XVIII.* . . . Frankfurt, 1741–44.

Guericke, Otto von. *Experimenta nova (ut vocantur) Magdeburgica de vacuo spatio.* Amsterdam, 1672.

———. *Die Belagerung, Eroberung und Zerstörung der Stadt Magdeburg am 10/20 Mai 1631.* Leipzig: Voigtlander, 1912.

———. *Neue (sogenannte) Magdeburger Versuche über den leeren Raum. Nebst Briefen, Urkunden und anderen Zeugnissen seiner Lebens- und Schaffensgeschichte.* Translated and edited by Hans Schimank. Düsseldorf: VDI-Verlag, 1968.

———. *The New (So-Called) Magdeburg Experiments of Otto von Guericke.* Translated by Margaret Ames. Dordrecht: Kluwer, 1994.

Hardy, Francis, S.J. *Theses philosophicae.* Würzburg, 1728.

Hartung, Johannes, S.J. *Iter peripateticum triennali labore feliciter confectum, festivo carmine celebratum....* Würzburg, 1713.

Haunold, Christophor, S.J. *Acroamata physica exotericis medicis permixta....* Dillingen, 1645.

———. *De ortu et interitu theoremata physica mathematicis permixta....* Dillingen, 1645.

Hauser, Berthold, S.J. *Elementa philosophiae ad rationis et experientiae ductum conscripta, atque usibus scholasticis accomodata....* Augsburg, 1755–64.

Henner, Blasius. *Conatus physico-experimentales de corporum affectionibus tum generalibus tum specialibus, ad usum philosophiae candidatorum suscepti....* 2 vols. Würzburg, 1756–69.

Henner, Georg, S.J. *Selectae propositiones physico-experimentales experimentis et usibus illustratae de quatuor elementis....* Würzburg, 1762.

———. *Propositiones selectae ex praecipuis partibus physicae experiementalis....* Würzburg, 1766.

———. *Positiones selectae ex physica experimentali....* Würzburg, 1769.

Hermann, Georg, S.J. *Lapis offensionis atomisticae a peripatetico motus, sive Dialogi inter peripateticum, & atomistam veritatis studiosum circa generationem formarum peripateticarum....* Ingolstadt, 1730.

Heyss, Ferdinand, S.J. *Disputatio philosophica de corpore simplici elementari et meteoro....* Dillingen, 1641.

Hinlang, Adam, S.J. *Theses ex universa philosophia.* Düsseldorf, 1760.

Höchtl, Christopher, S.J. *Natura, et proprietatis aeris....* Ingolstadt, 1730.

Huben, Franz, S.J. *Mega-cosmus primum in sua origine, natura, elementis, & quibusdam phoenomenis per hanc trieteridem observatis, demum in microcosmo consideratus....* Mainz, 1719.

Hubert, Franz, S.J. *Observationes meteorologico-thermometricae ad annos MDCCLXV et MDCCLXXVI. Cum nonnullis miscellaneis philosophico-literariis....* Würzburg, 1768.

———. *Observationes meteorologico-thermometricae ad annos MDCCLXVII et MDCCLXVIII. Cum nonnullis miscellaneis philosophico-literariis....* Würzburg, 1768.

———. *Observationes meteorologico-thermometricae ad annum MDCCLXVIII. Cum nonnullis miscellaneis philosophico-literariis....* Würzburg, 1769.

———. *Observationes meteorologicae ad annum MDCCLXIX. Cum nonnullis miscellaneis philosophico-literariis....* Würzburg, 1770.

———. *Observationes meteorologicae ad annum MDCCLXX. Cum nonnullis miscellaneis philosophico-literariis....* Würzburg, 1771.

Ignatius of Loyola. *The Letters of St. Ignatius of Loyola.* Edited and translated by William J. Young, S.J. Chicago: Loyola University Press, 1959.

———. *The Constitutions of the Society of Jesus.* Translated with an introduction and a commentary by George E. Ganss, S.J. St. Louis: Institute of Jesuit Sources, 1970.

———. *The Autobiography of St. Ignatius Loyola with Related Documents.* Edited with introduction and notes by John C. Olin. Translated by Joseph F. O'Callaghan. New York: Harper and Row, 1974.

———. *Counsels for Jesuits: Selected Letters and Instructions of Saint Ignatius Loyola.* Edited by Joseph N. Tylenda, S.J. Chicago: Loyola University Press, 1985.

———. *Briefe und Unterweisungen.* Translated by Peter Knauer. Würzburg: Echter, 1993.

———. *Personal Writings.* Edited by Joseph A. Munitz, S.J., and Philip Endean. Harmondsworth: Penguin, 1996.

———. *The Spiritual Exercises of Saint Ignatius.* Translated by Louis J. Puhl. New York: Vintage Books, 2000.

Institutum Societatis Jesu. Florence, 1893.

Instructio pro visitatoribus Societatis Iesu. Naples, 1608.

Kilber, Heinrich, S.J. *Theses ex universa philosophia, quas cum adjuncta dissertatione, praecipua Cartesianae doctrinae capita examinante. . . .* Mainz, 1747.

Kircher, Athanasius, S.J. *Romani Collegii Societatis JESU Musaeum Celeberrimum. . . .* Amsterdam, n.d.

———. *Musurgia universalis, sive Ars magna consoni et dissoni in X. libros. . . .* Rome, 1650.

———. *Iter exstaticum coeleste. . . .* Edited by Kaspar Schott. 1660. Würzburg, 1671.

Kleinbrodt, Anton, S.J. *Physica particularis.* 3 vols. Ingolstadt, 1704.

———. *Mundus elementaris disputationi subjectus.* 3 vols. Ingolstadt, 1704.

Kleiner, Joseph, S.J. *Meditatio de eo, quod justum esse videtur circa crisin de rebus philosophicis moderate ferendam. . . .* Würzburg, 1760.

———. *De transitu Veneris per discum solis, prout is observatus fuit Wirceburgi ad diem astron. 5, vulgatum 6. Junii, anno M. DCC. LXI., expositio.* Würzburg, 1761.

Kluntzingk, Johannes. *Placita philosophorum logica, physica, metaphysica. . . .* Mainz, 1610.

Knittel, Kaspar, S.J. *Aristoteles curiosus et utilis in quo centum praecipuae quaestiones peripateticae problematicè disputantur, ac utriusque partis argumenta dissolvuntur. . . .* Prague, 1682.

———. *Theses curiosiores ex universa Aristotelis philosophia . . . propugnatae. . . .* Prague, 1682.

Kretz, Marquard, S.J. *Magiae artificialis phaenomena quaedam magis curiosa accuratius discussa & explicata. . . .* Würzburg, 1750.

Le Cazre, Pierre, S.J. *Physica demonstratio qua ratio, mensura, modus ac potentia accelerationis motus in naturali decensu gravium determinatur. . . .* Paris, 1645.

———. "Clarissimo Sapientissimóque Viro D. D. Petro Gassendo Diniensis Ecclesiae Praeposito dignissimo, Petrus Cazreus Soc. Iesu." In Pierre Gassendi, *Opera omnia,* vol. 7. Lyon, 1698.

Lechner, Kaspar, S.J. *Disputatio philosophica de proprietatibus compositi physici. . . .* Ingolstadt, 1616.

Leibniz, Gottfried Wilhelm. *Die philosophischen Schriften von Gottfried Wilhelm Leibniz.* Vol. 2. Edited by C. J. Gerhardt. Berlin, 1879.

———. *Philosophical Papers and Letters: A Selection Translated and Edited, with an Introduction by Leroy E. Loemker.* Dordrecht: D. Reidel, 1969.

Leinberer, Wolfgang, S.J. *Disputatio philosophica de natura et perfectione mundi. . . .* Ingolstadt, 1670.

Lessius, Leonard. *De gratis efficaci, decretis divinis libertate arbitrarii et praescientia Dei conditionata.* Antwerp, 1610.

Line, Francis, S.J. *Tractatus de corporum inseperabilitate; in quo experimenta de vacuo, tam Torricelliana, quam Magdeburgica, & Boyliana, examinatur, veraque eorum causa detecta, ostenditur, vacuum naturaliter non dari posse. . . .* London, 1661.

Linz, Valentinus, S.J. *Certamen inaugurale ex philosophia universa . . . Micae philosophicae de mensa physicorum.* Würzburg, 1764.

Loder, Georg, S.J. *Theses de mundo. . . .* Würzburg, 1693.

Lohnmüller, Andreas, S.J. *Philosophiae antiquae utilitas & jucunditas eruditae crisi proposita. . . .* Mainz, 1750.

———. *Physiognomia philosophiae peripateticae pars utilis et jucunda a novis philosophis neglecta. . . .* Mainz, 1750.

Lorini, Bonaiuto. *Fünf Bücher von Vestung Bawen . . . ubergesetzt durch David Wormbser.* Frankfurt am Main, 1621.

Maaß, Ferdinand. *Der Josephinismus: Quellen zu seiner Geschichte in Österreich, 1760–1790: Amtliche Dokumente aus dem Wiener Haus-, Hof- und Staatsarchiv.* 5 vols. Vienna: Herold, 1951–61.

Magirus, Johann. *Physiologiae peripateticae. Libri sex: Cum commentariis. . . .* Cambridge, 1642.

Mangold, Joseph, S.J. *Philosophia rationalis et experimentalis hodiernis discentium studiis accomodata auctore. . . .* Ingolstadt, 1755.

Mangold, Maximus, S.J. *Positiones ex philosophia rationali et experimentali selectae. . . .* Munich, 1757.

———. *Philosophia recentior praelectionibus publicis accomodata a Patre Maximo Mangold Soc. Jesu. . . .* Vol. 1. Mainz,: 1763–64.

Manz, Caspar. *Iudicium super illa quaestione: Utrum dari possit melior et Christianae pietati conformior modus docendi philosophiam, quam sit vulgaris?* Ingolstadt, 1648. Reprinted in Arno Seiffert, "Der jesuitische Bildungskanon im Lichte Zeitgenossischer Kritik." *Zeitschrift für bayerische Landesgeschichte* 47 (1984): 43–75.

Mayer, Johannes. *Theses ex universis philosophiae partibus. . . .* Mainz, 1616.

Mayr, Anton, S.J. *Philosophia peripatetica antiquorum principiis, et recentiorum experimentis conformata. . . .* Ingolstadt, 1739.

Mederer, Joannes Nepomuk, S.J. *Annales Ingolstadiensis Academiae.* Pt. 3. *Ab anno 1672 ad annum 1772.* Ingolstadt, 1782.

Mohr, Nicolas, S.J. *Flores ex universa Aristotelis philosophia. . . .* Würzburg, 1658.

Monumenta Ignatiana, ex autographis vel ex antiquioribus exemplis collecta. Series prima. Sancti Ignatii de Loyola Societatis Jesu fundatoris epistolae et instructiones. 12 vols. Madrid: Typis G. Lopez del Horno, 1903–11.

Monumenta Paedagogica Societatis Jesu. 7 vols. Edited by L. Lukacs, S.J. Rome: Institutum Historicum Societatis Iesu, 1965–92.

Morasch, Johann Adam. *Philosophica atomistica in alma electorali Universitate Ingolstadiensi disputationi subjecta. . . .* Ingolstadt: 1727.

———. *Atomismus ad injustis peripateticorum censuris et imposturis vindicatus. . . .* N.p., 1733.

Natalis, Stephan, S.J. *Physica vetus et nova.* Paris, 1648.

Niderndorff, Heinrich, S.J. *Fundamenta geographiae, sive Propositiones geographiae mathematicae. . . .* Würzburg, 1735.

Nebel, Anton, S.J. *Disputatio I. Menstrua de principiis intrinsecis corporis naturalis in genere et in specie, cum resolutione quaestionis: An dentur accidentia absoluta?* Würzburg, 1747.

———. *Disputatio II. Menstrua de composito, natura, & arte cum resolutione quaestionis, an calor sit accidens absolutum?* Würzburg, 1747.

———. *Disputatio III. Menstrua de causis in genere, cum resolutione quaestionis: An frigus sit accidens absolutum?* Würzburg, 1747.

———. *Disputatio IV. Menstrua de causis in specie, cum resolutione quaestionis: An siccitas & humiditas sint accidentia absoluta?* Würzburg, 1747.

———. *Phaenomena barometri. Quatuor dissertationibus illustrata. Dissertatio I. De aeris gravitate, elasticitate & pressione, ratione et experimentis-hydro- & aero-technicis comprobata. . . .* Würzburg, 1748.

———. *Phaenomena barometri. Quatuor dissertationibus illustrata. Dissertatio II. De barometri inventione, progressu, & aliquibus circa hoc phaenomenon eruditorum quaestionibus. . . .* Würzburg, 1748.

———. *Phaenomena barometri. Quatuor dissertationibus illustrata. Dissertatio III. De causa suspensionis mercurii in barometro. . . .* Würzburg, 1748.

———. *Phaenomena barometri. Quatuor dissertationibus illustrata. Dissertatio IV. De phaenomenis barometri et prognosi ex illis deducenda. . . .* Würzburg, 1748.

Neubauer, Ignatius, S.J. *Influxus astrorum in sublunaria, quem una selectis corollariis ex universa philosophia. . . .* Heidelberg, 1763.

Neuff, Franz, S.J. *Authoritatis in philosophia pondus aequa examinatum trutina, ac philosophicis praefixum thesibus. . . .* Mainz, 1763.

Ockham, William of. *Tractatus de quantitate et Tractatus de corpore Christi.* Edited by Carolus A. Grassi. St. Bonaventure, N.Y.: St. Bonaventure University, 1986.

———. *Quodlibetal Questions.* 2 vols. Translated by Alfred J. Freddoso and Francis E. Kelley. New Haven: Yale University Press, 1991.

Offermann, Paul, S.J. *Theses selectae ex universa philosophia ad majorem dei optimi maximi gloriam, virginis sine labe conceptae honorem. . . .* Mainz, 1761.

Pachtler, G. M., ed. *Ratio Studiorum et Institutiones Scholasticae Societatis Jesu per Germaniam olim vigentes collectae concinnatae dilucidatae.* 4 vols. Berlin: A. Hoffman, 1887–94.

Padberg, John W., S.J., Martin D. O'Keefe, S.J., and John L. McCarthy, S.J. *For Matters of Greater Moment: The First Thirty Jesuit Congregations.* St. Louis: Institute of Jesuit Sources, 1994.

Pallavicino, Sforza, S.J. *Vindicationes Societatis Jesu quibus multorum accusationes in eius institutum, leges, gymnasia, mores refelluntur.* Rome, 1649.

———. *Vera Concilii Tridentini historia contra falsam Petri Suavis Polani narrationem. . . .* Antwerp, 1670.

Parnassus Boicus, oder neu-eröffneter Musen-Berg worauf verschidene Denck-und Leswürdigkeiten auß der gelehrten Welt zumahlen aber auß denen Landen zu Bayrn abgehandelt werden. Vols. 1–4. Munich, 1722–26. 2d ser., Ingolstadt, 1737–40.

Pfister, Adam, S.J. *Universa philosophia eclectica contentiosa & experimentalis in assertiones inaugurales selectas digesta. . . .* Würzburg, 1747.

Pfriemb, Joseph, S.J. *Apologia qua errores R.P. Andreae Gordon, O.S.B. contra philosophiam scholasticam. . . .* Mainz, 1748. Reprinted in Andreas Gordon, O. S. B., *Varia philosophiae mutationem spectantia . . .*, 35–60. Erfurt, 1749.

———. *Connubium felix philosophiae theoreticae cum experimentali. . . .* Mainz, 1749.

Picinelli, Philippo. *Mundus symbolicus, in emblematum universitate formatus, explicatus, et tam sacris, quam profanis eruditionibus ac sententiis illustratus. . . .* Cologne, 1694.

Plutarch. *The Lives of the Noble Grecians and Romans.* Translated by John Dryden. New York, n.d.

Porta, Giovanni Batista della. *Natural Magick.* London, 1658.

Ramelli, Agostino de. *Schatzkammer mechanischer Künste.* N.p., 1620.

Rassler, Franz. *Problema physicum inter Aristotelem et Empedoclem controversum de primis initiis corporis generalis.* Dillingen, 1685.

Redlhamer, Josef, S.J. *Philosophiae naturalis. Pars prima, seu Physica generalis ad praefixam in scholis nostris normam concinnata....* Vienna, 1755.

Reinzer, Franz, S.J. *Meteorologia philosophico-politica in duodecim dissertationes per quaestiones meteorologicas et conclusiones politicas divisa, appositisque symbolis illustrata....* Augsburg, 1709.

Remscheidt, Johannes. *Assertiae theologicae ex universa theologia....* Mainz, 1663.

Robert, Jacob, S.J. *Hortus philosophicus triennali labore cultus....* Mainz, 1674.

Roth, Johannes, S.J. *Theses ex universa philosophia selectae quas ad majorem Dei Tri-Unius gloriam sub patrocinio immaculatae semper virginis Mariae....* Mainz, 1743.

Rottenberger, Philipp Wolffgang, S.J. *Summa vitae S. Francisci Xaverii... assertionibus ex universa philosophia depromtis, illustrata....* Würzburg, 1648.

Salmanticenses. *Collegii Salmanticensis... Cursus theologicus juxta miram Divi Thomae Praeceptoris Angelici doctrinam. Tomus decimus octavus.* Paris: Palmé, 1882.

Sardagna, Carlo, S.J. *Theologia dogmatico-polemica, qua adversus veteres novasque haereses ex Scripturis, patribus, atque ecclesiastica historia catholica veritas propugnatur.* 8 vols. Regensburg, 1770–71.

Sarpi, Paolo [Pietro Soave Polano, pseud.]. *The History of the Council of Trent.* Translated by Nathanael Brent. London, 1676.

Sartorius, Eucharius, S.J. *Elementum aquae. Hoc est disputatio de elementis ingenere, de aqua in communi, de mari, fontibus, fluminibus, thermis alijsque aqueis....* Würzburg, 1646.

———. *Conclusionum peripateticorum centuria, quam post triennialem in sapientiae agone exercitationem pro suprema philosophiae laurea in arenam educet....* Würzburg, 1647.

Saur, Georg, S.J. *Difficultates physicae Cartesianae, thesibus inauguralibus philosophicis propositae, subjuncta doctrina Aristotelica contraria magis elucidatae & aggravatae, disquisitionibus peripateticis subjectae....* Würzburg, 1705.

———. *Relatio historica judiciorum et censurarum adversus philosophorum anti-peripateticum; subjunctis thesibus ex universa philosophia peripatetica.* Würzburg, 1708.

Scheidsach, Fructuosus. *Tractatus de accidentibus absolutis, sive Sacro-Sanctum Eucharistiae Sacramentum in principiis philosophiae peripateticae impugnatum, et ex placitis philosophiae corpuscularíae propugnatum.* Paderborn, 1718.

Scherffer, Karl, S.J. *Institutionum physicae.* Vienna: Trattner, 1752–53.

Schmid, Maximilian, S.J. *Veritatum peripateticarum pentadecas physica....* Würzburg, 1623.

Schmitz, Josef, S.J. *Theses ex universa philosophia.* Cologne, 1752.

Scholl, Jakob, et al. *Physica disputatio de quantitate, loco, et tempore....* Würzburg, 1594.

Schott, Kaspar, S.J. *Magia universalis naturae et artis, sive Recondita naturalium & artificialium rerum scientia....* 2 vols. Bamberg, 1657.

———. *Mechanica hydraulico-pneumatica: Qua praeterquam quod aquei elementi natura proprietas, vis motrix, atque occultas cum aere conflictus, a primis fundamentis demonstratur....* Würzburg, 1657.

———. *Cursus mathematicus, sive Absoluta omnium mathematicum disciplinarum encyclopaedia, in libros XXVIII. digesta.* . . . Würzburg, 1661.

———. *Mathesis Caesarea, sive Amussis Ferdinandea . . . ad problemata universae matheseos, praesertim verò architecturae militaris explicata jussu & auctoritate Augustissimi Imperatoris Ferdinandi III.* . . . Würzburg, 1662.

———. *Technica curiosa, sive Mirabilia artis.* Würzburg, 1664.

———. *Joco-seriorum naturae et artis, sive Magiae naturalis centuriae tres.* Würzburg, 1666.

Schwenter, Daniel, and Georg Philipp Harsdörffer. *Deliciae Physico-Mathematicae, oder Mathematische und Philosophische Erquickstunden.* Nuremberg: 1636–53.

Seitz, Ignatius, S.J. *Conclusiones ex universa philosophia contentiosa et experimentali.* . . . Würzburg, 1742.

———. *Disquisitio physico-experimentalis de cometarum causis & effectibus exhibens insuper seriem chronologicam cometarum insigniorum a Christo nato usque cometam hoc anno observatam.* . . . Würzburg, 1742.

Seitz, Wolfgang, and Bernard Schemmel, eds. *Die graphischen Thesen- und Promotionsblätter in Bamberg.* Wiesbaden: Harrassowitz, 2001.

Semery, Andreas, S.J. *Triennium philosophicum.* . . . Cologne, 1688.

[Sex Religiosi Societatis Jesu]. *Materia tentaminis publici quod ex praelectionibus philosophicis.* . . . Ingolstadt, 1770.

[Sex Societatis JESU Religiosi]. *Theses ex universa philosophia rationali et experimentali.* Ingolstadt, 1762.

Seyfrid, Joannes, S.J. *Theses ex universa philosophia.* . . . Würzburg, 1712.

Spindler, Max. *Electoralis Academiae Scientiarum Boicae Primordia: Briefe aus der Gründungszeit der Bayerischen Akademie der Wissenschaften.* Munich: C. H. Beck, 1959.

Sprat, Thomas. *History of the Royal Society.* London, 1667.

Stadler, Franz Sales, S.J. *Theses ex universa philosophia.* Munich, 1768.

Stattler, Benedikt, S. J, *Theses ex universa philosophia.* . . . Munich, 1763.

———. *Philosophia methodo scientiis propria explanata.* . . . Augsburg, 1769–72.

———. *Freudschaftliche Verteidigung der Gesellschaft Jesu.* Berlin, 1773.

Staudenhecht, Friederich, S.J. *Summa vitae S. Francisci Xavierii.* . . . Würzburg, 1648.

———. *Positiones ex universa desumptae philosophia.* . . . Würzburg, 1650.

Stebler, Franz Anton Ferdinand. *Aristoteles atomista, seu Explicari solitis apud atomistas principiis non difformis, sed conformis omnino Aristotelis doctrina.* Ingolstadt, 1740.

Steinhauser, Joseph, S.J. *Gassendus de gravitate corporum naturalium quid opinatus fuerit? Prolusio critica in Disputationem III. Physicam.* . . . Würzburg, 1764.

Stenmans, Peter. *Litterae annuae. Die Jahresberichte des Neusser Jesuitenkollegs, 1616–1773.* Neuss: Schriftenreihe des Stadtarchivs Neuss, 1966.

Sturm, Johann Christophor. *Collegium experimentale, sive Curiosum.* . . . Nuremberg, 1676.

Suárez, Francisco, S.J. *Metaphysicarum disputationum in quibus et universa naturalis theologia ordinate traditur.* . . . 2 vols. 1597. Mainz, 1606. Venice, 1610.

———. *Theologiae R.P. Fr. Suarez, summa, seu Compendium.* . . . Paris, 1858.

Tanner, Adam, S.J. *Dissertatio peripatetico-theologica. De coelis. In qua de coelorum ortu, interitu, substantia, accidentibus, novis phaenomenis, ac numero.* . . . Ingolstadt, 1621.

Temmik, Aloysius [Caspar Kuemmet?]. *Philosophia vera theologiae et medicinae ministra.* Cologne, 1706.

Thoelen, Heinrich, S.J. *Menologium oder Lebensbilder aus der Geschichte der deutschen Ordensprovinz der Gesellschaft Jesu.* Roermond, 1901.

Toledo, Francisco, S.J. *Commentaria una cum quaestionibus in universam Aristotelis logicam.* Rome, 1572.

Toledo, Francisco, S.J. *Commentaria una cum quaestionibus, in VIII libros, de physica auscultatione.* Cologne, 1574.

Tolomei, Giovanni Baptista, S.J. *Philosophia mentis et sensuum secundum utramque Aristotelis methodum pertracta metaphysice, et empirice.* Augsburg, 1698.

Trentel, Franz, S.J. *Positiones ex universa philosophia.* Würzburg, 1765.

Udr, Peter, S.J. *Stellae vigiliis astronomicis elucubratae . . . apud Friburgenses Helvet: . . . .* Bern, 1635.

Vaeth, Georg, S.J. *Propositiones ex universa philosophia. . . .* Würzburg, 1771.

———. *Concordia veritatis ontologicae de accidentibus cum veritate theologica de speciebus Eucharisticis brevi dissertatione stabilitata.* Würzburg, 1772.

Vogler, Joseph, S.J. *Corpus humanum, sive Disputatio physica de fabrica, nutritione, et vita partium corporis humani. . . .* Dillingen, 1696.

Voit, Edmund, S.J. *Veritas philosophiae peripatetico-Christianae ex contrariis doctrinis breviter appositis magis elucescens. . . .* Würzburg, 1740.

Weber, Adam, S.J. *Universa philosophia, quam assertionum peripateticorum tribus centuriis complexus. . . .* Würzburg, 1653.

Weiler, Heinrich, S.J. *Religio prudentum per sanioris philosophiae principia inprudentiae convicta, dissertatio philosophicae, quam cum annexis corollaris ex universa philosophia. . . .* Würzburg, 1763.

Westenrider, Lorenz. *Geschichte der Baierischen Akademie der Wissenschaften, auf Verlangen derselben verfertigte.* Vol. 1. *Von 1759–1777.* Munich, 1784.

Windtweh, Nicolas, S.J. *Peripatetica sapientia ramalia. . . .* Würzburg, 1660.

Wirceburgenses. *Theologia Wirceburgensis: Theologia dogmatica, polemica, scholastica et moralis praelectionibus publicis in alma universitate Wirceburgensi accommodata.* 5 vols. Paris, 1852–54.

Wolff, Christian. *Preliminary Discourse on Philosophy in General.* Translated by Richard J. Blackwell. Indianapolis: Bobbs-Merrill, 1963.

———. *Gesammelte Werke.* Hildesheim: Georg Olms, 1962–2001.

Wolff, Ignatius, S.J. *Wolffius de inhabitatione corporum naturalium quaenam sensa habuerit? Prolusio critica in Disputationem IV. Physicam. . . .* Würzburg, 1764.

Wysing, S.J. *Scientia physica actualis ad amussim Aristotelis demonstrativam exacta. . . .* Ingolstadt, 1631.

Zallinger, Jacob von, S.J. *Lex gravitatis universalis ac mutuae cum theoria de sectione coni, potissimum elliptica proposita.* Munich, 1769.

———. *Anmerkungen über den Auszug, und die Kritik eines berlinischen Herrn Recensenten des Boscovichische System betreffend.* Freiburg im Br., 1772.

———. *De expositione physica demonstrationum mathematicarum in philosophia naturali dissertatio.* Dillingen, 1772.

———. *Interpretatio naturae, seu Philosophia Newtoniana methodo exposita, atque academicis usibus adcommodata.* 3 vols. Augsburg, 1773–75.

Zeder, Georg, S.J. *De praecipuis megacosmi phaenomenis dissertationes inaugurales, exercitationibus scholasticis illustratae. Ad partem I. Phys. special.* . . . Mainz, 1766.

Zetl, Paul, S.J. *Praecipua barometri phaenomena rationibus physicis illustrata.* . . . Ingolstadt, 1718.

Zinck, Ignatius, S.J. *Positiones ex universa philosophiae Aristotelis tum contemplativa, tum politica.* . . . Würzburg, 1700.

Zucchi, Nicolò. *Experimenta vulgata non vacuum probare, sed plenum et antiperistasim instabilire.* Rome, 1648.

———. *Nova de machinis philosophia.* Rome, 1649.

Secondary Sources

Ariew, Roger. "Descartes and Scholasticism: The Intellectual Background to Descartes' Thought." In *The Cambridge Companion to Descartes,* edited by John Cottingham. Cambridge: Cambridge University Press, 1992.

———. "Damned If You Do: Cartesians and Censorship, 1663–1706." *Perspectives on Science* 2 (1994): 255–74.

———. "Quelques condamnations du cartèsianisme: 1662–1706." *Bulletin cartésien XXII, Archives de Philosophie* 57 (1994): 1–6.

———. *Descartes and the Last Scholastics.* Ithaca: Cornell University Press, 1999.

Ariew, Roger, and Marjorie Greene, eds. *Descartes and His Contemporaries: Mediations, Objections, and Replies.* Chicago: University of Chicago Press, 1995.

Armogathe, J.-R. *Theologia cartesiana: L'explication physique de l'Eucharistie chez Descartes et dom Desgabets.* The Hague: Martinus Nijhoff, 1977.

Ashworth, William B. "Light of Reason, Light of Nature: Catholic and Protestant Metaphors of Scientific Knowledge." *Science in Context* 3 (1989): 89–107.

Attlmayr, Ernst. "Die fünf gelehrten Naturwissenschaftler von Zallinger aus Bozen." *Beiträge zur Technikgeschichte Tirols* 2 (1970): 36–43.

Baldini, Ugo. "Additamenta Galilaeana: I. Galileo, le nuova astronomia e la critica all'Aristotelismo nel dialogo epistolare tra Giuseppe Biancani e i revisori romani della Compagnia de Gesu." *Annali dell'Instituto e Museo di Storia della Scienza di Firenze* 9 (1984): 13–43.

______. "Una fonte poco utilizzata per la storia intellettuale: Le 'censurae librorum' e 'opinionum' nell'antica Compagnia di Gesù." *Annali dell'Istituto storico italo-germanico in Trento* 9 (1985): 19–67.

———. *Legem impone subactis: Studi su filosofia e scienza dei gesuiti in Italia, 1540–1632.* Rome: Bulzoni, 1992.

———, ed. *Cristoph Clavius e l'attivita scientifica dei Gesuiti nell'eta di Galileo: Atti del Convegio Internationale, Chieti, 28–30 aprile 1993.* Rome: Bulzoni, 1995.

Baldini, Ugo, and George V. Coyne, S.J. *The Louvain Lectures (Lectiones Lovanienses) of Bellarmine and the Autograph Copy of His 1616 Declaration to Galileo.* Studi Galileiani, 1/2. Vatican City: Specola Vaticana, 1984.

Baldwin, Martha. "Alchemy and the Society of Jesus in the Seventeenth Century: Strange Bedfellows?" *Ambix* 40 (1993): 41–64.

Bangert, William V., S.J. *Jerome Nadal, S.J., 1507–1580: Tracking the First Generation of Jesuits.* Edited and completed by Thomas M. McCoog, S.J. Chicago: Loyola University Press, 1992.

Bartlett, Dennis A. "The Evolution of the Philosophical and Theological Elements of the Jesuit *Ratio Studiorum:* An Historical Survey, 1540–1599." Ed. D. diss., University of San Francisco, 1984.

Bennett, J. A. "The Challenge of Practical Mathematics." In *Science, Culture and Popular Belief in Renaissance Europe,* edited by Stephen Pumfrey, Paolo L. Rossi, and Maurice Slawinski. Manchester: Manchester University Press, 1991.

Bernard, Paul P. *Jesuits and Jacobins: Enlightenment and Enlightened Despotism in Austria.* Urbana: University of Illinois Press, 1971.

Biagioli, Mario. *Galileo, Courtier: The Practice of Science in the Culture of Absolutism.* Chicago: Chicago University Press, 1993.

Bianco, Bruno. "Le Wolffianisme et les lumières catholiques: L'anthropologie de Storchenau." *Archives de Philosophie* 56 (1993): 353–88.

Blackwell, Richard J. "The Structure of Wolffian Philosophy." *Modern Schoolman* 38 (1961): 203–18.

———. *Galileo, Bellarmine, and the Bible: Including a Translation of Foscarini's "Letter on the Motion of the Earth."* Notre Dame: University of Notre Dame Press, 1991.

Blanning, T. C. W. *Reform and Revolution in Mainz, 1743–1803.* Cambridge: Cambridge University Press, 1974.

Blum, Paul Richard. "Sentiendum cum paucis, loquendum cum multis: Die Aristotelische Schulphilosophie und die Versuchungen der Naturwissenschaften bei Melchior Cornaeus SJ." In *Aristoteles Werk und Wirkung,* vol. 2, *Kommentierung, Überlieferung, Nachleben,* edited by Jürgen Wiesner. Berlin: de Gruyter, 1987.

———. "Der Standardkursus der katholischen Schulphilosophie im 17. Jahrhundert." In *Aristotelismus und Renaissance,* edited by Eckhard Keßler, Charles H. Lohr, and Walter Sparn. Wiesbaden: Harrassowitz, 1988.

Boehm, A. "L'Aristotélisme d'Honoré Fabri." *Revue des Sciences Religieuses* 39 (1965): 305–60.

Boehm, Laetitia. "Das Hochschulwesen in seiner organisatorischen Entwicklung." In *Handbuch der bayerischen Geschichte,* vol. 2, edited by Max Spindler. Munich: C. H. Beck, 1988.

———. "Universität in der Krise? Aus der Forschungsgeschichte zu katholischen Universitäten in der Aufklärung am Beispiel der Reformen in Ingolstadt und Dillingen." *Zeitschrift für bayerische Landesgeschichte* 54 (1991): 107–57.

———, ed. *Biographisches Lexikon der Ludwig-Maximilians-Universität München.* Berlin: Duncker & Humboldt, 1998.

Boehm, Laetitia, and Johannes Spörl, eds. *Die Ludwig-Maximilians-Universität in ihren Fakultäten.* Berlin: Duncker & Humboldt, 1980.

Böhm, Wilhelm. *Die Wiener Universität. Geschichte, Sendung und Zukunft.* Vienna: Regina, 1952.

Bönicke, Christian. *Grundriß einer Geschichte von der Universität zu Wirzburg.* Würzburg: F. E. Nitribitt, 1782.

Bosl, Karl. "Stellung und Funktionen der Jesuiten in den Universitätsstädten Würzburg, Ingolstadt und Dillingen." In *Bischofs- und Kathedralstädte des Mittelalters und der Frühen Neuzeit,* edited by Franz Petri. Köln: Böhlau, 1976.

Brachner, Alto, ed. *G. F. Brander, 1713–1783, wissenschaftliche Instrumente aus seiner Werkstatt.* Munich: Deutsches Museum, 1983.

Brechka, Frank T. *Gerard van Swieten and His World, 1700–1772.* The Hague: Martinus Nijhoff, 1970.

Brinitzer, Carl. *A Reasonable Rebel: Georg Christoph Lichtenberg.* Translated by Bernard Smith. London: Allen and Unwin, 1960.

Brockliss, Laurence W. B. *French Higher Education in the Seventeenth and Eighteenth Centuries: A Cultural History.* Oxford: Clarendon Press, 1987.

———. "Pierre Gautruche et l'enseignement de la philosophie de la nature dans les collèges jésuites français vers 1650." In *Les jésuites à la Renaissance: Système éducatif et production du savoir,* edited by Luce Giard. Paris: Presses Universitaires de France, 1995.

Brodrick, James, S.J. *Saint Peter Canisius, S.J., 1521–1597.* London: Sheed and Ward, 1935.

Brooke, John Hedley. *Science and Religion: Some Historical Perspectives.* Cambridge: Cambridge University Press, 1991.

Brunschwig, Henri. *Enlightenment and Romanticism in Eighteenth-Century Prussia.* 1947. Reprint, Chicago: University of Chicago Press, 1974.

Buescher, Gabriel N., O. F. M. *The Eucharistic Teaching of William Ockham.* Washington, D.C.: Catholic University of America Press, 1950.

Burr, David. "Quantity and the Eucharistic Presence: The Debate from Olivi through Ockham." *Collectanea Franciscana* 44 (1974): 5–44.

———. *Eucharistic Presence and Conversion in Late Thirteenth-Century Franciscan Thought.* Philadelphia: American Philosophical Society, 1984.

Calinger, Ronald. "The German Classical *Weltanschauung* in the Physical Sciences." In *The Influence of Early Enlightenment Thought upon German Classical Science and Letters: Problems for Future Discussion.* New York: Science History Publications, 1972.

Camenietzki, Carlos Ziller. "L'harmonie du monde au XVIIe siècle: Essai sur la pensée scientifique d'Athanasius Kircher." Ph.D. diss, Université de Paris IV-Sorbonne, 1995.

Carraud, Vincent. "Arnauld: From Ockhamism to Cartesianism." In *Descartes and His Contemporaries: Meditations, Objections, and Replies,* edited by Roger Ariew and Marjorie Greene. Chicago: University of Chicago Press, 1995.

Casanovas, J. "Boscovich as an Astronomer." In *Bicentennial Commemoration of R. G. Boscovich,* edited by M. Bossi and P. Tucci, 57–70. Milan: Edizioni Unicopli, 1988.

Cesareo, Francesco C. "The Collegium Germanicum and the Ignatian View of Education." *Sixteenth Century Journal* 24 (1993): 829–41.

Chamberlain, Cecil H. "Leonard Lessius." In *Jesuit Thinkers of the Renaissance,* edited by Gerard Smith. Milwaukee: Marquette University Press, 1939.

Clark, William. "From the Medieval Universitas Scholarium to the German Research University: A Sociogenesis of the Germanic Academic." Ph.D. diss., University of California, Los Angeles, 1986.

———. "On the Dialectical Origins of the Research Seminar." *History of Science* 27 (1989): 111–54.

———. "On the Ironic Specimen of the Doctor of Philosophy." *Science in Context* 5 (1992): 97–137.

Clericuzio, Antonio. "Notes on Corpuscular Philosophy and Pneumatical Experiments in Robert Boyle's *New Experiments Physico-mechanical, touching the Spring of the Air.*" In *Die Schwere der Luft in der Diskussion des 17. Jahrhunderts,* edited by Wim Klever, 109–16. Wiesbaden: Harrassowitz, 1997.

Cohen, H. Floris. *The Scientific Revolution: A Historiographical Inquiry.* Chicago: University of Chicago Press, 1994.

Cohen, I. Bernard. *Revolution in Science.* Cambridge, Mass.: Harvard University Press, 1985.

Conant, James B. *On Understanding Science: An Historical Approach.* New Haven: Yale University Press, 1947.

Conti, Lino. "Galileo and the Ancient Dispute about the Weight of the Air." In *Die Schwere der Luft in der Diskussion des 17. Jahrhunderts,* edited by Wim Klever. Wiesbaden: Harrassowitz, 1997.

Costantini, Claudio. *Baliani e i gesuiti: Annotazioni in margine alla corrispondenza del Baliani con Gio Luigi Confalonieri e Orazio Grassi.* Florence: Giunti, 1969.

Crombie, A. C. "Mathematics and Platonism in the Sixteenth-Century Italian Universities and in Jesuit Educational Policy." In *Prismata: Naturwissenschaftsgeschichtliche Studien. Festschrift für Willy Hartner,* edited by Y. Maeyama and W. G. Saltzer. Wiesbaden: Franz Steiner, 1977.

Cunningham, Andrew. "How the *Principia* Got Its Name: Or, Taking Natural Philosophy Seriously." *History of Science* 29 (1991): 377–92.

Cunningham, Andrew, and Perry Williams. "De-centring the 'Big Picture': The Origins of Modern Science and the Modern Origins of Science." *British Journal for the History of Science* 26 (1993): 407–32.

Dadic, Zarko. "Boskovic and the Question of the Earth's Motion." In *The Philosophy of Science of Ruder Boskovic: Proceedings of the Symposium of the Institute of Philosophy and Theology,* edited by Ivan Macan and Valentin Pozaic. Zagreb: Institute of Philosophy and Theology, 1987.

Dainville, François de. *L'éducation des jésuites: xvie–xviiie siècles.* Paris: Editions de Minuit, 1978.

Daumas, Maurice. *Scientific Instruments of the Seventeenth and Eighteenth Centuries and Their Makers.* London: Batsford, 1972.

De Backer, Augustin, S.J., Aloys de Backer, S.J., and Carlos Sommervogel, S.J. *Bibliothèque de la Compagnie de Jesus.* Paris, 1890. Reprint, Louvain, 1960.

De Oliveira Dinis, Alfredo. "The Cosmology of Giovanni Battista Riccioli (1598–1671)." Ph.D. diss, University of Cambridge, 1989.

De Vries, Josef, S.J. "Die Erkenntnislehre des Franz Suarez und der Nominalismus." *Scholastik: Zeitschrift für Theologie und Philosophie* 20–24 (1949): 321–44.

———. "Zur Aristotelisch-scholastischen Problematik von Materie und Form." *Scholastik: Zeitschrift für Theologie und Philosophie* 32 (1957): 161–85.

———. *Grundbegriffe der Scholastik.* Darmstadt: Wissenschaftliche Buchgesellschaft, 1980.

———. "Zur Geschichte und Problematik der Barockscholastik in Deutschland." *Theologie und Philosophie* 57 (1982): 1–20.

Dear, Peter. "Jesuit Mathematical Science and the Reconstitution of Experience in the Early Seventeenth Century." *Studies in the History and Philosophy of Science* 18 (1987): 133–75.

———. "Miracles, Experiments, and the Ordinary Course of Nature." *Isis* 81 (1990): 663–83.

———. "The Church and the New Philosophy." In *Science, Culture and Popular Belief in Renaissance Europe,* edited by Stephen Pumfrey, Paolo L. Rossi, and Maurice Slawinski. Manchester: Manchester University Press, 1991.

———. "Narratives, Anecdotes, and Experiments: Turning Experience into Science in the Seventeenth Century." In *The Literary Structure of Scientific Argument,* edited by Peter Dear. Philadelphia: University of Pennsylvania Press, 1991.

———. *Discipline and Experience: The Mathematical Way in the Scientific Revolution.* Chicago: University of Chicago Press, 1995.

Demoustier, Adrien, S.J. "La distinction des fonctions et l'exercise du pouvoir selon les règles de la Compagnie de Jésus." In *Les jésuites a la Renaissance: Système éducatif et production du savoir,* edited by Luce Giard. Paris: Presses Universitaires de France, 1995.

Des Chene, Dennis. *Physiologia: Natural Philosophy in Late Aristotelian and Cartesian Thought.* Ithaca: Cornell University Press, 1996.

Dictionnaire de la théologie catholique. 16 vols. Paris: Letouzey et Ané, 1903–72.

Dimler, G. Richard, S.J. "A Bibliographical Survey of Jesuit Emblem Authors in German-Speaking Territories: Topography and Themes." *AHSI* 89 (1976): 129–37.

———. "The *Imago Primi Saeculi:* Jesuit Emblem Books and the Secular Tradition." *Thought* 56 (1981): 433–48.

Dolinar, France Martin. *Das Jesuitenkolleg in Laibach und die Residenz Pleterje, 1597–1704.* Ljubljana: Theologische Fakultät Laibach, 1977.

Donahue, William H. "The Solid Planetary Spheres in Post-Copernican Natural Philosophy." In *The Copernican Achievement,* edited by Robert S. Westman. Berkeley: University of California Press, 1975.

Donelly, John Patrick, S.J. "The Jesuit College at Padua: Growth, Suppression, Attempts at Restoration, 1552–1606." *AHSI* 51 (1982): 45–79.

Duffy, Christopher. *Siege Warfare: The Fortress in the Early Modern World, 1494–1660.* London: Routledge, 1979.

Duhr, Bernhard, S.J. *Geschichte der Jesuiten in den Ländern deutscher Zunge in der ersten Hälfte des XVII. Jahrhunderts.* Freiburg i. Br.: Herder, 1913.

———. *Geschichte der Jesuiten in den Ländern deutscher Zunge in der zweiten Hälfte des XVII. Jahrhunderts.* Munich-Regensburg: Manz, 1921.

———. *Geschichte der Jesuiten in den Ländern deutscher Zunge im 18. Jahrhundert.* Pts. 1 and 2. Munich-Regensburg: Manz, 1928.

"Eine Anklageschrift gegen die deutschen Jesuiten des 18. Jahrhunderts." *Stimmen aus Maria-Laach* 77 (1909): 343–56.

Erdmannsdörffer, Bernhard. *Deutsche Geschichte vom Westfälischen Frieden bis zum Regierungsantritt Friedrich's des Grossen, 1648–1740.* Berlin, 1892.

Erman, Wilhelm, and Ewald Horn. *Bibliographie der deutschen Universitäten.* 3 vols. Leipzig, 1904–5.

Eschweiler, Karl. "Die Philosophie der spanischen Spätscholastik auf den deutschen Universitäten des siebzehten Jahrhunderts." In *Gesammelte Aufsätze zur Kulturgeschichte Spaniens,* edited by

Johannes Vincke, Spanische Forschungen der Görresgesellschaft, 1st ser., vol. 1. Münster: Aschendorf, 1928.

———. "Roderigo de Arriaga S.J.: Ein Beitrag zur Geschichte der Barockscholastik." In *Gesammelte Aufsätze zur Kulturgeschichte Spaniens,* edited by Johannes Vincke, Spanische Forschungen der Görresgesellschaft, 1st ser., vol. 3. Münster: Aschendorf, 1931, 253–85.

Espinosa Pólit, Manuel María, S.J. *Perfect Obedience: Commentary on the Letter of Obedience of Saint Ignatius of Loyola.* Westminster: Newman Bookshop, 1947.

Ettelt, Beatrix, ed. *Die Jesuiten in Ingolstadt 1549–1773.* Ingolstadt: Stadtarchiv Ingolstadt, 1991–92.

Eulenberg, Franz. *Die Frequenzen der deutschen Universitäten von ihrer Gründung bis zur Gegenwart.* Leipzig: Teubner, 1904.

Evans, G. R. *Old Arts and New Theology: The Beginnings of Theology as an Academic Discipline.* Oxford: Clarendon, 1980.

———. *Philosophy and Theology in the Middle Ages.* New York: Routledge, 1993.

———. "Theology: The Vocabulary of Teaching and Research 1300–1600: Words and Concepts." In *Vocabulary of Teaching and Research Between Middle Ages and Renaissance,* edited by Olga Weijers. Turnhout, Belgium: Brepols, 1995.

Evans, G. R., Alister E. McGrath, and Allan D. Galloway. *The Science of Theology.* Grand Rapids, Mich.: Eerdmans, 1986.

Evans, R. J. W. "German Universities after the Thirty Years War." *History of Universities* 1 (1981): 169–90.

Fabre, Pierre-Antoine. "Dépouilles d'Egypte: L'expurgation des auteurs latins dans les collèges jésuites." In *Les jésuites à la Renaissance: Système éducatif et production du savoir,* edited by Luce Giard. Paris: Presses Universitaires de France, 1995.

Fantoli, Annibale. *Galileo: For Copernicanism and for the Church.* Vatican City: Specola Vaticana, 1994.

Farrell, Allan P., S.J. *The Jesuit Code of Liberal Education. Development and Scope of the Ratio Studiorum.* Milwaukee: Bruce, 1938.

Fazzari, Michela. "Robert Boyle: Elasticity of Air and Philosophical Polemics." In *Die Schwere der Luft in der Diskussion des 17. Jahrhunderts,* edited by Wim Klever, 117–34. Wiesbaden: Harrassowitz, 1997.

Feldhay, Rivka. "Knowledge and Salvation in Jesuit Culture," *Science in Context* 1 (1987): 195–213.

———. *Galileo and the Church: Political Inquisition or Critical Dialogue?* Cambridge: Cambridge University Press, 1995.

Feldhay, Rivka, and Michael Heyd. "The Discourse of Pious Science." *Science in Context* 3 (1989): 109–42.

Findlen, Paula. *Possessing Nature: Museums, Collecting, and Scientific Culture in Early Modern Italy.* Berkeley: University of California Press, 1994.

Finnochiaro, Maurice A. *The Galileo Affair: A Documentary History.* Berkeley: University of California Press, 1989.

Fischer, Karl Adolf Franz. "Jesuiten Mathematiker in der deutschen Assistenz bis 1773." *AHSI* 47 (1978): 159–224.

———. "Newton's Ideas at the Jesuit Universities of Slovakia." *Centaurus* 31 (1988): 164–67.

Flasch, Kurt. *Einführung in die Philosophie des Mittelalters.* Darmstadt: WBG, 1987.

Frängsmyr, Tore. "Christian Wolff's Mathematical Method and its Impact on the Eighteenth Century." *Journal of the History of Ideas* 36 (1975): 653–68.

Freisen, Joseph. *Die Universität Paderborn.* Vol. 1. *Quellen und Abhandlungen von 1614–1808.* Paderborn: Jungfermann, 1898.

———. *Die Matrikel der Universität Paderborn: Matricula Universitatis Theodorianae Padibornae, 1614–1844.* Würzburg, 1931.

French, Roger, and Andrew Cunningham. *Before Science: The Invention of the Friars' Natural Philosophy.* Aldershot: Ashgate, 1996.

Funkenstein, Amos. *Theology and the Scientific Imagination from the Middle Ages to the Seventeenth Century.* Princeton: Princeton University Press, 1986.

Gabbey, Alan. "Between *Ars* and *Philosophia Naturalis:* Reflections on the Historiography of Early Modern Mechanics." In *Renaissance and Revolution: Humanists, Scholars, Craftsmen and Natural Philosophers in Early Modern Europe,* edited by J. V. Field and Frank A. J. L. James. Cambridge: Cambridge University Press, 1993.

Garcia Villoslada, Ricardo. *Storia del Collegio romano dal suo inizio (1551) alla soppressione della Compagnia di Gesù (1773).* Rome: Typis Pontificiae Universitatis Gregorianae, 1954.

Gascoigne, John. "A Reappraisal of the Role of the Universities in the Scientific Revolution." In *Reappraisals of the Scientific Revolution,* edited by David C. Lindberg and Robert S. Westman. Cambridge: Cambridge University Press, 1990.

Gatto, Romano. *Tra scienza e immaginazione: Le matematiche presso il collegio gesuitico napoletano (1552–1670 ca.).* Florence: Olschki, 1994.

Gericke, Helmuth. *Zur Geschichte der Mathematik an der Universität Freiburg i. Br.* Freiburg i. Br.: Eberhard Albert Universität, 1955.

Giard, Luce. "Remapping Knowledge, Reshaping Institutions." In *Science, Culture and Popular Belief in Renaissance Europe,* edited by Stephen Pumfrey, Paolo L. Rossi, and Maurice Slawinski. Manchester: Manchester University Press, 1991.

———. "Le Collège romain: La diffusion de la science (1570–1620) dans le réseau jésuite." In *XIXth International Congress of History of Science. Symposia Survey Papers,* edited by Jean Dhombres, Mariano Hormigan, and Elena Ausejo. Zarazoga, Spain: International Union of History and Philosophy of Sciences, 1993.

———, ed. *Les jésuites à la Renaissance: Système éducatif et production du savoir.* Paris: Presses Universitaires de France, 1995.

———. "Le devoir d'intelligence, ou l'insertion des jésuites dans le monde du savoir." In *Les jésuites a la Renaissance: Système éducatif et production du savoir,* edited by Luce Giard. Paris: Presses Universitaires de France, 1995.

Gillispie, Charles Coulston, ed. *Dictionary of Scientific Biography.* New York: Scribner, 1981.

Goddu, André. "The Dialectic of Certitude and Demonstrability According to William of Ockham and the Conceptual Relation of His Account to Later Developments." In *Studies in Medieval Natural Philosophy,* edited by Stefano Caroti. Biblioteca di Nuncius, Studi e Testi 1. Florence, 1989.

Gorman, Michael John. "Jesuit Explorations of the Torricellian Space: Carp Bladders and Sulphurous Fumes." *Melanges de l'Ecole Française de Rome* 106 (1994): 7–32.

———. "A Matter of Faith? Christoph Scheiner, Jesuit Censorship, and the Trial of Galileo." *Perspectives on Science* 4 (1996): 283–320.

Grabmann, Martin. "Die Disputationes metaphysicae des Franz Suarez in ihrer methodischen Eigenart und Fortwirkung." In *Mittelalterliches Geistesleben: Abhandlungen zur Geschichte der Scholastik und Mystik.* Munich: Hueber, 1926.

Grant, Edward. *A Source Book in Medieval Science.* Cambridge, Mass.: Harvard University Press, 1974.

———. *Much Ado about Nothing: Theories of Space and Vacuum from the Middle Ages to the Scientific Revolution.* Cambridge: Cambridge University Press, 1981.

———. *In Defense of the Earth's Centrality and Immobility: Scholastic Reaction to Copernicanism in the Seventeenth Century.* Philadelphia: American Philosophical Society, 1984.

———. *Planets, Stars, and Orbs: The Medieval Cosmos, 1200–1687.* Cambridge: Cambridge University Press, 1994.

Grass, Franz. "Der Innsbrucker Professor Ignaz von Weinhart und sein Testament." In *Festschrift Hans Lentze,* edited by Nikolaus Grass and Werner Ogris. Innsbrück: Wagner, 1969.

Greene, Marjorie. *Descartes among the Scholastics.* Milwaukee, Wisc.: Marquette University Press, 1991.

———. "Aristotelico-Cartesian Themes in Natural Philosophy: Some Seventeenth-Century Cases." *Perspectives on Science* 1 (1993): 66–87.

Guibert, Joseph de, S.J. *The Jesuits: Their Spiritual Doctrine and Practice. A Historical Study.* Translated by William J. Young, S.J. Chicago: Institute for Jesuit Sources, 1964.

Gurr, John Edwin. *The Principle of Sufficient Reason in Some Scholastic Systems, 1750–1900.* Milwaukee: Marquette University Press, 1959.

Haaß, Robert. *Die geistige Haltung der katholischen Universitäten Deutschlands im 18. Jahrhundert.* Freiburg i Br.: Herder, 1952.

Haberzettl, Hermann. *Die Stellung der Exjesuiten in Politik und Kulturleben Österreichs zu Ende des 18. Jahrhunderts.* Vienna: Verband der wissenschaftlichen Gesellschaften Österreichs Verlag, 1973.

Hammermayer, Ludwig. "Aufklärung im katholischen Deutschland des 18. Jahrhunderts: Werk und Wirkung von Andreas Gordon O. S. B. (1712–1751), Professor der Philosophie an der Universität Erfurt." *Jahrbuch des Instituts für deutsche Geschichte* 4 (1975): 53–109.

———. *Geschichte der Bayerischen Akademie der Wissenschaften 1759–1807.* 2 vols. Munich: C. H. Beck, 1983.

Hammerstein, Notker. *Aufklärung und katholisches Reich: Untersuchungen zur Universitätsreform und Politik katholischer Territorien des Heiligen Römischen Reichs deutscher Nation im 18. Jahrhundert.* Berlin: Duncker & Humboldt, 1977.

———. "Christian Wolff und die Universitäten: Zur Wirkungsgeschichte des Wolffianismus im 18. Jahrhundert." In *Christian Wolff, 1679–1754: Interpretationen zu seiner Philosophie und deren Wirkung,* edited by Werner Schneiders. Hamburg: Felix Meiner, 1983.

Hankins, Thomas L., and Robert J. Silverman. *Instruments and the Imagination.* Princeton: Princeton University Press, 1995.

Harris, Steven J. "Jesuit Ideology and Jesuit Science: Scientific Activity in the Society of Jesus, 1540–1773." Ph.D. diss., University of Wisconsin–Madison, 1988.

———. "Transposing the Merton Thesis: Apostolic Spirituality and the Establishment of the Jesuit Scientific Tradition." *Science in Context* 3 (1989): 29–65.

———. "Boscovich, the 'Boscovich Circle' and the Revival of the [sic] Jesuit Science." In *R.J. Boscovich: Vita e Attività Scientifica. His Life and Work,* edited by Piers Bursill-Hall. Rome: Istituto della Enciclopedia italiana, 1993.

———. "Les chaires de mathématiques." In *Les jésuites à la Renaissance: Système éducatif et production du savoir,* edited by Luce Giard. Paris: Presses Universitaires de France, 1995.

Haupt, Herman, ed. "Chronik der Universität Gießen von 1607 bis 1907." In *Die Universität Gießen von 1607 bis 1907.* Gießen: Töpelmann, 1907.

Heigel, Theodor. "Zur Geschichte des Censurwesens in der Gesellschaft Jesu." *Archiv für Geschichte des deutschen Buchhandels* 6 (1891): 162–67.

Heilbron, John L. *Electricity in the Seventeenth and Eighteenth Centuries: A Study in Early Modern Physics.* Berkeley: University of California Press, 1979.

———. "Science in the Church." *Science in Context* 3 (1989): 9–28.

Helden, Anne C. van. "The Age of the Air-Pump." *Tractix* 3 (1991): 149–72.

Hellyer, Marcus. "'Because the Authority of My Superiors Commands': Censorship, Physics and the German Jesuits." *Early Science and Medicine* 1 (1996): 319–54.

———. "Jesuit Physics in Eighteenth-Century Germany: Some Important Continuities." In *The Jesuits: Culture, Sciences, and the Arts, 1540–1773,* edited by John W. O'Malley, Gauvin Alexander Bailey, Steven J. Harris, and T. Frank Kennedy. Toronto: University of Toronto Press, 1999.

———. "The Construction of the Ordinatio pro Studiis Superioribus of 1651." *AHSI* 72 (2003): 3–44.

Henckel, Arthur, and Albrecht Schöne. *Emblemata: Handbuch zur Sinnbildkunst des XVI. und XVII. Jahrhunderts.* Stuttgart: Metzler, 1967.

Hengst, Karl, S.J. *Jesuiten an Universitäten und Jesuitenuniversitäten: Zur Geschichte der Universitäten in der Oberdeutschen und Rheinischen Provinz der Gesellschaft Jesu im Zeitalter der konfessionellen Auseinandersetzung.* Paderborn: Ferdinand Schöningh, 1981.

Hill, Elizabeth. "Biographical Essay." In *Roger Joseph Boscovich, S.J., F.R.S., 1711–1787,* edited by Lancelot Law Whyte. New York: Fordham University Press, 1961.

Hoffmann, Hermann. *Friedrich II. von Preussen und die Aufhebung der Gesellschaft Jesu.* Rome: Institutum Historicum Societatis Iesu, 1969.

Hsia, R. Po-Chia. *The World of Catholic Renewal, 1540–1770.* Cambridge: Cambridge University Press, 1998.

Huter, Franz. "Zur Einführung." In *Die Fächer Mathematik, Physik, Chemie an der Philosophischen Fakultät zu Innsbruck bis 1945,* edited by Franz Huter. Innsbruck: Kommissionsbuchh. in Komm., 1971.

Huter, Franz, and Anton Haidacher, eds. *Die Matrikel der Universität Innsbruck.* 3 vols. Innsbruck: Wagner, 1952–61.

Inauen, Andreas, S.J. "Stellung der Gesellschaft Jesu zur Lehre des Aristoteles und des hl. Thomas vor 1583." *Zeitschrift für katholische Theologie* 40 (1916): 201–37.

Jacob, Margaret C. *Scientific Culture and the Making of the Industrial West.* Oxford: Oxford University Press, 1997.

Jaenicke, Walter. "Naturwissenschaften und Naturwissenschaftler in Erlangen, 1743–1993." In *250 Jahre Friedrich-Alexander-Universität Erlangen-Nürnberg. Festschrift,* edited by Henning Kössler. Erlangen: Universitätsbund Erlangen-Nürnberg, 1993.

Jansen, Bernhard, S.J. "Deutsche Jesuiten-Philosophen des 18. Jahrhunderts in ihrer Stellung zur neuzeitlichen Naturauffassung." *Zeitschrift für katholische Theologie* 57 (1933): 384–410.

———. "Philosophen katholischen Bekenntnisses in ihrer Stellung zur Philosophie der Aufklärung." *Scholastik: Vierteljahresschrift für Theologie und Philosophie* 11 (1936): 1–51.

———. "Die scholastische Philosophie des 17. Jahrhunderts." *Philosophisches Jahrbuch der Görres-Gesellschaft* 50 (1937): 401–44.

———. "Die Pflege der Philosophie im Jesuitenorden während des 17./18. Jahrhunderts." *Philosophisches Jahrbuch der Görres-Gesellschaft* 51 (1938): 172–215, 344–66, 436–56.

Jansen, F. "Eucharistiques (accidents)." In *Dictionnaire de la théologie catholique,* vol. 5, cols. 1368–1452. Paris: Letouzey et Ané, 1903–72.

Johnson, Trevor. "Blood, Tears and Xavier-Water: Jesuit Missionaries and Popular Religion in the Eighteenth-Century Upper Palatinate." *In Popular Religion in Germany and Central Europe, 1400–1800,* edited by Robert Scribner and Trevor Johnson. New York: St. Martin's, 1996.

Jürgensmeier, Friedhelm. *Johann Philipp von Schönborn (1605–1672) und die römische Kurie: Ein Beitrag zur Kirchengeschichte des 17. Jahrhunderts.* Mainz: Selbstverlag der Gesellschaft für Mittelrheinische Kirchengeschichte, 1977.

Just, Leo, and Helmut Mathy. *Die Universität Mainz: Grundzüge ihrer Geschichte.* Maulheim: Mushak, 1965.

Kallinen, Maija. *Change and Stability: Natural Philosophy at the Academy of Turku, 1640–1713.* Helsinki: Suomen Historiallinen Seura, 1995.

Kelter, Irving A. "The Refusal to Accommodate: Jesuit Exegetes and the Copernican System." *Sixteenth-Century Journal* 26 (1995): 273–83.

Kern, Anton. "Die Promotionsschriften der Jesuiten-Universitäten in der Zeit des Barocks." In *Festschrift Julius Franz Schütz,* edited by Berthold Sutter. Graz-Köln: Böhlaus, 1954.

Keßler, Eckhard, Charles H. Lohr, and Walter Sparn, eds. *Aristotelismus und Renaissance.* Wiesbaden: Harrassowitz, 1988.

Kink, Rudolf. *Geschichte der kaiserlichen Universität zu Wien.* Vienna: C. Gerold, 1854.

Kistner, Adolf. *Die Pflege der Naturwissenschaften in Mannheim zur Zeit Karl Theodors.* Mannheim: Selbstverlag des Mannheimer Altertumsvereins, 1930.

Klein, Michael. "Das Schottenkloster St. Jakob." In *Im Turm, im Kabinett, im Labor: Streifzüge durch die Regensburger Wissenschaftsgeschichte,* edited by Martina Lorenz. Regensburg: Universitätsverlag, 1995.

Kleineidam, Erich. *Universitas Studii Erffordensis. Überblick über die Geschichte der Universität Erfurt.* Vol. 4. *Die Universität Erfurt und ihre theologische Fakultät von 1633 bis zum Untergang 1816.* Leipzig: St. Benno, 1981.

Klingenstein, Grete. "Despotismus und Wissenschaft: Zur Kritik norddeutscher Aufklärer an der österreichischen Universität, 1750–1790." In *Formen der europäischen Aufklärung: Untersuchungen zur Situation von Christentum, Bildung und Wissenschaft im 18. Jahrhundert,* edited by Friedrich Engel-Janosi et al. Munich: R. Oldenbourg, 1976.

Klueting, Harm, ed. *Katholische Aufklärung: Aufklärung im katholischen Deutschland.* Hamburg: Felix Meiner, 1993.

Knobloch, Eberhard. "Klassifikationen." In *Maß, Zahl und Gewicht: Mathematik als Schlüssel zu Weltverständnis und Weltbeherrschung,* edited by Menso Folkerts. Weinheim: VCH, Acta Humaniora, 1989.

Komorowski, Manfred. "Die Hochschulschriften des 17. Jahrhunderts und ihre bibliographische Erfassung." *Wolfenbütteler Barock-Nachrichten* 24 (1997): 19–42.

Koutná-Karg, Dana. "Experientia fuit, mathematicum paucos discipulos habere . . . Zu den Naturwissenschaften und der Mathematik an der Universität Dillingen zwischen 1563 und 1632." In *Das andere Wahrnehmen: Beiträge zur europäischen Geschichte,* edited by Wolfgang Stürner, Martin Kintzinger, and Johannes Zalten. Cologne: Böhlau, 1991.

Krafft, Fritz. "Jesuiten als Lehrer an Gymnasium und Universität Mainz und ihre Lehrfächer: Eine chronologisch-synoptische Übersicht 1561–1773." In *Tradition und Gegenwart: Studien und Quellen zur Geschichte der Universität Mainz,* vol. 1, *Aus der Zeit der Kurfürstlichen Universität,* edited by Helmut Mathy and Ludwig Petry. Wiesbaden: Franz Steiner, 1977.

———. *Otto von Guericke.* Darmstadt: Wissenschaftliche Buchgesellschaft, 1978.

———. *Das Selbstverständnis der Physik im Wandel der Zeit.* Weinheim: Chemie Verlag, 1982.

———. "Die Schwere der Luft in der Diskussion des 17. Jahrhunderts: Otto von Guericke." In *Die Schwere der Luft in der Diskussion des 17. Jahrhunderts,* edited by Wim Klever. Wiesbaden: Harrassowitz, 1997.

Krammer, Otto. *Bildungswesen und Gegenreformation: Die Hohen Schulen der Jesuiten im katholischen Teil Deutschlands von 16. bis zum 18. Jahrhundert.* Würzburg: Vögel, 1988.

Kratz, Wilhelm, S.J. "Aus den Frühtagen der bayerischen Akademie der Wissenschaften: Zur Vorgeschichte der Aufhebung des Jesuitenordens." *AHSI* 7 (1938): 181–219.

Kraus, Andreas. *Die naturwissenschaftliche Forschung an der Bayerischen Akademie der Wissenschaften im Zeitalter der Aufklärung.* Munich: C. H. Beck, 1978.

———. "Bayerische Wissenschaft in der Barockzeit (1579–1750)." In *Handbuch der bayerischen Geschichte,* vol. 2, edited by Max Spindler. Munich: C. H. Beck, 1988.

Krayer, Albert. *Mathematik im Studienplan der Jesuiten: Die Vorlesungen von Otto Cattenius an der Universität Mainz (1610/11).* Stuttgart: Franz Steiner, 1991.

———. *Mathematik in Mainz: Handschriften und Drucke zur Mathematikausbildung an Schule und Universität, 1500–1800.* Mainz, 1992.

Kreh, Fritz. *Leben und Werk des Reichsfreiherrn Johann Adam von Ickstatt (1702–1776).* Paderborn: Ferdinand Schöningh, 1974.

Kretzmann, Norman, Anthony Kenny, and Jan Pinborg, eds. *Cambridge History of Later Medieval Philosophy.* Cambridge: Cambridge University Press, 1982.

Kröll, Helmut. "Die Auswirkungen der Aufhebung des Jesuitenordens in Wien und Niederösterreich." *Zeitschrift für bayerische Landesgeschichte* 34 (1971): 567–617.

Krones, Franz von. *Geschichte der Karl Franzens-Universität in Graz.* Graz: Verlag der Karl Franzens-Universität, 1886.

Kuckhoff, Josef. *Die Geschichte des Gymnasiums Tricoronatum.* Cologne, 1931.

Kurrus, Theodor. *Die Jesuiten an der Unversität Freiburg i. Br.* Freiburg i. Br.: Albert, 1963.

———. "Zur Einführung der Experimentalphysik durch die Jesuiten an der Universität Freiburg i. Br." In *Zur Geschichte der Universität Freiburg i. Br.,* edited by Johannes Vincke. Freiburg i. Br.: 1966.

Kusukawa, Sachiko. *The Transformation of Natural Philosophy: The Case of Philip Melanchthon.* Cambridge: Cambridge University Press, 1995.

Laeven, A. H. *The "Acta Eruditorum" under the Editorship of Otto Mencke (1644–1707): The History of an International Learned Journal between 1682 and 1707.* Amsterdam: APA-Holland University Press, 1990.

Laird, W. R. "The Scope of Renaissance Mechanics." *Osiris,* 2d ser., 2 (1986): 43–68.

Landwehr, John. *German Emblem Books, 1531–1888.* Utrecht: Haentjens Dekker & Gumbert, 1972.

Lang, Helen S. *Aristotle's Physics and Its Medieval Varieties.* Albany: SUNY Press, 1992.

Lattis, James M. *Between Copernicus and Galileo: Christoph Clavius and the Collapse of Ptolemaic Cosmology.* Chicago: Chicago University Press, 1994.

Lawn, Brian. *The Rise and Decline of the Scholastic "Quaestio Disputata": With Special Emphasis on Its Use in the Teaching of Medicine and Science.* Leiden: E. J. Brill, 1993.

Leaman, Oliver. *Averroes and His Philosophy.* Oxford: Clarendon, 1988.

Leff, Gordon. "Ockham and Wyclif on the Eucharist." *Reading Medieval Studies* 2 (1976): 1–13.

Leijenhorst, Cees. "Jesuit Conceptions of *Spatium Imaginarium* and Thomas Hobbes' Doctrine of Space." *Early Science and Medicine* 1 (1996): 355–80.

Leinkauf, Thomas. "Athanasius Kircher und Aristoteles: Ein Beispiel für das fortleben Aristotelischen Denkens in fremden Kontexten." In *Aristotelismus und Renaissance,* edited by Eckhard Keßler, Charles H. Lohr, and Walter Sparn. Wiesbaden: Harrassowitz, 1988.

———. *Mundus Combinatus: Studien zur Struktur der barocken Universalwissenschaft am Beispiel Athanasius Kirchers SJ (1602–1680).* Berlin: Akademie, 1993.

Lerner, Michel-Pierre. "L'entrée de Tycho Brahe chez les jésuites ou le chant du cygne de Clavius." In *Les jésuites à la Renaissance: Système éducatif et production du savoir,* edited by Luce Giard. Paris: Presses Universitaires de France, 1995.

Lesch, Karl Joseph. *Neuorientierung der Theologie im 18. Jahrhundert in Würzburg und Bamberg.* Würzburg: Echter, 1978.

Lewalter, Ernst. *Spanisch-jesuitische und deutsch-Lutherische Metaphysik des 17. Jahrhunderts. Ein Beitrag zur Geschichte der iberisch-deutschen Kulturbeziehungen und zur Vorgeschichte des deutschen Idealismus.* Hamburg: Ibero-Amerikanisches Institut, 1935.

Liess, Albrecht, ed. *Von der Academia Ottonia zur Otto-Friedrich-Universität Bamberg: Eine Ausstellung des Staatsarchivs Bamberg anlässlich des 37. Deutschen Historikertages.* Munich: Selbstverlag der Generaldirektion der Staatlichen Archive Bayerns, 1988.

Lind, Gunter. *Physik im Lehrbuch, 1700–1850: Zur Geschichte der Physik und ihrer Didaktik in Deutschland.* Berlin: Springer, 1992.

Lindberg, David C., ed. *Science in the Middle Ages.* Chicago: University of Chicago Press, 1978.

Lindberg, David C. "Conceptions of the Scientific Revolution from Bacon to Butterfield." In *Reappraisals of the Scientific Revolution,* edited by David C. Lindberg and Robert S. Westman. Cambridge: Cambridge University Press, 1990.

Lindberg, David C., and Ronald L. Numbers, eds. *God and Nature: Historical Essays on the Encounter between Christianity and Science.* Berkeley: University of California Press, 1986.

Lindberg, David C., and Robert S. Westman, eds. *Reappraisals of the Scientific Revolution.* Cambridge: Cambridge University Press, 1990.

Logermann, Karl-Heinz. "Personalbibliographien von Professoren der Philosophischen Fakultät der Alma Mater Julia Wirceburgensis:. Mathematik, Experimentalphysik, Theoretische Physik, Botanik, Chemie, Naturgeschichte von 1582–1803." Ph.D. diss., Erlangen-Nürnberg, 1970.

Lohr, Charles H., S.J. "Jesuit Aristotelianism and Sixteenth-Century Metaphysics." In *Paradosis: Studies in Memory of Edwin A. Quain,* edited by H. G. Fletcher III and M. B. Scholtel. New York: Fordham University Press, 1976.

———. "The Medieval Interpretation of Aristotle." In *The Cambridge History of Later Medieval Philosophy: From the Rediscovery of Aristotle to the Disintegration of Scholasticism, 1100–1600,* edited by Norman Kretzmann, Anthony Kenny, and Jan Pinborg. Cambridge: Cambridge University Press, 1982.

———. "The Sixteenth-Century Transformation of the Aristotelian Natural Philosophy." In *Aristotelismus und Renaissance: In Memoriam Charles B. Schmitt,* edited by Eckhard Kessler, Charles H. Lohr, and Walter Sparn. Wiesbaden: Harrassowitz, 1988.

———. *Latin Aristotle Commentaries.* Vol. 2. *Renaissance Authors.* Florence: Loe S. Olschki, 1988.

———. "Les jésuites et l'aristotélianisme du XVIe siècle." In *Les jésuites à la Renaissance: Système éducatif et production du savoir,* edited by Luce Giard. Paris: Presses Universitaires de France, 1995.

Lorenz, Martina. "Der Beitrag Christian Wolffs zur Rezeption von Grundprinzipien der Mechanik Newtons in Deutschland zu Begin des 18. Jahrhunderts." *Historia Scientiarum* 31 (1986): 87–100.

Lucas, Thomas M., S.J. *Landmarking: City, Church, and Jesuit Urban Strategy.* Chicago: Jesuit Way, 1997.

Lukács, Ladislaus, S.J., "Introductio Generalis." In *Monumenta Paedagogica Societatis Iesu,* vol. 2, edited by Ladislaus Lukács, S.J. Rome: Institutum Historicum Societatis Iesu, 1974.

———. "A History of the Jesuit *Ratio Studiorum.*" In *Church, Culture, and Curriculum: Theology and Mathematics in the Ratio Studiorum,* edited by Frederick A. Homan. Philadelphia: St. Joseph's University Press, 1999.

Lukens, David Clough. "An Aristotelian Response to Galileo: Honoré Fabri, S.J. (1608–1688) on the Causal Analysis of Motion." Ph.D. diss., University of Toronto, 1979.

M***, M. l'Abbé [Barthelemy Mercier]. *Notice raisonnée des ouvrage de Gaspar Schott, jésuite; contenant des observations curieuses sur la physique expérimentale, l'histoire naturelle & les arts.* Paris: Lagrange, 1785.

MacDonnell, Joseph, S.J. "Jesuit Mathematicians before the Suppression." *Archivum Historicum Societatis Iesu* 89 (1976): 139–47.

Macy, Gary. "The Dogma of Transubstantiation in the Middle Ages." *Journal of Ecclesiastical History* 45 (1994): 11–41.

Mälzer, Gottfried, ed. *Würzburger Hochschulschriften, 1581–1803 Bestandsverzeichnis.* Würzburg: Universitäts-Bibliothek, 1992.

Mancosu, Paolo. *Philosophy of Mathematics and Mathematical Practice in the Seventeenth Century.* Oxford: Oxford University Press, 1996.

Mangenot, E. "Eucharistie du XIIIe au XVe siècle." *Dictionaire de Theologie Catholique,* vol 5, cols. 1302–26.

Marchand, James W. "The Reception of Science among German Men of Letters in the Late Eighteenth Century." In *The Influence of Early Enlightenment Thought upon German Classical Science and Letters: Problems for Future Discussion.* New York: Science History Publications, 1972.

Marenbon, John. *Later Medieval Philosophy (1150–1350).* New York: Routledge, 1987.

Markovic, Zeljko. "Boscovich's *Theoria.*" In *Roger Joseph Boscovich, S.J., F.R.S. 1711–1787,* edited by Lancelot Law Whyte. New York: Fordham University Press, 1961.

Martin, A. Lynn. *The Jesuit Mind: The Mentality of an Elite in Early Modern France.* Ithaca: Cornell University Press, 1988.

Mathy, Helmut. *Die Universität Mainz, 1477–1977.* Mainz: Hanns Krach, 1977.

Mattiussi, Guido, S.J. *Le XXIV Tesi della Filosofia di S. Tommaso d'Aquino.* Rome: Università Gregoriana, 1925.

McClelland, Charles E. *State, Society and University in Germany, 1700–1714.* Cambridge: Cambridge University Press, 1980.

McCue, James F. "The Doctrine of Transubstantiation from Berengar through Trent: The Point at Issue." *Harvard Theological Review* 61 (1968): 385–430.

Meinel, Christoph. "Natur als moralische Anstalt: Die 'Meteorologia philosophico-politica' des Franz Reinzer, S.J.: Ein naturwissenschaftliches Emblembuch aus dem Jahre 1698." *Nuncius* 1 (1987): 37–94.

Menn, Stephen. "The Greatest Stumbling Block: Descartes' Denial of Real Qualities." In *Descartes and His Contemporaries: Meditations, Objections, and Replies,* edited by Roger Ariew and Marjorie Greene. Chicago: University of Chicago Press, 1995.

Methuen, Charlotte. "The Role of the Heavens in the Thought of Philip Melanchthon." *Journal of the History of Ideas* 57 (1996): 385–403.

Methuen, Charlotte. *Kepler's Tübingen: Stimulus to a Theological Mathematics.* Brookfield, Vt.: Ashgate, 1998.

Middleton, W. E. Knowles. *The History of the Barometer.* Baltimore: Johns Hopkins University Press, 1968.

Minnigerode, Gunther von. "250 Jahre Demonstrationsversuche in der Physik." In *Naturwissenschaften in Göttingen: Eine Vortragsreihe,* edited by Hans-Heinrich Vogt. Göttingen: Vandenhock & Rupprecht, 1988.

Mir, Gabriel Codina, S.J. *Aux sources de la pédagogie des jésuites: Le "modus Parisiensis."* Rome: Institutum Historicum Societatis Iesu, 1968.

Moran, Bruce, ed. *Patronage and Institutions.* Rochester: Boydell, 1980.

Moss, Jean Dietz. "Newton and the Jesuits in the *Philosophical Transactions.*" In *Newton and the New Direction in Science,* edited by G. V. Coyne and J. Aycinski. Vatican City: Specola Vaticana, 1988.

Müller, Winfried. "Die Aufhebung des Jesuitenordens in Bayern: Vorgeschichte, Durchführung, Administrative Bewältigung." *Zeitschrift für bayerische Landesgeschichte* 48 (1985): 285–352.

———. *Universität und Orden: Die bayerische Landesuniversität Ingolstadt zwischen der Aufhebung des Jesuitenordens und der Säkularisation 1773–1803.* Berlin: Duncker & Humboldt, 1986.

Mugnai, Massimo. *Leibniz' Theory of Relations.* Stuttgart: Franz Steiner, 1992.

Nauck, E. T. *Zur Vorgeschichte der naturwissenschaftlich-mathematischen Fakultät der Albert-Ludwigs-Universität Freiburg i. Br. Die Vertretung der Naturwissenschaften durch Freiburger Medizinprofessoren.* Freiburg i. Br.: Eberhard Albert Universität, 1954.

Neunheuser, Burkhard, O. S. B. *Eucharistie in Mittelalter und Neuzeit.* Freiburg i. Br.: Herder, 1963.

Oberkofler, Gerhard. "Zur Geschichte der Innsbrucker Mathematikerschule." In *Die Fächer Mathematik, Physik, Chemie an der Philosophischen Fakultät zu Innsbruck bis 1945,* edited by Franz Huter. Innsbruck: Kommissionsbuchh. in Komm., 1971.

Olesko, Kathryn M. *Physics as a Calling: Discipline and Practice in the Königsberg Seminar for Physics.* Ithaca: Cornell University Press, 1991.

O'Malley, John W. *The First Jesuits.* Cambridge, Mass.: Harvard University Press, 1993.

———. "The Historiography of the Society of Jesus: Where Does It Stand Today?" In *The Jesuits: Culture, Sciences, and the Arts, 1540–1773,* edited by John W. O'Malley, Gauvin Alexander Bailey, Steven J. Harris, and T. Frank Kennedy. Toronto: University of Toronto Press, 1999.

O'Malley, John W., Gauvin Alexander Bailey, Steven J. Harris, and T. Frank Kennedy, eds. *The Jesuits: Culture, Sciences and the Arts, 1540–1773.* Toronto: University of Toronto Press, 1999.

Oorschot, Theo G. M. van, S.J. "Die Lebensdaten." In *Friedrich Spee im Licht der Wissenschaften: Beiträge und Untersuchungen,* ed. Anton Arens. Mainz: Selbstverlag der Gesellschaft für Mittelrheinische Kirchengeschichte, 1984.

Ornstein, Martha. *The Role of Scientific Societies in the Seventeenth Century.* Chicago: University of Chicago Press, 1928.

Palmerino, Carla Rita. "Infinite Degrees of Speed: Marin Mersenne and the Debate over Galileo's Law of Free Fall." *Early Science and Medicine* 4 (1999): 269–328.

Park, David. *The Fire within the Eye: A Historical Essay on the Nature and Meaning of Light.* Princeton: Princeton University Press, 1997.

Paulsen, Friedrich. *Geschichte des gelehrten Unterrichts auf den deutschen Schulen und Universitäten vom Ausgang des Mittelalters bis zur Gegenwart.* 2 vols. Leipzig: Veit, 1919.

Pelikan, Jaroslav. *The Christian Tradition: A History of the Development of Doctrine.* Vol. 4. *Reformation of Church and Dogma (1300–1700).* Chicago: University of Chicago Press, 1984.

Polgar, Laszlo, S.J. *Bibliographie sur l'histoire de la Compagnie de Jesus.* Rome: Institutum Historicum Societatis Iesu, 1981–90.

Powers, Joseph M. *Eucharistic Theology.* New York: Seabury, 1967.

Prantl, Carl. *Geschichte der Ludwig-Maximilians-Universität in Ingolstadt, Landshut, München.* Munich, 1872.

Press, Volker. *Korbinian von Prielmair (1643–1707).* Ottenhofen: Kliempt, 1978.

Probst, Jacob. *Geschichte der Universität Innsbruck seit ihrer Entstehung bis zum Jahre 1860.* Innsbruck, 1869.

Randles, W. G. L. "Le ciel chez les jésuites espagnols et portugais (1590–1651)." In *Les jésuites à la Renaissance: Système éducatif et production du savoir,* edited by Luce Giard. Paris: Presses Universitaires de France, 1995.

Redondi, Pietro. *Galileo Heretic.* Princeton: Princeton University Press, 1987.

Reif, Mary Richard. "Natural Philosophy in Some Early Seventeenth Century Scholastic Textbooks." Ph.D. diss., Saint Louis University, 1962.

———. "The Textbook Tradition in Natural Philosophy, 1600–1650." *Journal of the History of Ideas* 30 (1969): 17–32.

Reindl, Maria. *Lehre und Forschung in Mathematik und Naturwissenschaften, insbesondere Astronomie, an der Universität Würzburg von der Gründung bis zum Beginn des 20. Jahrhundert.* Neustadt: Degener, 1966.

Rice, Louise. "Jesuit Thesis Prints and the Festive Academic Defence at the Collegio Romano." In *The Jesuits: Culture, Sciences, and the Arts, 1540–1773,* edited by John W. O'Malley, Gauvin Alexander Bailey, Steven J. Harris, and T. Frank Kennedy. Toronto: University of Toronto Press, 1999.

Riedl, Clare, C. "Suárez and the Organization of Learning." In *Jesuit Thinkers of the Renaissance,* edited by Gerard Smith, S.J. Milwaukee: Marquette University Press, 1939.

Riepe, Christian. *Geschichte der Universitat Osnabrück.* Osnabrück: A. Fromm, 1965.

Romano, Antonella. *La contre-réforme mathématique: Constitution et diffusion d'une culture mathématique jésuite à la Renaissance (1540–1640).* Rome: Ecole française de Rome, 1999.

Romstöck, Franz Sales. *Die Jesuitennullen Prantl's an der Universität Ingolstadt und ihre Leidensgenossen: Ein biobibliographische Studie.* Eichstatt, 1898.

Rose, Paul Lawrence, and Stillman Drake. "The Pseudo-Aristotelian *Questions of Mechanics* in Renaissance Culture." *Studies in the Renaissance* 18 (1971): 65–104.

Russell, John L., S.J. "Catholic Astronomers and the Copernican System after the Condemnation of Galileo." *Annals of Science* 46 (1989): 365–86.

Schaff, Josef. *Geschichte der Physik an der Universität Ingolstadt.* Erlangen: Wiedemann, 1912.

Schillebeeckx, Edward, O.P. *The Eucharist.* New York: Sheed and Ward, 1968.

Schilling, Klaus. *Die Kirchenlehre der Theologia Wirceburgensis.* Paderborn: Ferdinand Schöningh, 1969.

Schimank, Hans. "Zur Geschichte der Physik an der Universität Göttingen vor Wilhelm Weber (1734–1830)." *Rete: Strukturgeschichte der Naturwissenschaften* 2 (1974): 207–52.

Schindling, Anton. *Bildung und Wissenschaft in der frühen Neuzeit, 1650–1800.* Munich: R. Oldenbourg, 1994.

Schmid, Alois. "Geschichtsschreibung am Hofe Kurfürst Maximilians I. von Bayern." In *Um Glauben und Reich: Kurfürst Maximilian I,* edited by Hubert Glaser. Munich: Himer, 1980.

Schmidt, Wilhelm. "Heron von Alexandria im 17. Jahrhundert." *Abhandlungen zur Geschichte der Mathematik* 8 (1898): 195–214.

Schmidt-Biggeman, Wilhelm. "Aristoteles im Barock: Über den Wandel der Wissenschaften." In *Res Publica Litteraria: Die Institutionen der Gelehrsamkeit in der frühen Neuzeit,* vol. 1, edited by Sebastian Neumeister and Conrad Wiedemann. Wiesbaden: Harrassowitz, 1987.

Schmitt, Charles B. "Experimental Evidence for and against a Void: The Sixteenth-Century Arguments." *Isis* 58 (1967): 352–66.

———. "Experience and Experiment: A Comparison of Zabarella's View with Galileo's in 'De Motu.'" *Studies in the Renaissance* 16 (1969): 80–138.

———. "Towards a Reassessment of Renaissance Aristotelianism." *History of Science* 11 (1973): 159–93.

———. "Reappraisals in Renaissance Science." *History of Science* 16 (1978): 200–214.

———. *Aristotle and the Renaissance.* Cambridge, Mass.: Harvard University Press, 1983.

———. "Galilei and the Seventeenth-Century Text-Book Tradition." In *Novità celesti e crisi del sapere: Atti del Convegno Internazionale di Studi Galileiani,* edited by Paolo Galluzzi. Florence: Barbèra, 1984.

———. "The Rise of the Philosophical Textbook." In *The Cambridge History of Renaissance Philosophy,* edited by Charles B. Schmitt, Quentin Skinner, Eckhard Kessler, and Jill Kraye. Cambridge: Cambridge University Press, 1988.

Schmitt, Charles B., Quentin Skinner, Eckhard Kessler, and Jill Kraye, eds. *The Cambridge History of Renaissance Philosophy.* Cambridge: Cambridge University Press, 1988.

Schneider, Burkhart, S.J. "Der Konflikt zwischen Claudius Aquaviva und Paul Hoffaeus." *AHSI* 26 (1957): 3–56.

Schubert, Ernst. *Materielle und organisatorische Grundlagen der Würzburger Universitätsentwicklung.* Neustadt an der Aisch: Degener, 1973.

Schuppener, Georg. *Jesuitische Mathematik in Prag im 16. und 17. Jahrhundert (1556–1654).* Leipzig: Universitätsverlag, 1999.

Seifert, Arno. "Die jesuitische Reform: Geschichte der Artistenfakultät im Zeitraum 1570 bis 1650." In *Die Ludwig-Maximilians-Universität in ihren Fakultäten,* edited by Laetitia Boehm and Johannes Spörl. Berlin: Duncker & Humboldt, 1980.

———. "Der jesuitische Bildungskanon im Lichte zeitgenossischer Kritik." *Zeitschrift für bayerische Landesgeschichte* 47 (1984): 43–75.

Seitz, Wolfgang. "Die graphischen Thesenblätter des 17. und 18. Jahrhunderts: Ein Forschungsvorhaben über ein Spezialgebiet barocker Graphik." *Wolfenbütteler Barock-Nachrichten* 11 (1984): 105–14.

Shapin, Steven. "Pump and Circumstance: Robert Boyle's Literary Technology." *Social Studies of Science* 14 (1984): 481–520.

———. *A Social History of Truth: Civility and Science in Seventeenth-Century England.* Chicago: University of Chicago Press, 1994.

———. *The Scientific Revolution.* Chicago: University of Chicago Press, 1996.

Shapin, Steven, and Simon Schaffer. *Leviathan and the Air Pump: Hobbes, Boyle, and the Experimental Life.* Princeton: Princeton University Press, 1989.

Sievernich, Michael, S.J. *Friedrich von Spee, Priester-Poet-Prophet.* Frankfurt a. M.: Josef Knecht, 1986.

Sortais, G., S.J. *Le Cartésianisme chez les jésuites français au XVIIe et au XVIIIe siècle.* Paris: Beauchesne, 1929.

Specht, Thomas. *Geschichte der ehemaligen Universität Dillingen (1549–1804) und der mit ihr verbundenen Lehr- und Erziehungsanstalten.* Freiburg i. Br.: Herder, 1902.

Steiner, Jürgen. *Die Artistenfakultät der Universität Mainz 1477–1562: Ein Beitrag zur vergleichenden Universitätsgeschichte.* Stuttgart: Franz Steiner, 1989.

Steinmaurer, Rudolf. "Die Lehrkanzel für Experimentalphysik." In *Die Fächer Mathematik, Physik, Chemie an der Philosophischen Fakultät zu Innsbruck bus 1945,* edited by Franz Huter. Innsbruck: Kommissionsbuchh. in Komm., 1971.

Stölzle, Remigius. "Der philosophische Unterricht an der Universität Würzburg 1762–63 im Urteil eines ehemaligen Jesuitenzöglings." *Zeitschrift für Geschichte der Erziehung und des Unterrichts* 5 (1915): 235–38.

Stötter, Peter. "Vom Barock zur Aufklärung: Die Philosophische Fakultät der Universität Ingolstadt in der zweiten Hälfte des 17. und im 18. Jahrhundert." In *Die Ludwig-Maximilians-Universität in ihren Fakultäten,* edited by Laetitia Boehm and Johannes Spörl. Berlin: Duncker & Humboldt, 1980.

Stone, Darwell. *A History of the Doctrine of the Holy Eucharist.* London: Longmans, 1909.

Stosiek, Winfried. "Die Personalbibliographien der Professoren der Aristotelischen Physik in der Philosophischen Fakultät der Alma Mater Julia Wirceburgensis von 1582–1773." Ph.D. diss., Universität Erlangen-Nürnberg, 1972.

Stump, Eleonore. "Theology and Physics in *De sacramento altaris:* Ockham's Theory of Indivisibles." In *Infinity and Continuity in Ancient and Medieval Thought,* edited by Normann Kretzmann. Ithaca: Cornell University Press, 1982.

Sylla, Edith Dudley. "Autonomous and Handmaiden Science: St. Thomas Aquinas and William of Ockham on the Physics of the Eucharist." In *The Cultural Context of Medieval Learning,* edited by John Murdoch and Edith Dudley Sylla. Dordrecht: D. Reidel, 1975.

Synopsis historiae Societatis Jesu. Louvain, 1950.

Taft, Robert. "The Frequency of the Eucharist throughout History." *Concilium* 152 (1982): 13–24.

Telesko, Werner. *Thesenblätter österreichischer Universitäten.* Salzburg: Das Museum, 1996.

Thorndike, Lynn. *A History of Magic and Experimental Science.* New York: Macmillan, 1923–58.

———. "The 'Cursus Philosophicus' before Descartes." *Archives Internationales d'Histoire des Sciences* 4 (1951): 16–24.

Tonelli, Giorgio. "Einleitung." In Christian August Crusius, *Die philosophischen Hauptwerke,* edited by Giorgio Tonelli. Hildesheim: Georg Olms, 1969.

Trapnell, William H. *Voltaire and the Eucharist.* Oxford: Voltaire Foundation, 1981.

Turner, Anthony. *Early Scientific Instruments: Europe, 1400–1800.* London: Sotheby's, 1987.

Turner, R. Steven. "University Reformers and Professorial Scholarship in Germany 1760–1806." In *The University in Society,* vol. 2, *Europe, Scotland and the United States from the Sixteenth to the Twentieth Century,* edited by Lawrence Stone. Princeton: Princeton University Press, 1974.

Turner, S. "The Prussian Professoriate and the Research Imperative, 1790–1840." In *Epistemological and Social Problems of the Sciences in the Early Nineteenth Century,* edited by H. N. Jahnke and M. Otte. Dordrecht: D. Riedel, 1981.

Uhlenbrock, Gudrun. "Personalbibliographien der Professoren der Philosophischen Fakultät der Alma Mater Julia Wirceburgensis von 1582–1803." Vol. 3. Ph.D. diss., Universität Erlangen-Nürnberg, 1973.

Van Dülmen, Richard. "Antijesuitismus und katholische Aufklärung in Deutschland." *Historisches Jahrbuch* 89 (1969): 52–80.

———. *The Society of the Enlightenment: The Rise of the Middle Class and Enlightenment Culture in Germany.* Cambridge: Polity Press, 1992.

Veit, Andreas Ludwig. "Aus der Geschichte der Universität zu Mainz, 1477–1731." *Historisches Jahrbuch* 40 (1920): 106–36.

Verlet, Loup. *La malle de Newton.* Paris: Gallimard, 1993.

Viller, Marcel, ed. *Dictionnaire de spiritualité ascétique et mystique, doctrine et histoire.* 17 vols. Paris: Beauchesne, 1932–95.

Villoslada, Ricardo Garcia. *Storia del Collegio romano dal suo inizio (1551) alla soppressione della Compagnia di Gesù (1773).* Rome: Gregorian University, 1954.

Wallace, William A. "Causes and Forces in Sixteenth-Century Physics." *Isis* 69 (1978): 400–412.

———. "The Problem of Apodictic Proof in Early Seventeenth Century Mechanics: Galileo, Guevara and the Jesuits." *Science in Context* 3 (1989): 67–87.

———. *Galileo and His Sources: The Heritage of the Collegio Romano in Galileo's Science.* Princeton: Princeton University Press, 1984.

———. *Galileo, the Jesuits and the Medieval Aristotle.* Brookfield, Vt.: Ashgate, 1991.

Wegele, Franz X. *Geschichte der Universität Wirzburg.* Würzburg: Stahel, 1882.

Weijers, Olga. *Terminologie des Universités au XIIIe. Siècle.* Rome: Ateneo, 1987.

Westfall, Richard S. *Force in Newton's Physics: The Science of Dynamics in the Seventeenth Century.* New York: American Elsevier, 1971.

———. *The Construction of Modern Science: Mechanisms and Mechanics.* Cambridge: Cambridge University Press, 1977.

Westman, Robert S. "The Melanchthon Circle, Rheticus, and the Wittenberg Interpretation of the Copernican Theory." *Isis* 66 (1975): 164–93.

———. "The Astronomer's Role in the Sixteenth Century: A Preliminary Study." *History of Science* 18 (1980): 105–47.

———. "The Copernicans and the Churches." In *God and Nature: Historical Essays on the Encounter between Christianity and Science,* edited by David C. Lindberg and Ronald L. Numbers. Berkeley: University of California Press, 1986.

White, Andrew D. *A History of the Warfare of Science with Theology in Christendom.* 1896. Reprint, New York: Braziller, 1955.

Whyte, Lancelot Law. "Boscovich's Atomism." In *Roger Joseph Boscovich, S.J., F.R.S., 1711–1787,* edited by Lancelot Law White. New York: Fordham University Press, 1961.

Wulf, Maurice de. *An Introduction to Scholastic Philosophy Medieval and Modern.* New York: Dover, 1956.

Wundt, Max. *Die deutsche Schulphilosophie im Zeitalter der Aufklärung.* Hildesheim: Georg Olms, 1964.

Zeeden, Ernst Walter. "Die Freiburger Philosophische Fakultät im Umbruch des 18. Jahrhunderts." In *Beiträge zur Geschichte der Freiburger Philosophischen Fakultät* by Clemens Bauer, Ernst Walter Zeeden, and Hans-Günter Zmarzlik. Freiburg i. Br.: Albert, 1957.

Ziggelaar, August, S.J. "Jesuit Astronomy North of the Alps: Four Unpublished Jesuit Letters, 1611–1620." In *Cristoph Clavius e l'attivita scientifica dei Gesuiti nell'eta di Galileo: atti del Convegio Internationale, Chieti, 28–30 aprile 1993,* edited by Ugo Baldini. Rome: Bulzoni, 1995.

Zinner, Ernst. *Entstehung und Ausbreitung der copernicanischen Lehre.* Erlangen: Mencke, 1943.

INDEX

Marcus Hellyer

received his Ph.D. in the history of science at the University of California, San Diego. He is a senior research officer at Parliament House in Canberra, Australia.